Conference Proceedings of the Society for Experimental Mechanics Series

Series Editor

Kristin B. Zimmerman, Ph.D.
Society for Experimental Mechanics, Inc.,
Bethel, CT, USA

More information about this series at http://www.springer.com/series/8922

Piyush R. Thakre • Raman Singh • Geoff Slipher
Editors

Mechanics of Composite and Multi-functional Materials, Volume 6

Proceedings of the 2017 Annual Conference on Experimental and Applied Mechanics

Springer

Editors
Piyush R. Thakre
Dow Chemical Company
Freeport, TX, USA

Geoff Slipher
U.S. Army Research Laboratory
Aberdeen, MD, USA

Raman Singh
School of Mechanical and Aerospace Engineering
Helmerich Advanced Technology Research Center
Oklahoma State University
Tulsa, OK, USA

School of Materials Science and Engineering
Helmerich Advanced Technology Research Center
Oklahoma State University
Tulsa, OK, USA

ISSN 2191-5644 ISSN 2191-5652 (electronic)
Conference Proceedings of the Society for Experimental Mechanics Series
ISBN 978-3-319-63407-4 ISBN 978-3-319-63408-1 (eBook)
DOI 10.1007/978-3-319-63408-1

Library of Congress Control Number: 2016951852

Printed on acid-free paper

This Springer imprint is published by Springer Nature
The registered company is Springer International Publishing AG
The registered company address is: Gewerbestrasse 11, 6330 Cham, Switzerland

Preface

Mechanics of Composite and Multifunctional Materials represents one of nine volumes of technical papers presented at the 2017 SEM Annual Conference & Exposition on Experimental and Applied Mechanics organized by the Society for Experimental Mechanics and held in Indianapolis, IN, June 12–15, 2017. The complete Proceedings also includes volumes on: *Dynamic Behavior of Materials; Challenges In Mechanics of Time-Dependent Materials; Advancement of Optical Methods in Experimental Mechanics; Mechanics of Biological Systems, Materials and Other Topics in Experimental and Applied Mechanics; Micro-and Nanomechanics; Fracture, Fatigue, Failure and Damage Evolution; Residual Stress, Thermomechanics & Infrared Imaging, Hybrid Techniques and Inverse Problems; and Mechanics of Additive and Advanced Manufacturing*.

The commercial market for composites continues to expand with a wide range of applications from sporting equipment to aerospace vehicles. This growth has been fueled by new material developments, greater understanding of material behaviors, novel design solutions, and improved manufacturing techniques. The broad range of applications and the associated technical challenges require an increasingly multidisciplinary and collaborative approach between the mechanical, chemical, and physical sciences to sustain and enhance the positive impact of composites on the commercial and military sectors.

New materials are being developed from recycled source materials, leading to composites with unique properties and more sustainable sources. Existing materials are also being used in new and critical applications, which require a deeper understanding of material behaviors and failure mechanisms on multiple length and time scales. In addition, the unique properties of composites present many challenges in manufacturing and in joining with other materials. New testing methods must be developed to characterize the novel composite properties, to evaluate application and product life cycle performance, as well as to evaluate impacts and merits of new manufacturing methods.

This volume presents early research findings from experimental and computational investigations related to the processing, characterization, and testing of composite, hybrid, and multifunctional materials.

Freeport, TX, USA — Piyush R. Thakre
Tulsa, OK, USA — Raman Singh
Aberdeen, MD, USA — Geoff Slipher

Contents

Chapter 1
Scrap-Rubber Based Composites Reinforced with Boron and Alumina

A.B. Irez, Jennifer Hay, Ibrahim Miskioglu, and Emin Bayraktar

Abstract Composites made of reinforced scrap rubber are generating interest in transportation industries due to their unique combination of high strength, low density, and limitless availability. Whether the vehicle is an airplane, truck, car, or boat, weight reduction leads directly to reduced fuel consumption and operating costs. In the present study, different composite formulations are prepared by means of a low-cost production process. In this process, fresh scrap rubber is combined with varying amounts of boron and alumina, with the aim of optimizing strength through control of both composition and constituent bonding. Mechanical properties are evaluated by impact testing, bend testing, and nanoindentation. Microstructure is analyzed by scanning-electron microscopy (SEM).

Keywords Recycled composites • Ceramic reinforcements • High-speed nanoindentation, SEM

1.1 Introduction

Polymer-matrix composites are used increasingly for engineering applications. Among the numerous polymers that can be used as matrix material, epoxy resins are very popular. Epoxy- resin-based composites are used in many engineering applications, especially in the aeronautical and automotive industries. Superior advantages of epoxy include high specific strength and stiffness, chemical resistance, ease of processing, environmental stability and relatively low cost [1]. However, the brittleness of epoxy limits its usage in some areas. This low fracture resistance arises from highly cross-linked network structure. Therefore, for a few decades, engineers have been working on the improvement of the fracture toughness of epoxies in order to widen their applications [2, 3]. Generally, secondary-phase particles in the epoxy matrix can be incorporated to toughen epoxy resins. Secondary-phase particles can be dispersed in the matrix and they can either be soft fillers such as thermoplastic particles, rubber, or rigid fillers such as silica beads, titania, or alumina [4].

In order to make epoxy-resin-based composites resistant to high mechanical and tribological loads, reinforcements of the matrix with certain fillers are absolutely necessary. In the literature, certain fillers are chosen for the sake of simplicity in processing. However, new composites (lightweight and high-performance) need special fillers in the matrix in order to increase ductility and stiffness of the matrix materials. In reality, scrap rubber particles are ideal, because they are easy to incorporate and have the effect of increasing the toughness of epoxy-based composites. However, scrap rubber in its virgin form is not compatible with epoxy due to poor adhesion between the rubber and epoxy. Surface treatment of the rubber is needed in order to improve its adhesion to the matrix. In the literature [5–14], some of the methods proposed for chemical surface treatments, such as plasma-surface modification and other de-vulcanizations, are time consuming and expensive for the recycling process. In this research, surface treatment of the recycled scrap rubber was achieved by a simple chemical treatment with silane. Hence, the low-cost epoxy-resin-based composites were designed with scrap rubber chemically treated with silane with the addition of boron, alumina and glass bubbles. After researching the

Jennifer Hay and Ibrahim Miskioglu contributed equally to this work.

A.B. Irez • E. Bayraktar
Supmeca-Paris, School of Mechanical and Manufacturing Engineering, Saint-Ouen, France

J. Hay (✉)
Nanomechanics, Inc., 105 Meco Lane, Oak Ridge, TN 37830, USA
e-mail: jenny.hay@nanomechanicsinc.com

I. Miskioglu (✉)
Michigan Technological University ME-EM Department, Houghton, MI, USA
e-mail: imiski@mtu.edu

P.R. Thakre et al. (eds.), *Mechanics of Composite and Multi-functional Materials, Volume 6*,
Conference Proceedings of the Society for Experimental Mechanics Series, DOI 10.1007/978-3-319-63408-1_1

compatibility of various reinforcements with the epoxy matrix, we found that the most suitable recipe for aeronautical applications included the addition of fine-scrap-rubber powders (after simple chemical treatments with silane), and then additions of boron, alumina and glass. Further, alumina has good thermal conductivity, inertness to most acids and alkalis, high adsorption capacity, thermal stability and electrical insulation, and so on. Also, it is inexpensive, non-toxic and highly abrasive [15–17].

This work presents a method for processing scrap-rubber powder with epoxy resin (EpometF (SiO_2 filled), Buehler USA), reinforced with alumina, boron, and glass, to create novel composites in an economic way. The main objective of this research was to determine the mechanical and tribological properties of these composites. Properties of constituent materials were measured by means of nanoindentation. Impact behaviour was investigated by means of drop-weight testing. Scanning electron microscopy (SEM) was used to study the microstructure of these composites.

1.2 Experimental Conditions

1.2.1 Materials Processing

A new design of epoxy-based composite, reinforced with fine scrap-rubber powders, boron, alumina and glass bubbles, was prepared in several steps:

1. The fresh scrap rubber was milled with a fast-rotating toothed-wheel mill to obtain fine rubber powder.
2. Very fine solid-epoxy resin and the scrap-rubber powder (10 wt%) were mixed and chemically treated by using toluene and acrylic acid and vinyltriethoxysilane (2%). After treatment, the mixture was dried in the oven to eliminate any trace of the treatment chemicals. The mixture of epoxy resin and rubber powder was then milled 2 h to obtain a homogenous compound then heated at 80 °C for 2 h and this mixture was used as matrix.
3. Dry boron and alumina powders (micro-scale particles) were mixed together in predefined ratios and the boron-alumina mixture was heated at 80 °C for 2 h to prevent it from absorbing any moisture.
4. The rubber-epoxy and boron-alumina mixtures were combined in a blender and milled for 4 h.
5. Glass bubbles (GB-hollow glass microspheres manufactured by 3 M with a density of 0.227 g/cm^3, specified as S38HSS & K1) were added and mixed to obtain homogenous distribution of the constituents (ultrasound) for an additional 2 h.
6. The specimens were then manufactured by hot compacting (double uniaxial action) under a pressure of 70 MPa at a temperature of 180 °C for 30 min. All of the specimens (30 mm in diameter) were cooled slowly.
7. Specimens were cured isothermally at 80 °C for 48 h.

General compositions of all the composites manufactured are given in the Table 1.1 with specified weight ratios for each constituent.

In Fig. 1.1 two different specimen sizes are shown after manufacturing by hot compacting.

1.2.2 Mechanical Tests and Microstructure of the Compositions

Dynamic compression tests (drop weight tests) were carried out using a universal drop weight test device. A standard conical punch was released from a height of 900 mm. Impact behavior of the manufactured composites were observed over the test specimens with the help of force-time curves.

General microstructures in the transversal direction of two compositions are shown in the Fig. 1.2. All of the compositions have shown a considerably homogenous distribution of the reinforcements in the structure. Some small local agglomerations of reinforcement particles are observed in the structure, which may indicate the need for longer and more aggressive mixing. Also, relatively large black rubber particles are seen in the microstructure.

Table 1.1 Composition of the epoxy-rubber based composites

	Reinforcements (wt%)		
Epoxy-rubber based composition	Boron	Alumina	Glass bubbles
EBAL 1	5	5	–
EBAL 2	5	5	5
EBAL 3	10	5	5
EBAL 4	5	10	5

Fig. 1.1 Macrograph of the sample after compacting and post curing (**a**) d = 50 mm (**b**) d = 30 mm

1.2.3 Damage Analysis by Means of Scratch Test and 3D Optical Roughness Meter

Wear testing quantifies the tribological behavior of the composites. Each wear test involved sliding a 2.5 mm diameter zirconia probe over a 1.5 to 2 mm long track at a frequency of 15 Hz under a load of 30 N for either 50,000 cycles or 100,000 cycles. The damage zone was investigated by a 3D optical-surface scanner. The resistance to scratch deformation (damage) was quantified in terms of scratch depth and wear volume.

1.2.4 Nanoindentation to Measure Constituent Properties

Nanoindentation was used to measure the mechanical properties of the various constituents of EBAL 4. Nanoindentation was performed with an iNano fit with a Berkovich indenter (Nanomechanics, Inc., Oak Ridge, TN). EBAL 4 was tested with a new high-speed indentation technique in order to mechanically characterize all materials within a square area of 0.3 mm × 0.3 mm; results are presented as highly resolved contour maps and spectra of properties. Within the selected area, an array of 60 × 60 indents was performed (a total of 3600 indents). Each indent had a peak load of 2mN and individual indents were separated by 5 microns. Indentations were performed at a rate of approximately 1 indent per second; thus, all 3600 indents were completed in about 1 h.

1.3 Results and Discussions

1.3.1 Dynamic Compression (Impact) Testing

As mentioned earlier, reinforcing materials were incorporated into the epoxy matrix with the aim of improving its toughness. Figure 1.3 shows the impact force as a function of time for each composite. Further, the additional boron incorporated into

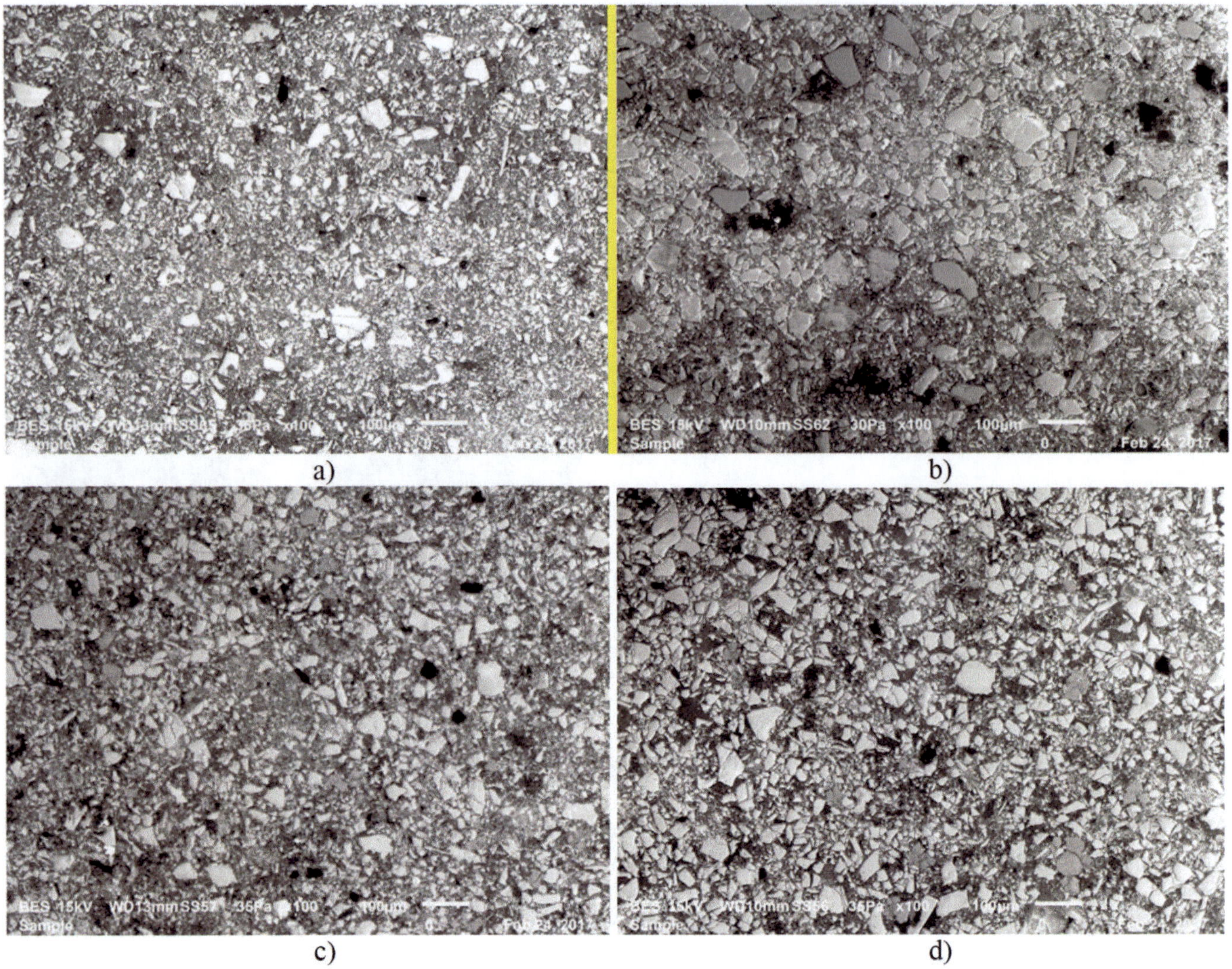

Fig. 1.2 Microstructure of EBAL specimens (**a**) EBAL 1, (**b**) EBAL 2, (**c**) EBAL 3, (**d**) EBAL 4

EBAL 3 seemed to weaken the resulting composite as evidenced by the reduced maximum impact force, relative to EBAL 2 and 4. However, the hollow glass bubbles had a toughening effect, which is consistent with previous studies [18].

1.3.2 Wear Testing

Figure 1.4 shows the profilometry scan for EBAL 2 following a 50,000-pass wear test. The wear tracks from all tests were scanned in the same way, and the depth and volume were measured; these results are provided in Fig. 1.5. High interfacial shear stress is the primary cause of tribological damage [19]. However, fluctuation in stress is also a contributing factor – ahead of the probe, the stress is compressive, but behind the probe, it is tensile. Based on these results, it seems that additional boron provides a greater resistance to wear damage than additional alumina. There is one more interesting observation from these tests. As shown in Fig. 1.6, it seems that some toughening particles were pulled out during wear testing. Such dislodged particles may have the effect of accelerating wear in an uneven and unpredictable way.

In Fig. 1.5 damaged areas were given as a function of lost volume during wear tests. By increasing the number of the cycles, surface damage increases as expected. On the other hand, it is difficult to establish a correlation about wear resistance depending on the reinforcement elements.

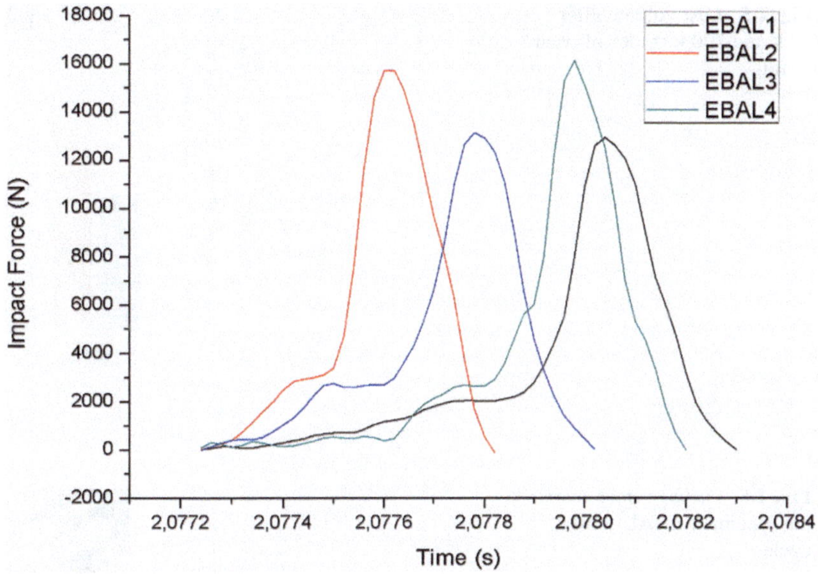

Fig. 1.3 Impact behavior of EBAL composites by dynamic compression test

Fig. 1.4 Three dimensional damage traces obtained in the direction of width and length for the specimen EBAL I for 50 k cycles

1.3.3 Nanoindentation

Figure 1.7 shows the microscopic image for the area of EBAL 4 selected for high-speed indentation. The same set of 3600 measurements of Young's modulus and hardness are presented both as contour plots (Figs. 1.8 and 1.9) and as histograms (Figs. 1.10 and 1.11). Each indentation manifests as one pixel in the contour plot and as one relative count in the histogram plot.

The contour plots of Young's modulus and hardness give a quick impression of the mechanical inhomogeneity of these materials: The rubber appears purple and the reinforcing ceramics appear red. The mechanical influence of the reinforcing

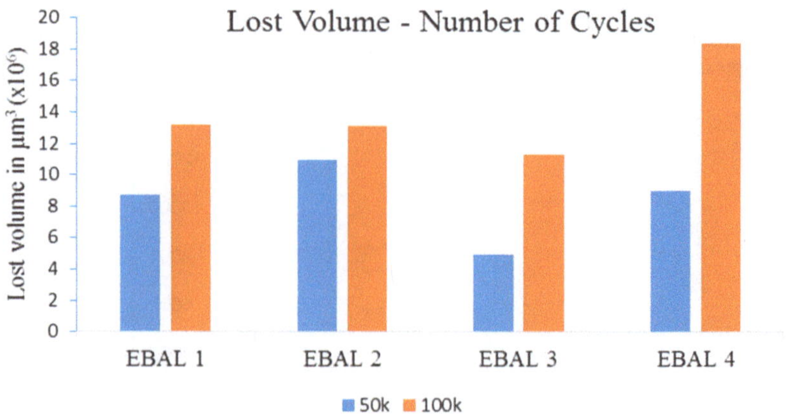

Fig. 1.5 Lost volume after 50 k and 100 k cycles of wear testing

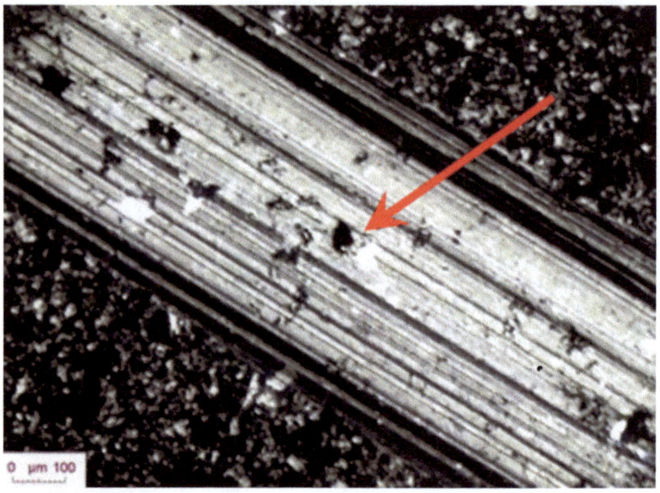

Fig. 1.6 Damage trace over the specimen EBAL- I 50 k cycles

Fig. 1.7 Surface of EBAL 4 subject to high-speed indentation array. Area is 0.3 mm on a side

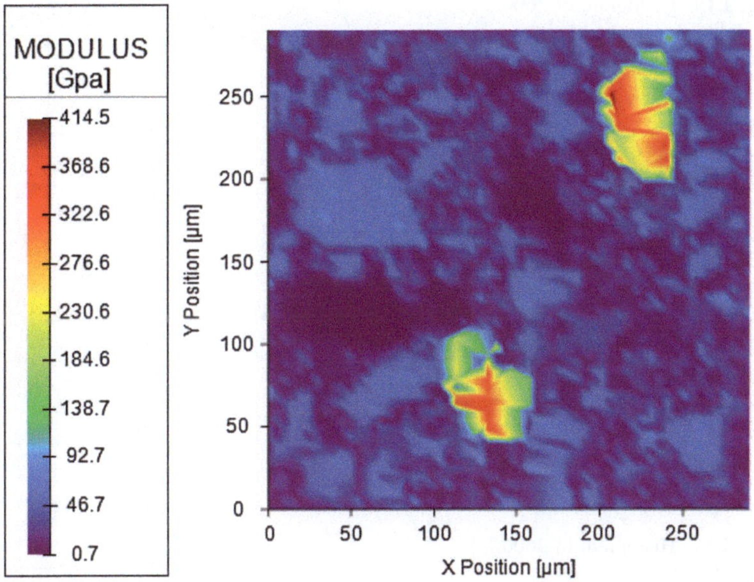

Fig. 1.8 Modulus contour plot for EBAL 4

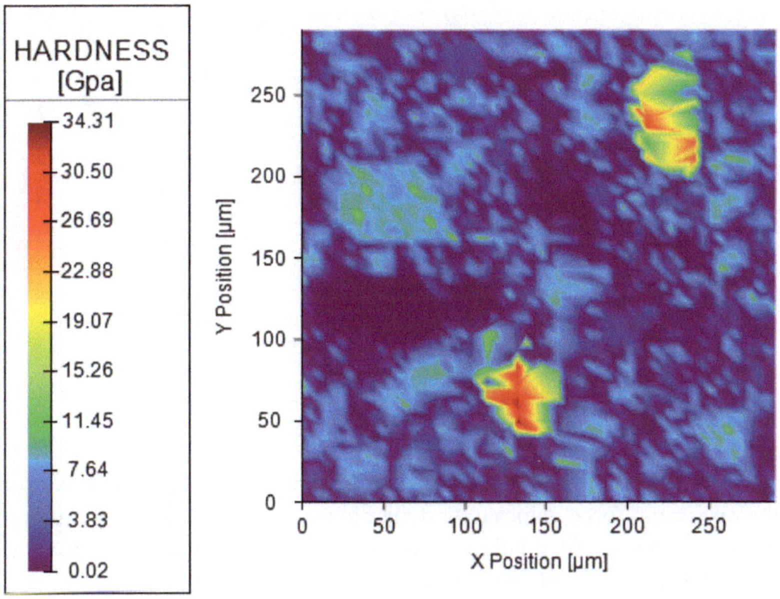

Fig. 1.9 Hardness contour plot for EBAL 4

particle does not extend beyond its boundaries. In other words, the presence of the reinforcing particle does not seem to mechanically alter nearby rubber or epoxy, as evidenced by the sharp contrast in color at material boundaries.

The histograms quickly show which materials are most influential in the tested area. There are clear peaks for the epoxy (E = 11.0GPa, H = 0.59GPa) and glass (E = 58.2GPa, H = 8.14GPa). However, these peaks are not sharp: most measurements fall between them, thus indicating a good mixture of materials with varied properties. Finally, although the ceramic particles draw the eye in the contour plots, their influence hardly registers in the histograms. As development of these materials continues, such histograms may be a useful mechanical characterization for predicting performance.

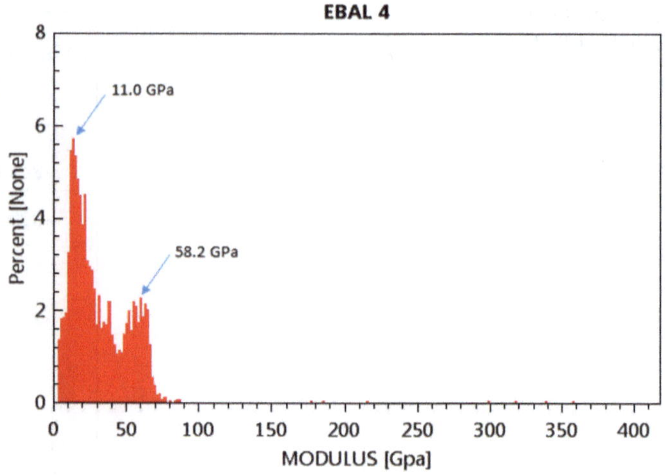

Fig. 1.10 Histogram of 3600 modulus measurements by nanoindentation on EBAL 4

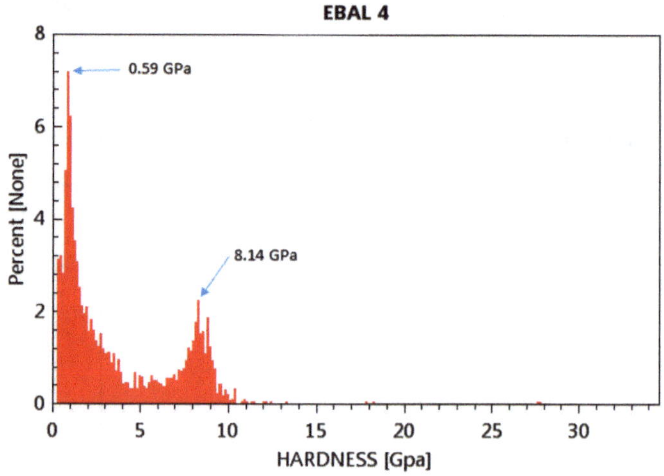

Fig. 1.11 Histogram of 3600 hardness measurements by nanoindentation on EBAL 4

1.4 Conclusions

In this study, hollow glass spheres were used to see their effect of damping. According to the drop weight test, they increased the maximum force values over dynamic compression curves. In other words, they improved the energy absorbing capacity of the composite. Also hollow glass spheres tend to promote decrease in density. From these perspectives, their use can be generalized after a detailed study.

In addition, the positive effect of boron particles on the surface wear resistance was seen after a comparative study. Results were found promising and this situation can be supported in the light of longer tests.

The histograms obtained as the result of the nanoindentation testing show which materials are most influential in the tested area. These histograms may be useful mechanical characterization for predicting performance.

References

1. May, C.A.: Epoxy Resins-Chemistry and Technology, 2nd edn. Marcel Dekker/Wiley, New York (1988)
2. Wetzel, B., Rosso, P., Haupert, F., Friedrich, K.: Epoxy nanocomposites–fracture and toughening mechanisms. Eng. Fract. Mech. **73**(16), 2375–2398 (2006)
3. Garg, A.C., Mai, Y.: Failure mechanisms in toughened epoxy resins-a review. Compos. Sci. Technol. **31**(3), 179–223 (1988)
4. Lee, J., Yee, A.: Inorganic particle toughening I: micro-mechanical deformations in the fracture of glass bead filled epoxies. Polymer. **42**(2), 577–588 (2001)
5. Manzione, L.T., Gillham, J.K.: Rubber-modified epoxies. I. transitions and morphology. J. Appl. Polym. Sci. **16**, 889–905 (1981)

6. Nakamura, Y., Yamaguchi, M., Okubo, M., Matsumoto, T.: Effects of particle size on mechanical and impact properties of epoxy resin filled with spherical silica. J. Appl. Polym. Sci. **45**(7), 1281–1289 (1992)
7. Okazaki, M., Murota, M., Kawaguchi, Y., Tsubokawa, N.: Curing of epoxy resin by ultrafine silica modified by grafting of hyperbranched polyamidoamine using dendrimer synthesis methodology. J. Appl. Polym. Sci. **80**(4), 573–579 (2001)
8. Wetzel, B., Haupert, F., Zhang, M.Q.: Epoxy nanocomposites with high mechanical and tribological performance. Compos. Sci. Technol. **63** (14), 2055–2067 (2003)
9. Zee, R.H., Huang, Y.H., Chen, J.J., Jang, B.Z.: Properties and processing characteristics of dielectric-filled epoxy resins. Polym. Compos. **10** (4), 205–214 (1989)
10. Jin Kim, D., Hyun Kang, P., Chang Nho, Y.: Characterization of mechanical properties of γAl_2O_3 dispersed epoxy resin cured by γ-ray radiation. J. Appl. Polym. Sci. **91**, 1898–1903 (2004)
11. Arayasantiparb, D., McKnight, S., Libera, M.: Compositional variation within the epoxy/adherend interphase. J. Adhes. Sci. Technol. **15**(12), 1463–1484 (2001)
12. Nielsen, L.E., Landel, R.F.: Mechanical Properties of Polymers and Composites, 2nd edn. Marcel Dekker, Inc., New York (1994)
13. Rothon, R. (ed.): Particulate-Filled Polymer Composites. Longman Scientific and Technical, London (1995)
14. Rothon, R.N.: Adv. Polym. Sci. **139**, 67–107 (1999)
15. Mirjalili, F., Mohamad, H., Chuah, L.: Preparation of nano-scale α-Al2O3 powder by the sol–gel method. Ceram-Silikáty. **55**, 378–383 (2011)
16. Amirshaghaghi, A., Kokabi, M.: Tailoring size of α-Al2O3 nanopowders via polymeric gel–net method. Iran. Polym. J. **19**, 615–624 (2010)
17. Zhang, Z., Lei, H.: Preparation of α-alumina/polymethacrylic acid composite abrasive and its CMP performance on glass substrate. Microelectron. Eng. **85**, 714–720 (2008)
18. Keivani, M., et al.: Synergistic toughening in ternary silica/hollow glass spheres/epoxy nanocomposites. Polym.-Plast. Technol. Eng. **54**(6), 581–593 (2015)
19. Zaimova, D., et al.: Manufacturing and damage analysis of epoxy resin-reinforced scraprubber composites for aeronautical applications. In: Experimental Mechanics of Composite, Hybrid, and Multifunctional Materials, vol. 6, pp. 65–76. Springer International Publishing, Cham Heidelberg\New York\Dordrecht\London (2014)

Chapter 2
Characterization of Thermoplastic Matrix Composite Joints for the Development of a Computational Framework

Joseph R. Newkirk, Cassandra M. Degen, and Albert Romkes

Abstract Many industries, notably the automotive and aerospace industries, are now utilizing thermoplastic matrix composites (TPMCs) for their improved strength and stiffness properties compared to pure thermoplastic polymers, as well as their manufacturability compared to traditional thermoset matrix composites. The increase in the utilization of TPMCs ushers in the need for the development and characterization of joining methods for these materials. A widely used technique for joining thermoplastics is ultrasonic spot welding (USSW). During USSW, high frequency, low amplitude vibrations are applied through an ultrasonic horn resting on the polymer surface. The vibrations induce frictional heat, producing a solid state joint between polymer sheets. Advantages such as short weld cycle time, fewer moving components and reproducibility make this technique attractive for automation and industrial use. Prior work showed USSW as a feasible, repeatable joining method for a polycarbonate matrix filled with chopped glass fibers. The mechanical properties required for full characterization of the TPMC used in this work were not provided by the manufacturer. As such, the constitutive behavior of both as-received and USSW thermoplastic composite material (polypropylene matrix filled with 30 wt% chopped glass fibers) was characterized. The fiber orientation and distribution in TPMCs has a direct impact on constitutive behavior. To characterize these qualities, optical techniques such as scanning electron microscopy (SEM) and micro computed tomography (micro-CT) were employed. Digital image correlation (DIC) was used to acquire full field strain measurements from the composite material under different loading scenarios. Because the constitutive behavior of polymers is greatly dependent on temperature, temperature measurements during the USSW process and measurement of mechanical properties as a function of temperature will be conducted through infrared (IR) imaging and dynamic mechanical analysis (DMA), respectively. Following the calibration of the constitutive model for the polypropylene matrix TPMC, the mechanical and thermal properties will be used to develop a computational framework for the purpose of predicting the structural response of a composite joint under various loadings.

Keywords Composite joining • Ultrasonic spot welding • X-ray micro computed tomography • Thermoplastic matrix composites • Fiber orientation

2.1 Introduction

With growing regulations in many industries, especially in the case of emissions, the demand for lightweight strong materials is increasing rapidly. As a result of this, many of these industries are turning to composite materials [1]. The increased usage of these materials drives the need for development of both the manufacturing and joining processes applied to them. Composite materials are generally joined using mechanical fasteners, adhesive bonding, or fusion welding processes. Each of these methods brings unique advantages and disadvantages. Mechanical fasteners are a common joining method, used traditionally with isotropic materials. Anisotropic composite materials are more sensitive to holes and cutouts which shear fibers and create large stress concentrations. Adhesive bonding eliminates the stress concentration associated with mechanical fasteners, but also requires more time intensive preparation work, involves unavoidable curing cycles, and the quality of the bond is dependent on the adherend materials [2]. Fusion welding relies on the melting of the materials to be joined, intermolecular diffusion across the joint, and re-solidification of the matrix material. Fusion bonding techniques generally eliminate, or at least mitigate, stress concentrations and result in joint efficiency which can approach the bulk properties of the joined materials [2].

J.R. Newkirk • C.M. Degen (✉) • A. Romkes
Department of Mechanical Engineering, South Dakota School of Mines and Technology, 501 E. Saint Joseph St., Rapid City, SD 57701, USA
e-mail: cassandra.degen@sdsmt.edu

P.R. Thakre et al. (eds.), *Mechanics of Composite and Multi-functional Materials, Volume 6*,
Conference Proceedings of the Society for Experimental Mechanics Series, DOI 10.1007/978-3-319-63408-1_2

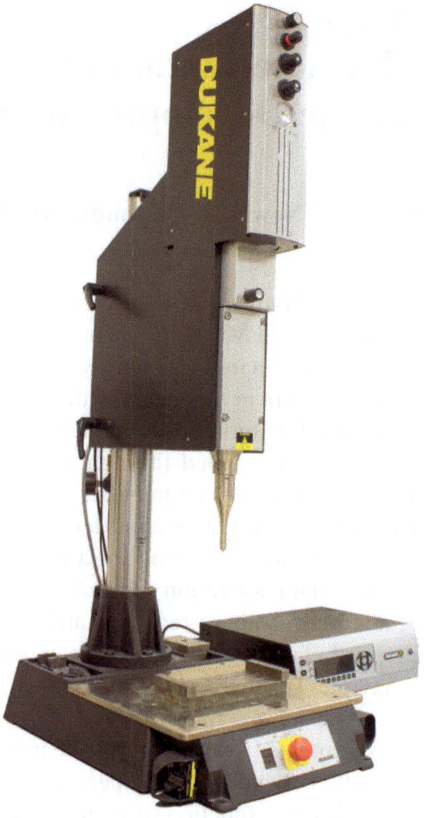

Fig. 2.1 20 kHz Dukane iQ series ultrasonic pneumatic press 43Q-220 used for this work

Historically, composites have been used in applications where cost is not a major design consideration. These composites have traditionally been thermosetting matrix composites (TSPCs), but the inherent limitations of TSPCs, most notably the inability to be joined by fusion welding and high manufacturing cost, have led to a surge in the usage of thermoplastic matrix composites (TPMCs) [2]. TPMCs have a number of advantages over TSPCs: higher fracture toughness, better resistance to wet environments, shorter processing cycles which lower manufacturing cost, recyclability [3]. Of consequence for this work is their ability to be melted and re-solidified which makes TPMCs suitable candidates for fusion welding processes.

Fusion welding can be broken into two categories: bulk heating, and frictional heating. Bulk heating involves common processes like autoclaving and compression molding, frictional heating includes spin welding, vibrational welding and ultrasonic welding [2]. The technique studied in this work is ultrasonic spot welding (USSW). In this joining process, the energy transferred to a thermoplastic material subjected to ultrasonic vibrations is dissipated through intermolecular friction, leading to heating of the material and localized melting. An ultrasonic welder typically consists of six main components: a base-plate, pneumatic press, control system, transducer, booster, and welding horn [4]. The ultrasonic spot welder used in this work is a 20 kHz Dukane iQ series Ultrasonic pneumatic press 43Q-220 (Rapid City, SD, USA) (Fig. 2.1).

When joining materials through USSW, a large range of weld qualities can be obtained by varying weld parameters. Taking this into consideration, a study was performed using varying weld parameters. The parameters used in the ultrasonic welding process were correlated to the quality of the welds created evaluated by lap shear testing in order to select suitable weld parameters. In general, the mechanical behavior of composite materials differs based upon the loading direction. The behavior of the TPMC used in this work was initially evaluated along two perpendicular loading directions. The mechanical response of the material under tensile loading was evaluated with respect to a fiber orientation that was assumed to be caused by the rolling step in the processing of the TPMC sheets.

The orientation of the fibers in a material is of consequence when evaluating mechanical properties, particularly for short fiber composites, a common type of TPMC. Changes in fiber orientation are related to a number of factors, such as the geometrical properties of the fibers, viscoelastic behavior of the matrix material, and the change in shape of the material produced by the processing operation [5]. With processes varying between manufacturers, a means of characterizing fiber orientation is necessary. To evaluate the fiber orientation in the TPMC used in this work, the optical methods of x-ray micro computed tomography (μCT) and scanning electron microscopy (SEM) were employed. These methods were used to characterize the fiber orientation of a TPMC before and after ultrasonic spot welding of the material.

The end goal of this work is to couple the experimental and optical characterization of ultrasonic spot welds in a chopped fiber-thermoplastic matrix composite with computational efforts to produce a physics-based model to guide the design of composite joints. The experimental work discussed here will guide the computational efforts, which will first focus on the development of a mathematical model that accurately describes the multi-physics of the welding process. This model will be based on principles of continuum mechanics and therefore will address both the mechanics and thermodynamics of the welding process, as well as the coupling between these. The development of this model will help guide the design of composite joints created with USSW. This predictive model will be used to derive optimal welding parameters based upon weld application and material properties. This will shorten the time necessary to apply USSW in different applications proving useful for many industries concerned with the manufacturing of novel, light-weight materials.

2.2 Mechanical Characterization of Composite Material

The use of USSW on polymer matrix composites has been studied and characterized previously [6]. In this work, initial weld characterizations were performed on pure polypropylene matrix material, followed by further characterization of polypropylene matrix composite materials welds. The composite used in this work was ESP 105 CC, a polypropylene matrix TPMC, produced by RTP Company (Winona, MN, USA). The ESP 105 CC composite is composed of 30 wt% chopped glass fibers and was provided in 1.52 mm thick extruded sheets [7]. The mechanical properties of this material provided by the manufacturer can be found in Table 2.1.

The ultrasonic spot welding process varies based upon the input parameter settings used during the weld. As such, a study was necessary to investigate the effects of weld parameters. The parameters that were varied in this study were amplitude, time and weld pressure. In the interest of limiting the use of the composite material, the study was performed on pure polypropylene material. Since the composite material is composed of 70 wt% pure polypropylene, the response to the welding parameters was assumed to be similar to that of plain polypropylene sheeting. The quality of the welds resulting was characterized both optically, by observing bond area, and mechanically, by measuring peak load during lap shear testing.

The criteria used to characterize the welds were total bonding area (shear area), and the peak load under lap shear. The bonding area was measured optically prior to lap shear testing using the open source image analysis software ImageJ. In the preparation of the lap shear samples, tabs were adhered to the ends of each test specimen to avoid the production of a bending moment during testing and the samples were tested at a rate of 1.3 mm/min as outlined in ASTM 3164-03 [8]. Samples with a broad range of weld parameters were investigated in order to isolate the parameters with the greatest impact on the quality of the weld. It was found that these parameters were the weld time and weld amplitude. The failure mode exhibited by the samples was almost immediate fracture, at the base of the weld, when the maximum load was reached. The location of the failure in the samples is most likely due to bending stresses, and subsequent stress concentrations that are inherent in lap shear testing, which are also highlighted by Degen et al. [6]. The results of the optical and mechanical characterization of plain polypropylene sheeting are shown in Table 2.2.

From this study, the following welding parameters were found to create welds with the largest bond area and highest peak load during lap shear testing in pure polypropylene sheeting: an amplitude of 24 μm, a weld time of 2 s and a weld pressure of 483 kPa. The chosen parameters are not immediately apparent upon inspection of Table 2.2, but these values were found to create the largest shear weld areas while exhibiting the least standard deviation and therefore creating the most repeatable welds. Upon finding suitable welding parameters for the pure polypropylene sheeting, the same parameters were used to weld the polypropylene matrix composite, ESP 105 CC. As a reference, a representative 25.4 mm × 25.4 mm USSW sample of the composite material is shown in Fig. 2.2. Five lap shear samples of both the plain polypropylene and composite materials were welded using the above parameters and evaluated in terms of bond area, maximum load during lap shear, and maximum shear stress. From the results shown in Fig. 2.3, it is apparent that the pure polypropylene samples exhibited better performance in terms of both average shear area and average failure load in lap shear than their composite material counterparts.

Table 2.1 Mechanical properties of ESP 105 CC [7]

Property	Value (Metric units)
Young's modulus	6205 MPa
Ultimate tensile strength	55 MPa
Elongation at break	1.5–2.5%
Impact strength (IZOD, unnotched)	0.32 J/mm

Table 2.2 Welding parameter combinations and results of both optical and mechanical analysis of welds

Amplitude	Time	Weld pressure	Energy	Area	Peak load (Lap Shear)
m	*s*	*kPa*	*J*	*mm*^{2}	*N*
24	1.50	345	168	142	903
30			161	82	459
24	2.00		192	112	671
30			189	90	510
24	1.50	414	160	121	743
30			161	100	511
24	2.00		195	141	722
30			189	93	567
24	1.50	483	177	144	825
30			158	94	637
24	2.00		208	141	863
30			196	111	588

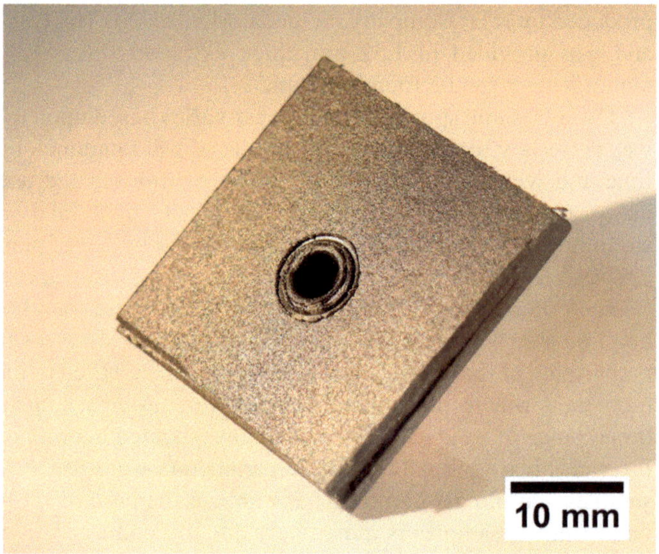

Fig. 2.2 Representative ESP 105 CC USSW sample

The maximum load measured in lap shear for the composite samples was directly correlated to the smaller measured shear area. When normalizing the average load by corresponding shear area, the results for the shear stress at failure are more closely aligned for both materials, with the composite showing slightly higher maximum stress. The results of the lap shear testing indicated that the welding parameters found to create suitable welds in the pure polypropylene sheeting may not have been suitable for creating welds in the polypropylene TPMC. The change in the apparent viscosity of the polymer melt present with the addition of the glass fibers in the polypropylene composite was one noted difference in the joining of the two different materials. This effect will be further studied for ESP 105 CC.

Rolled sheets of thermoplastic matrix composite often show a directional fiber orientation in the rolling direction. In fact, the manufacturing process is often manipulated to achieve desired fiber orientation to achieve optimal strength properties in the loading direction. Short fiber composites manufactured with conventional molding techniques possess axial mechanical properties approaching those obtained from continuous-fiber composites of equal fiber volume fraction [9]. Taking this into consideration, the mechanical response of the polypropylene composite when loaded both parallel to, and perpendicular to the orientation of the fibers must be characterized. To achieve an initial measure of this behavior, dog bone tensile samples were prepared and tested according to ASTM D638-08 [10]. Samples were cut out in perpendicular and parallel orientations relative to the observed fiber orientation. Representative results of the behavior observed in this testing are shown in Fig. 2.4. All samples tested exhibited the same brittle failure mode. A higher maximum tensile load was achieved in the samples with the loading direction parallel to the fiber orientation when compared to with the loading direction oriented perpendicular to the fiber orientation. This loading behavior indicates that the material is roughly orthotropic in nature.

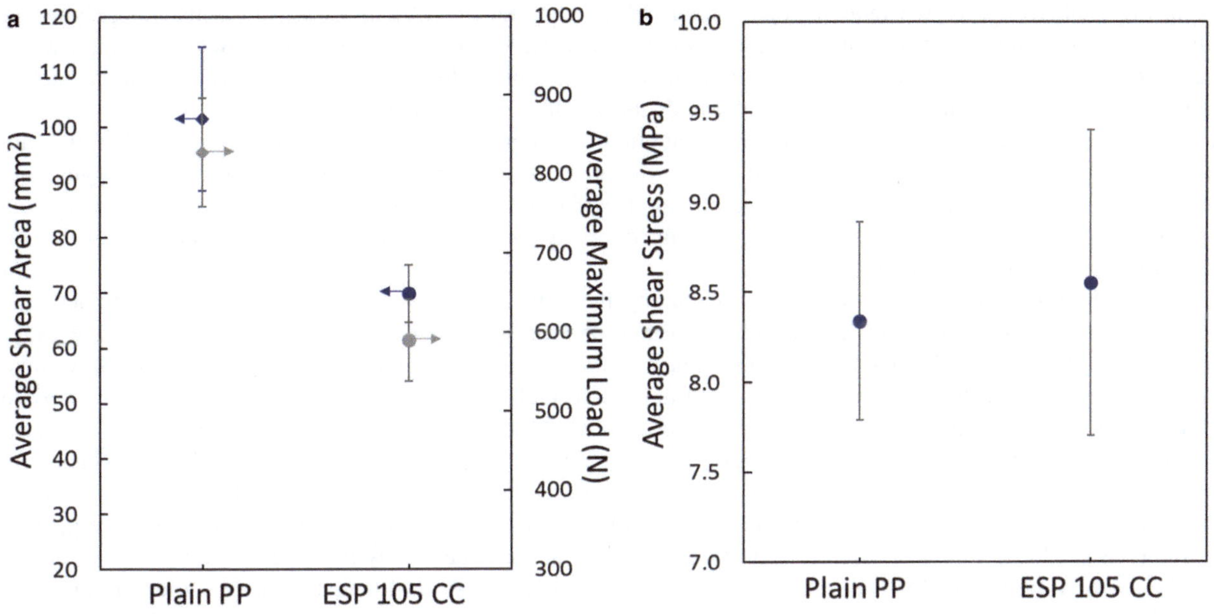

Fig. 2.3 Average Shear area and average maximum load in lap shear (**a**) and average shear stress (**b**) for pure polypropylene and ESP 105 CC composite lap shear Samples

Fig. 2.4 Stress-strain diagrams for ESP 105 CC composite tensile samples tested perpendicular and parallel to the fiber orientation tested at 5.08 mm/min

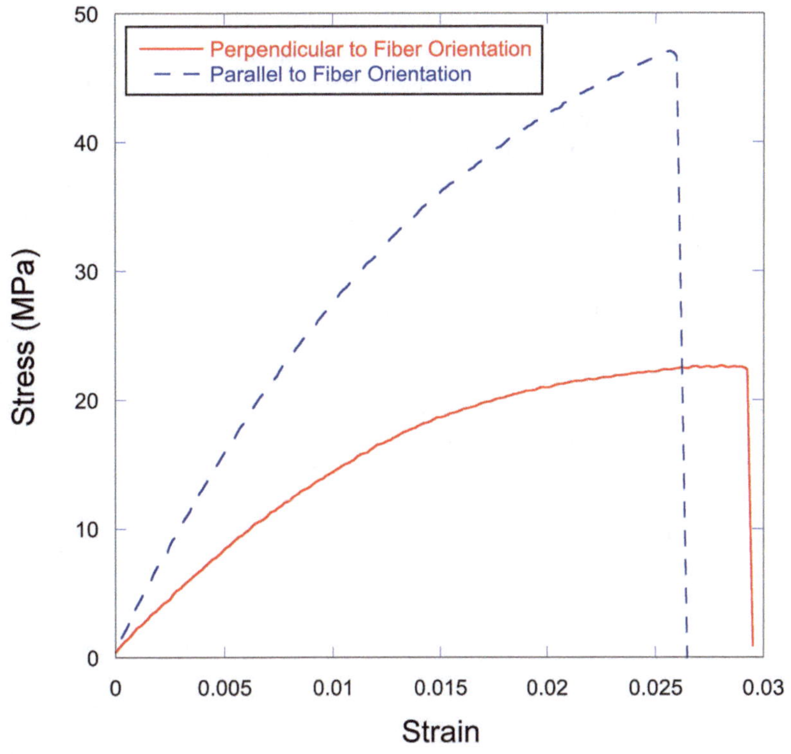

The material properties of ESP 105 CC must be characterized, in order to provide the necessary input parameters for the development of a computational model. Several approaches have been explored to characterize these properties. The first method employed was nanoindentation, a method in which the load and displacement of a Berkovich indenter are correlated to the material properties of a sample [11]. Nanoindentation allows for the investigation of the mechanical properties of a material at a submicron scale [11]. This capability is intriguing when applied to composite materials as the properties of the matrix and the fiber can be evaluated separately, in addition to the bulk properties of the material. This method relies on estimating the contact area of the indenter under load, which can lead to difficulties when applied to polymers. These difficulties are a result of the load resolution of the nanoindenter which is often too coarse for use with thermoplastics. Softer materials, like thermoplastics deform more than traditional materials in the loading range typically used in this method, and this leads to poor correlation in the curve fitting methods used with this technique [12].

Another method that experiences limitations when applied to polymers is the use of strain gauges. When working with thin samples of composite materials, the material being evaluated is often more compliant than the strain gauge being used. This results in a phenomena known as the "reinforcement effect" [13]. As a result of the reinforcement effect, strain gauges measure lower strains compared with the strains experienced locally by the test sample which can lead to significant errors in determination of strain and elastic modulus [13].

To mitigate the inherent issues involved with evaluating the material properties of polymers, a non-contact evaluation method was employed. This method is digital image correlation (DIC) [14]. DIC works by applying a speckle pattern to a sample and comparing digital photographs of that sample at different stages of deformation. By tracking blocks of pixels, the system can measure surface displacement and build up full field 2D and 3D deformation vector fields and strain maps. In order to achieve effective correlation, the pixel blocks utilized need to be random and unique with a good range of contrast and intensity levels [14]. In this work, a pattern was applied to samples using spray paint, with the nozzle at a distance from the sample that allows for a sparse amount of coverage. This method was used to evaluate the full field stress and strain exhibited in ESP 105 CC. Initial DIC analysis was performed using GOM Correlate 2-D analysis software (Braunschweig, Germany). Early testing has produced values of extension, stress, and strain that correlate well with values calculated through traditional means. This correlation allows for material properties to be extrapolated from this testing with some certainty, aiding in characterizing mechanical properties, like Poisson's ratio, that are difficult to measure using the methods outlined above.

2.3 Optical Characterization of Composite Material

The orientation of fibers in a composite material directly affects the mechanical properties of the material. This is especially true when working with discontinuous fiber composites. It can be difficult to manufacture TPMCs with a consistent fiber orientation. Practical difficulties arise in using new or existing processes to align fibers with a high degree of precision because of the large number of factors affecting fiber orientation during processing [9]. In the polypropylene matrix composite studied in this work, chopped glass fibers in the material were randomly oriented through the extrusion screw, and some orientation of the fibers was thought to be achieved through the rolling of the material into sheets. No further details of the anisotropic properties of the material were provided by the manufacturer. Taking this into consideration, the fiber orientation of the as-received ESP 105 CC sheeting, as well as the effect of the USSW process on the fiber orientation was characterized and compared to the results of mechanical characterization.

The use of micro-computed tomography (μCT) as a fiber orientation characterization technique has been investigated previously by Alemdar et al. [15]. It was found that the μCT could successfully characterize fiber diameter, length and orientation in TPMC. In this work, a Xradia MicroXCT-400 was used for all μCT characterization. In order to establish a baseline fiber orientation, the fiber orientation of the as-received ESP 105 CC was qualitatively characterized. A μCT image of the as-received TPMC is shown in Fig. 2.5. From this image, it was apparent that there was a strong fiber orientation along

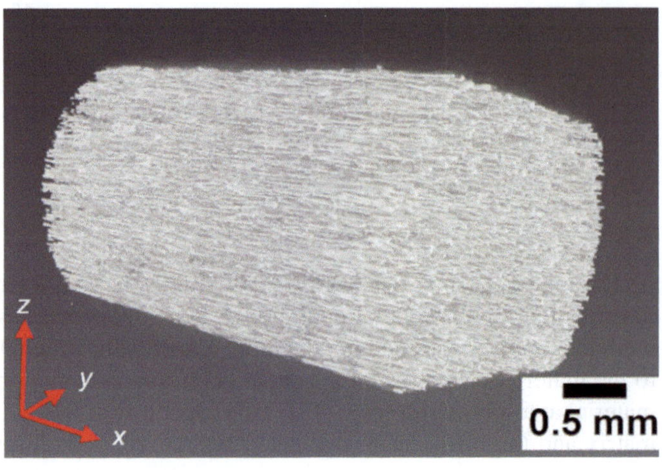

Fig. 2.5 ESP 105 CC as-received μCT image

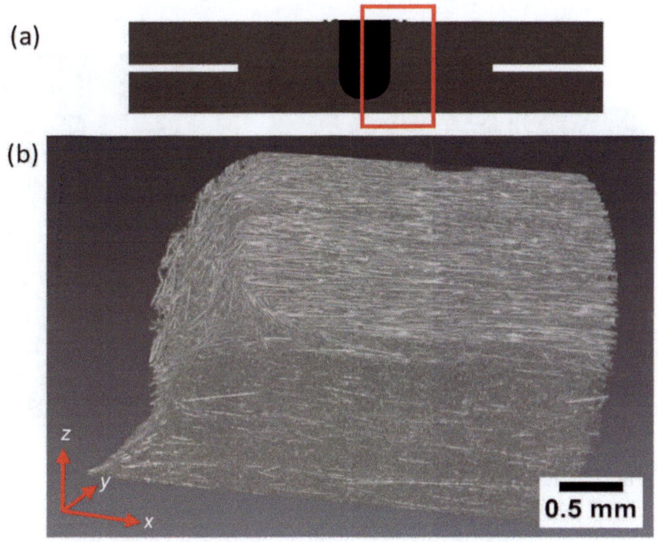

Fig. 2.6 Schematic of USSW cross section (**a**) with μCT field of view outlined in *red*, and USSW ESP 105 CC μCT image (**b**)

the x-axis of the sample, indicating the mechanical properties of the ESP 105 CC were roughly orthotropic. The orthotropic nature of the material under axial loading was observed in the study performed during mechanical characterization of the polypropylene composite and correlates well to these μCT findings (Fig. 2.4). From these scans, fiber diameter was approximately 13–15 microns and fiber length was approximately 270 microns.

After establishing a baseline of the fiber orientation for the as-received TPMC, the effect of the USSW process on the orientation of the fibers was investigated. This study was performed on 25.4 mm × 25.4 mm × 1.52 mm thick TPMC samples welded using the parameters chosen during the weld parameter study (Fig. 2.2). Figure 2.6a shows a schematic of the cross-section of the joint of two sheets. The area under investigation in the μCT image of Fig. 2.6b is highlighted with the box in Fig. 2.6a. This area encompasses the region just to the right of the void left where the weld horn was inserted. Here, the orientation of the fibers of each sheet was roughly perpendicular to one another, with the top sheet in Fig. 2.6b having fiber orientation along the x-axis and the bottom sheet having fiber orientation along the y-axis. In the left region of Fig. 2.6b, which was directly in contact with the weld horn, fiber folding in the direction of travel of the horn is present. In addition, the flow of the weld has a distinct, cone-shaped region converging towards the right of the sample at the interface of the two sheets. In this region, there are two distinguishable fiber orientations relative to the flow. At the edge of the flow, the fibers are aligned with the weld flow. In the core of the weld flow, the fibers are oriented perpendicular to the direction of the weld flow.

This behavior draws parallels to behavior observed in the study of injection molding of TPMCs [16]. When injection molding short fiber TPMCs, orientation of fibers is caused by differences in velocity between the two ends of a fiber. When the melt flow is homogenous, no change in the orientation of fibers will take place. When the flow is not homogenous, the variation in velocity can be perpendicular or parallel to the flow direction depending on whether the flow is convergent or divergent [16]. This behavior may lead to the differing alignment of fibers at the edge of the weld flow region. It was also apparent that the fiber orientations seen in the two pieces of the welded sample were oriented roughly perpendicular to one another, and this should be taken into consideration when analyzing the changes in fiber orientation throughout the weld affected region of the sample. From this analysis, it can be seen qualitatively that the fiber orientation is more randomly oriented in the region affected by the weld than it is in the rest of the sample and this will behavior will need to be further quantified.

The fiber orientation of the ESP 105 CC composite material was further characterized using scanning electron microscopy (SEM). As-received samples were polished along the short axis of the sheet in order to view the apparent fiber orientation throughout the cross section of the sheet. A SEM image of the cross section of as-received ESP 105 CC is shown in Fig. 2.7b. In this cross section, three distinct regions exist in which the fiber orientation is grouped.

In Fig. 2.7c, the fibers generally have an orientation in line with the length of the sheet, which is in agreement with the orientation observed through μCT. There is a region in the middle of the cross section, Fig. 2.7a, which exhibits a much more random fiber orientation than observed in the outer layers, Fig. 2.7c. It is thought that these zones are a product of the rolling manufacturing method. In terms of evaluating the mechanical response of this material under loading, the interaction between these three zones will need to be characterized for the constitutive behavior of the polypropylene to be modeled effectively throughout the entire cross section.

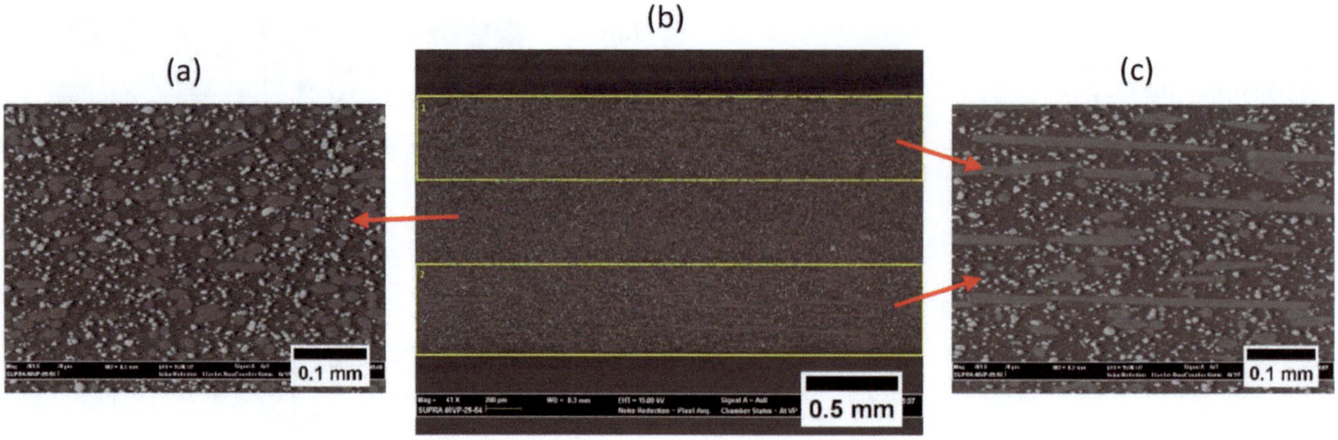

Fig. 2.7 ESP 105 CC cross section imaged by SEM

2.4 Conclusions

The scope of this work was centered around characterizing the mechanical properties of ESP 105 CC before and after USSW, with the end goal of obtaining the input parameters necessary for a computational model. To achieve this, a combination of mechanical and optical characterization techniques were employed.

An initial study was performed to obtain suitable welding parameters for use with this material. This was initially done with pure polypropylene sheeting, and parameters found for the pure polypropylene sheeting were also used for the ESP 105 CC composite material. It was found that the ESP 105 CC samples exhibited smaller bonded shear areas and carried less maximum load when tested in lap shear as compared to the pure polypropylene sheeting. The results indicated that the parameters found to create suitable welds in the pure polypropylene did not create ideal welds in the ESP 105 CC composite and further testing will be needed to find parameters suitable for use with the TPMC. To further the understanding of the directional behavior of ESP 105 CC, samples were cut both parallel and perpendicular to the fiber orientation and pulled in tension. From this, a sizeable difference in maximum strength was found between the samples tested at each respective orientation.

To further characterize the mechanical behavior of ESP 105 CC, several techniques were employed with varying degrees of success. The first technique, nanoindentation, was found to behave poorly when applied to polymers due to limitations in the instrumentation as well as the curve fitting method employed. Strain gauges were also employed, but found to require calibration for use with polymers as well as introduce edge effects around the gauge. To mitigate the issues inherent with these methods, DIC was employed. Initial results from this method have shown good correlation to results of displacement, stress, and strain found through traditional means.

Optical techniques were employed achieve a better understanding of the fiber orientation within the ESP 105 CC before and after USSW. The first technique employed for this purpose was x-ray μCT. Through this method, images of as-received ESP 105 CC and USSW ESP 105 CC were obtained. μCT images showed the as-received material to have a strongly directional fiber orientation. The USSW samples showed folding of fibers from the travel of the weld horn, as well as orientation of fibers relative to the weld flow, a behavior similar to that seen in injection molding. SEM was also employed to investigate the distribution of fiber orientation throughout the cross section of ESP 105 CC. This analysis showed that there are distinct regions of the cross section in which fibers are oriented by the rolling process, with a zone in the center of the cross section where fibers have maintained a more random orientation. The optical characterization performed has been largely qualitative in nature and further quantitative characterization will be necessary.

2.5 Future Work

The goal of this work is to develop a computational framework representative of the constitutive behavior of USSW thermoplastic matrix composites. To further characterize the material properties of the ESP 105 CC composite, the fiber orientation will be quantitatively analyzed using image analysis techniques in conjunction with μCT, and the temperature dependent properties of the material will be investigated using both infrared (IR) imaging and dynamic mechanical analysis (DMA).

The outcomes of these experimental efforts are to be incorporated into the definition of a constitutive model of the TPMC materials before, during, and after the USSW process. Principles of continuum mechanics will be subsequently invoked to address the coupling between the mechanics and thermodynamics of the USSW process. These concern the energy equation, conservation of linear momentum, and the 1st and 2nd laws of thermodynamics; leading to a system of time dependent partial differential equations (PDEs) governing the response features of temperature and displacement during the welding.

The physics of polymer flow must be resolved due to melting and subsequent flow of the matrix material during the USSW process. Since the corresponding velocities are small, Lagrangian descriptions can be used toward this purpose. To establish stable computer simulations, a new finite element analysis (FEA) methodology, the k-Version of FEA by Surana et al. [17, 18], is to be applied. It provides numerical and computational processes that are unconditionally stable and do not degenerate, even in the presence of convection as is encountered here due to the occurrence of flow of the melted polymer. The k-Version FEA is based on principles of variational calculus and employs a control/approximation parameter k in addition to the usual h (element size) and p (degree of polynomial approximation), which defines the global order of differentiability of the numerical approximation. Instead of classical piecewise continuous numerical approximations, the numerical prediction can have continuous derivatives up to order k. The k-Version of the FEA therefore provides the flexibility to obtain numerically accurate and stable simulations with continuous stresses, strains, heat fluxes, as well as stress- and strain-rates.

By developing a k-Version FEA framework of USSW the computational framework will allow the virtual simulation of the welding as well as the prediction of the structural strength and constitutive behavior of the resulting joints after welding has been completed.

Acknowledgements This work is funded by the South Dakota Board of Regents through a Competitive Research Grant (CRG) award. The authors would also like to thank Dr. William Cross and Dr. Edward Duke for their assistance in the optical characterization process, the Department of Mechanical Engineering at the South Dakota School of Mines and Technology (SDSM&T), the SDSM&T machine shop, especially Mr. Adam Kuenkel and Mr. Aaron Lalley, and our colleagues at the Additive Manufacturing Lab (AML) for their continuing guidance and assistance.

References

1. Mazumdar, S.: What will drive composites growth in 2015? Composites manufacturing magazine, 01 January 2015
2. Ageorges, C., Ye, L.: Fusion Bonding of Polymer Composites. Springer Science & Business Media, London (2002)
3. Friedrich, K., Almajid, A.A.: Manufacturing aspects of advanced polymer composites for automotive applications. Appl. Compos. Mater. **20**(2), 107–128 (2013)
4. Ultrasonic welding of injection moulded components – Part 1 – Process and equipment – Job Knowledge 61. [Online]. Available: http://www.twi-global.com/technical-knowledge/job-knowledge/ultrasonic-welding-of-injection-moulded-components-part-1-process-and-equipment-061/. Accessed 26 Feb 2017
5. Matthews, F.L., Rawlings, R.D.: Composite Materials: Engineering and Science. Elsevier, New York (1999)
6. Degen, C.M., Kjerengtroen, L., Valseth, E., Newkirk, J.R.: Impact and lap shear properties of ultrasonically spot welded composite lap joints. In: Cloud, G.L., Patterson, E., Backman, D. (eds.) Joining Technologies for Composites and Dissimilar Materials, Volume 10: Proceedings of the 2016 Annual Conference on Experimental and Applied Mechanics, pp. 59–69. Springer International Publishing, Cham (2017)
7. RTP 105 CC Polypropylene (PP) product data sheet – RTP Company. [Online]. Available: http://web.rtpcompany.com/info/data/0100/RTP105CC.htm. Accessed 22 Feb 2017
8. A. Standard. D3164–03
9. Papthanasiou, T., Guell, D.C.: Flow-Induced Alignment in Composite Materials. Elsevier, New York (1997)
10. A. Standard, D638-08. Standard test method for tensile properties of plastics (2008)
11. Oliver, W.C., Pharr, G.M.: An improved technique for determining hardness and elastic modulus using load and displacement sensing indentation experiments. J. Mater. Res. **7**(6), 1564–1583 (1992)
12. VanLandingham, M.R., Villarrubia, J.S., Guthrie, W.F., Meyers, G.F.: Nanoindentation of polymers: an overview. Macromol. Symp. **167**, 15–44 (2001)
13. Zike, S., Mikkelsen, L.P.: Correction of gauge factor for strain gauges used in polymer composite testing. Exp. Mech. **54**(3), 393–403 (2014)
14. McCormick, N., Lord, J.: Digital image correlation. Mater. Today. **13**(12), 52–54 (2010)
15. Alemdar, A., Zhang, H., Sain, M., Cescutti, G., Müssig, J.: Determination of fiber size distributions of injection moulded polypropylene/natural fibers using X-ray microtomography. Adv. Eng. Mater. **10**(1–2), 126–130 (2008)
16. Horst, J.J., others: Incluence of Fibre Orientation on Fatigue of Short Glassfibre Reinforced Polyamide. Delft University Press, Delft, Netherlands (1997)
17. Surana, K.S., Reddy, J.N., Romkes, A.: h,p,k Mathematical and computational finite element framework for boundary value and initial value problems. Acta Mech. Solida Sin. **23**, 12–25 (2010)
18. Surana, K.S., Ma, Y.T., Reddy, J.N., Romkes, A.: Development of mathematical models and mathematical, computational framework for multi-physics interaction processes. Mech. Adv. Mater. Struct. **17**, 488–508 (2010)

Chapter 3
Experimental Study of Laser Cutting Process of Titanium Aluminium (Ti-Al) Based Composites Designed Through Combined Method of Powder Metallurgy and Thixoforming

S. Ezeddini, G. Zambelis, E. Bayraktar, I. Miskioglu, and D. Katundi

Abstract Laser beam machining (LBM) is widely used as thermal energy based non-contact type advance machining process which can be applied for almost whole range of materials. It is suitable for geometrically complex profile cutting in the metals used in manufacturing engineering.

In the present work, high-power laser cutting process of Titanium Aluminium (Ti-Al) based intermetallic composites designed through combined method of powder metallurgy and thixoforming were carried out. The thermal effects of laser cutting and effects of main operating parameters such as laser power, and cutting speed on the cutting edge and on the cutting surfaces were examined. The evolution of the microhardness underneath the cutting surface due to laser power is also examined. The composite used in this study was produced through combined method of powder metallurgical (P/M) and thixoforming. Microstructure of cutting edge and cutting surfaces are also investigated by scanning electron microscopy (SEM). Cutting surfaces have been analyzed with 3D optical surface roughness-meter (3D–SurfaScan).

Roughness evaluations of the cutting surface depending on the cutting speed and cutting power were taken as optimization criteria that have been carried out by Taguchi method. A simple and useful tool was proposed for using in real manufacturing environment. Results exposed that good quality cuts can be produced in this Ti-Al based composite, at a window of laser cutting speed variable between 0.35 and 0.6 m/min and at a minimum heat input variable between 1400 and 2000 W.

Keywords Intermetallic composite • Recycled materials • Laser cutting • Taguchi method • SEM

3.1 Introduction

Titanium-aluminide alloys based on Ti-Al have an excellent potential to become one of the most important materials in aeronautic/aerospace applications in aircraft engines, industrial gas turbines and also automotive industry because of their low density, high melting temperature, high resistance at elevated-temperature, etc. They have excellent resistance to oxidation and hydrogen absorption, and excellent creep properties. However they show short ductility (very often brittle) at low to middle temperatures that give low fracture toughness and a fast fatigue-crack growth rate. Last 10 years, much effort has been spent to improve these ductile properties [1–9]. These research activities have found some success through physical-chemistry modification and microstructure control and at the same time improved mechanical properties essentially high creep resistance and other good wear properties [8–14].

Development and processing of high-temperature materials is the key to technological advancements in engineering areas where materials have to meet extreme requirements. Examples for such areas are the aerospace and spacecraft industry or the automotive industry. New structural materials have to be "stronger, stiffer, hotter, and lighter" to withstand the extremely demanding conditions in the next generation of aircraft engines, space vehicles, and automotive engines. For this reason, intermetallic γ-TiAl-based alloys show a great potential to fulfill these demands.

Recently, in the literature, an interesting work has been carried out on the optimization of multiple quality characteristics in laser cutting of titanium alloy sheets [4]. In this work, an important control factors related to the quality characteristics for

S. Ezeddini (✉) • E. Bayraktar (✉) • D. Katundi
Supmeca-Paris, School of Mechanical and Manufacturing Engineering, Saint-Ouen, France
e-mail: sonia.ezeddini@supmeca.fr; bayraktar@supmeca.fr

G. Zambelis
Airbus-Helicopter, Composite Design Department, La Courneuve-, Paris, France

I. Miskioglu
Michigan Technological University ME-EM Department, Houghton, MI, USA

P.R. Thakre et al. (eds.), *Mechanics of Composite and Multi-functional Materials, Volume 6*,
Conference Proceedings of the Society for Experimental Mechanics Series, DOI 10.1007/978-3-319-63408-1_3

surface roughness. Another optimization study has been carried out for the laser cutting process, incorporating the laser parameters with multi-performance characteristics [11]. It was shown in this study that the laser power had more effect on the cutting quality than that of the cutting speed. In our knowledge, there is a gap in the literature on the laser cutting process of $\alpha_2 + \gamma$ TiAl-based composites.

In this work, a new processing route through combined method of powder metallurgical (P/M) and thixoforming will be explained and laser cutting conditions will be discussed in point of view manufacturing feasibility of these alloys for different industrial applications. For this aim, microstructures of cutting edge and cutting surfaces have been evaluated by scanning electron microscopy (SEM). Cutting surfaces have been analyzed with 3D optical surface roughness-meter (3D–SurfaScan). This process is a commercially-viable route for the manufacturing of aeronautical components and optimizing mechanical properties.

3.2 Experimental Conditions

3.2.1 *Manufacturing of Intermetallic Composite: Composition of the Specimens Designed in this Work Was Arranged as Follows*

Ti: 53, Al: 27, Nb_2Al: 4, Nb15, Zr doped yttrium: 1, Mo: 0.1, B: 0.15, Zn-St: 1 (wt %). Processing route has been carried out through combined method of powder metallurgical (P/M) and thixoforming. Firstly, all of the element of this composition have been mixed and milled in a pulverise during 4 h by using a protection and successfully ball milled (4000 rpm) during 1 h and compacted by CIP (Cold Isostatic Pressure) at a pressure of 250 MPa and after that sinter thixoforming (very similar to sinter forging) after heating at the temperature of 1100 °C for 45 min – 1 h under He + Argon inert atmosphere, then it was cooled in the oven. After manufacturing of the composite, microstructural evaluation has been carried out and certain basic mechanical parameters (static bending, microhardness, etc.) have been given here as an indicative results of this composite.

3.2.2 *Wearing-Scratch Tests*

This type of damage analysis has been carried out by using Scratch Tester Device with a zirconium ball indenter. Test conditions were set as 1.9 N normal load with the frequency of 15 Hz and the numbers of cycles were set at 50000 and 100,000 cycles respectively. Tangential and normal forces were controlled by software developed in our laboratory (Supmeca-Paris) during the test their ratio was used to calculate the coefficient of friction. After scratch test, damaged zone were investigated by a 3D optical roughness meter (Surf-Scan) to characterize the damage that occurred during the scratch test. Volume loss/time, maximum depth, and friction coefficient were evaluated.

3.2.3 *Laser Cutting Process and Cutting Parameters*

All of the cutting processes have been carried out on a laser machine called “2000 RAPIDO-PRIMA INDUSTRY” Basic parameters of laser cutting machine has been given in the Table 3.1. In the present work, three levels of power (Pw) and three levels of cutting speed (S), have been chosen in order to examine their impact on the basic performance parameters. As the basic cutting process parameters, the control factors and their designated symbols and also four different levels have been shown respectively in the Table 3.2. After achievement of the cutting process, the widths of resulting cuts were measured

Table 3.1 Basic properties of laser cutting machine

Laser	CO_2
Marque	Rapido (prima industry)
IR	10,6 μm
Mode decape	TEM = Traverse electric magnetic
2000 W continue	
Work piece	Ti-al intermetallic composite.
Gas pressure	12.1 bar

Table 3.2 Laser cutting process parameters and their levels

			Levels			
Factor	Parameters	Unit	Level 1	Level 2	Level 3	Level 4
A	Power (pw)	Watts	1400	1600	1800	2000
B	Cutting speed (S)	m/min	0,35	0,4	0,5	0,6

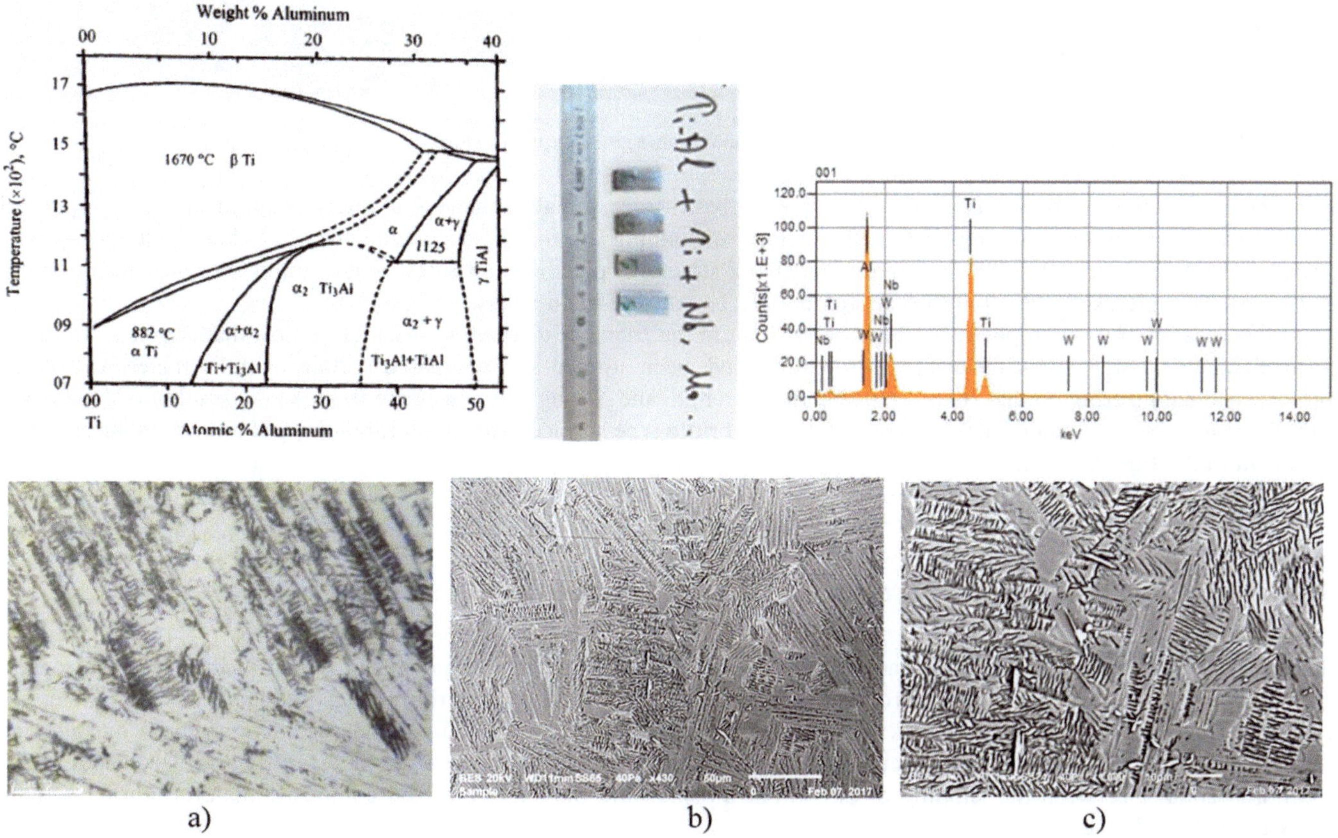

Fig. 3.1 Microstructural evaluation of Ti-Al intermetallic composite with binary phase diagram, at the surface (**a**) and thickness (transversal direction) (**b**) and (**c**)

and SEM micro-photography of the cutting surfaces was carried out. To measure the striation depths and widths, the surface roughness measurement was carried out using a 3D optical roughness meter (Surf-Scan).

The experimental matrix with variables was employed to develop 18 experimental tests, for each experimental test, one sample were used under time regulated condition. Taguchi method and L18 Orthogonal Array were used to reduce number of the experience. The design of experiments (DOE) and measured the average roughness Ra.

3.3 Results and Discussions

3.3.1 Microstructural Evaluation of the Intermetallic Composite

Optical microstructure was evaluated on the surface and thickness of the specimens as shown in Fig. 3.1 with binary diagram, respectively. Nearly, developed lamellar plate-structure was observed with a lamellar spacing of 3–5 μm. practically not so porosity was seen in the structure. According to our observation of the composition and binary phase diagram, typical microstructure of this composite was evaluated as a mixed "γ-TiAl and α_2-Ti_3Al" phases. Furthermore, any

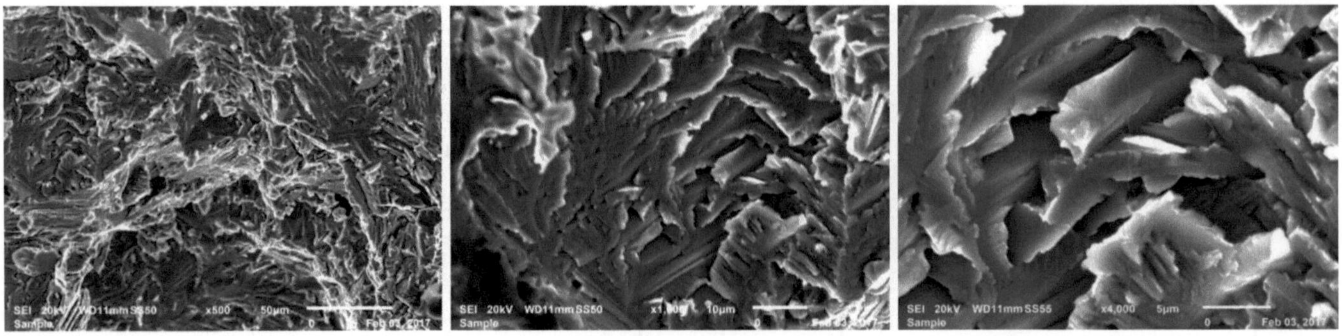

Fig. 3.2 Typical static bending fracture of the intermetallic composite that give a brittle fracture

grain boundary precipitation was observed in the microstructure. Finally, the major microstructure is nearly equiaxed lamellar plate-structure consisting of lamellar colony containing alternating γ ordered fcc (111)-based tetragonal (L_{10} structure) and very fine α2—(0001) ordered hcp ($D0_{19}$ structure) platelets and a few small grains of γ structure (around 8–12 μm). The microstructure comprise approximately 10 vol% of the α_2 phase.

In fact, design of this composite has been carried out in the frame of a common research project that is going on. In the former annual report, certain mechanical properties have been studied and compared certain metallic materials used in aeronautical engineering. In that study, static bending strength and Young Modulus were found at the level of 625 MPa and 145GPa respectively. Fracture surface show a typically brittle type surface; mixed intergranular and transgranular cleavage as shown in the Fig. 3.2 [10].

3.3.2 Evaluation of Surface Wearing by Scratch Test Results

Scratch tests results give useful information on the wear resistance of the composites designed in the current research. Tangential and normal forces were controlled by software – SUPMECA-Paris during the test that gives also coefficient of friction. After the test, damaged zone was investigated by 3D optical roughness meter to measure, scratch depth, and average scratch roughness. All of the data discussed here are controlled with special scratch test software-SUPMECA-Paris. Frictional contact between the slipping of zirconium indenter and the surface of the composite material during scratch test was evaluated here.

In fact, because of the high shear stress at the interfaces the interfacial shear stress should probably be the main reason for damage of the matrix and reinforced filler interfaces [10]. When the indenter is slipping, tangential tensile stress is caused on the surface behind the indenter, while in front of the indenter the tangential stress is compressive. In the frame of this present work wear resistance was evaluated only under two test conditions (50,000 and 100,000 number of cycles for the composite presented here. Figures 3.3 and 3.4 indicate scratch damaged zone and characteristic parameters obtained by 3D optical roughness meter after the test conditions carried up to 50,000 and 100,000 number of cycles respectively. Table 3.3 recaps damage results obtained by scratch test with two different number of cycle.

3.3.3 Evaluation of the Results of Laser Cutting Process

Laser gas assisted cutting process of Ti-Al intermetallic composite pieces has been carried out. The effect of laser output power and cutting speed on the surface striations and surface roughness evolution of the cutting edge surfaces were analyzed. Profiles of the laser cutting surfaces were also evaluated depending on the process parameters (power and cutting speed, etc.).

Figure 3.5a, b show surface roughness evolution of the workpieces for two laser output power settings (1400 and 2000 Watts) at a constant cutting speed (0.4 m/min) respectively. In the same way, Fig. 3.6a, b show 3D surface topographies and SEM micrographs for the cutting edge surfaces of the at two different power settings, 1400 and 2000 Watts at a constant cutting speed (0.4 m/min) respectively. The cutting edge surfaces seem similar with little differences the width of striations with flow of liquid metal.

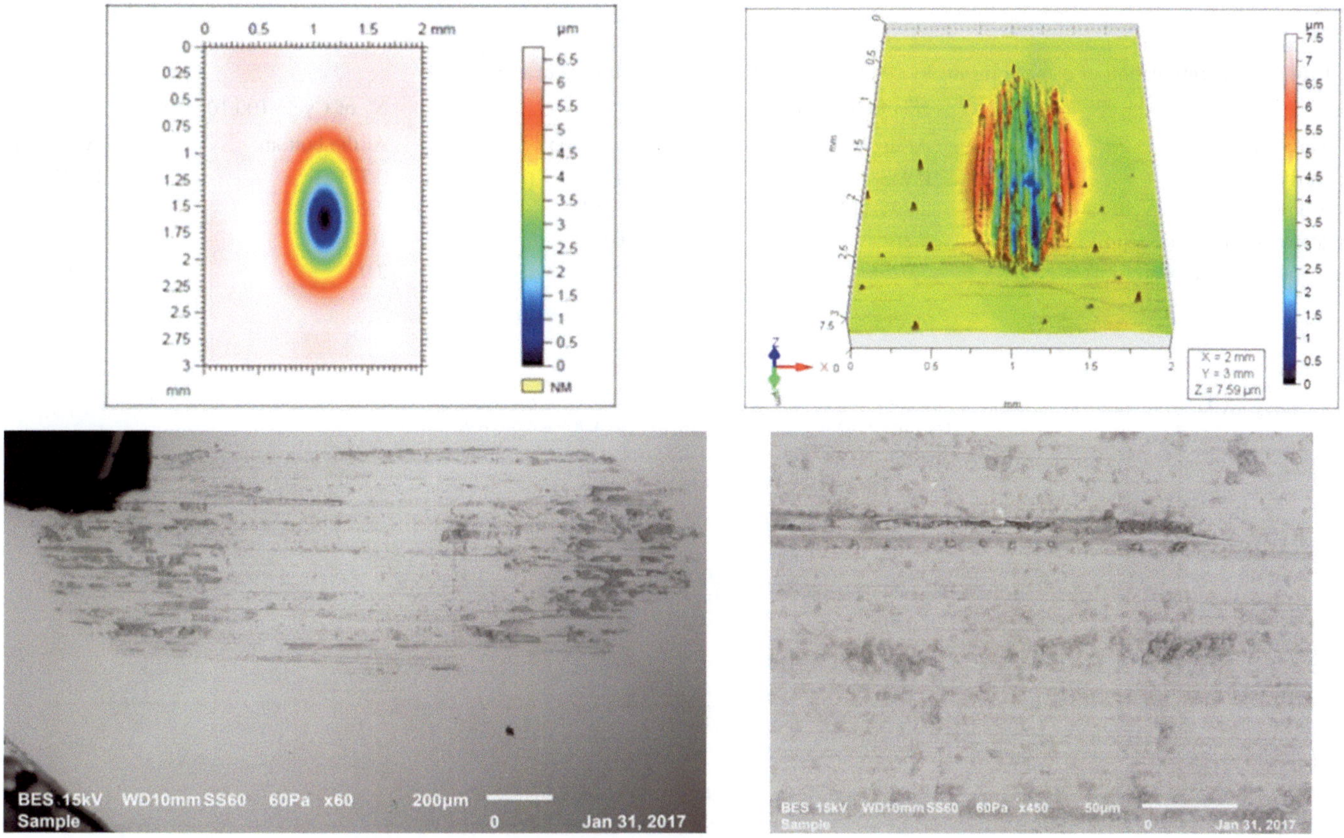

Fig. 3.3 Damage traces obtained in the direction of width and length by scratch test after 50,000 cycles and maximum depth profile (direction X-axis) obtained by scratch test determined with SEM microscopy

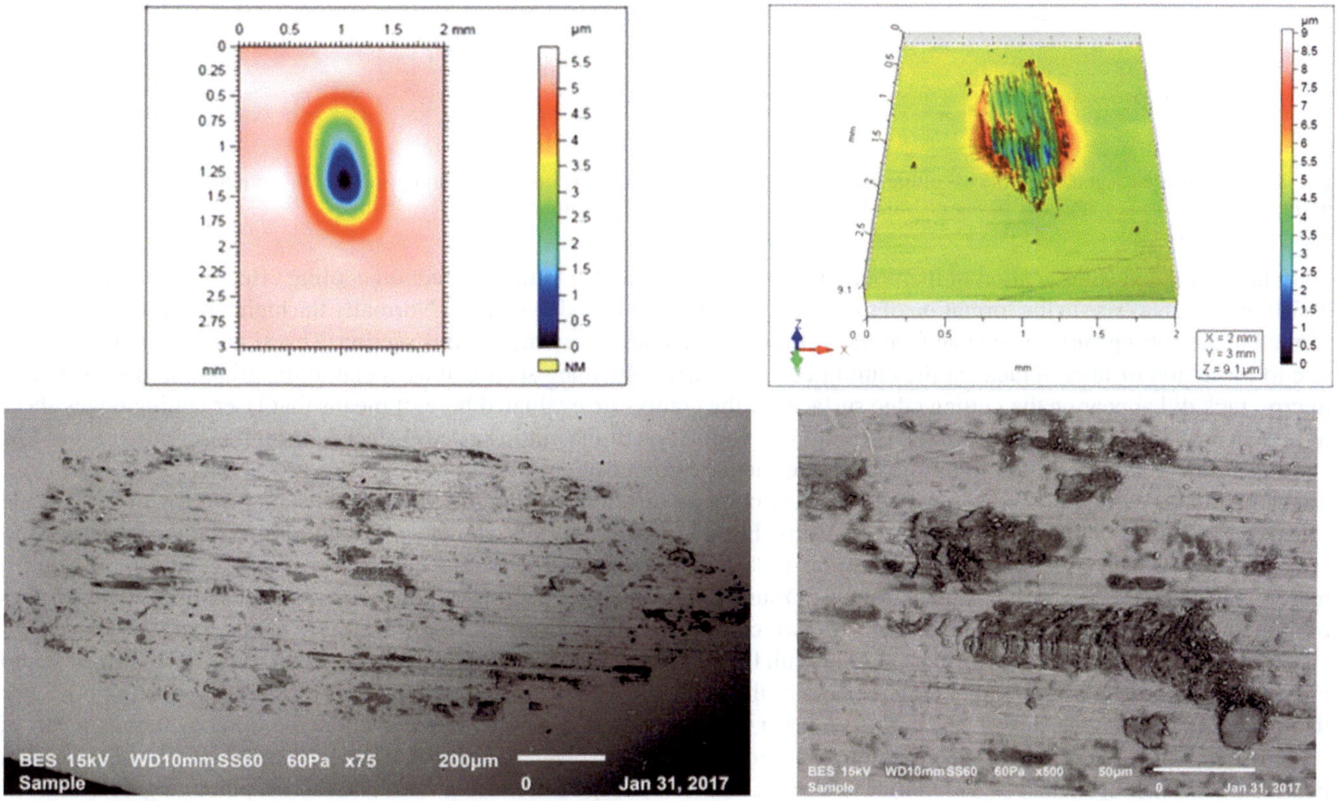

Fig. 3.4 Damage traces obtained in the direction of width and length by scratch test after 100,000 cycles, and maximum depth profile (direction X-axis) obtained by scratch test determined with SEM microscopy

Table 3.3 Identification of damage results by scratch test for two different number of cycle

	N° of Cycle 50.10^3		N° of Cycle 100.10^3	
Parameters	At the bottom	At the top	At the bottom	At the top
Surface (mm^2)	0.953	0.175	0.771	0.147
Volume (μm^3)	2,877,112	51,881	2,248,293	43,818
Depth.- height max. (μm)	9.86	2.04	10.7	2.66
Depth.- height mean (μm)	3.02	0.297	2.92	0.297

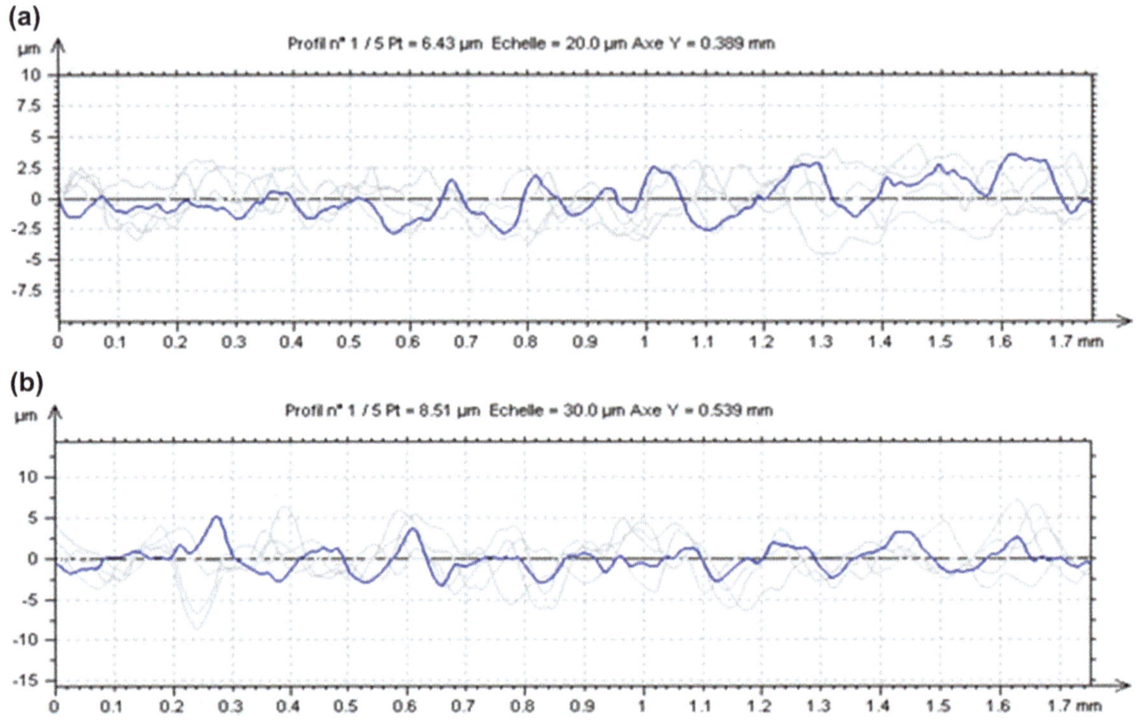

Fig. 3.5 (**a**) Surface roughness profile of cutting surface at a laser power of 1400 Watts with a cutting speed of 0.4 m/min. (**b**) Surface roughness profile of cutting surface at a laser power of 2000 Watts with a cutting speed of 0.4 m/min

In fact, the rapid solidification of the liquid metal at the surface, due to the convection cooling effect of the high pressure assisting gas, gives rise to the formation of the cast layer at the cutting edge surface. Normally the high temperature gradients formed in the cutting surface layer and naturally, thermal strain level stay high in this section. This situation very often gives rise to formation of micro-cracks at the cutting edge surfaces, which are seen mainly metallic materials. In this work, any micro-crack did appear on the cutting edge surface for the composite evaluated here. It means that laser cutting edges show in regular way increment of width size of striations – undulation at the cutting edge surfaces for high laser powers and low cutting speeds that were observed in this TiAl intermetallic composite.

Another laser cutting processes have been carried out with different test conditions (at a constant power with two cutting speeds). Figures 3.7 and 3.8 show these test results. In these figures, surface roughness evolutions of the workpieces were shown at a constant laser output power setting (2000 Watt) with two different cutting speeds (0.35 and 0.50 m/min) respectively. In the same way, Fig. 3.8a, b show 3D surface topographies and SEM micrographs for the cutting edge surfaces of at a constant power setting (2000 Watt) with two cutting speeds (0.35 and 0.50 m/min) respectively.

The cutting edge surfaces seem quasi similar with little differences the width of striations-undulations with flow of liquid metal. This difference is ascribed to the thickness of the melting layer, which should increase with the laser output power or the low cutting speeds. Therefore, the cooling rates at the surface region of the melted layer are modified by the main part of heated layer below the surface. In the literature, this zone is very often called as a heat affected zone (HAZ). For this reason, heat transfer from the melted layer towards to surface area can change the cooling rates and the temperature gradient in the surface vicinity of the melted layer reduces.

a

ISO25178	
Parameters for Height	
Sa	1.98 μm
Sq	2.46 μm
Space parameters	
Str	0.0746

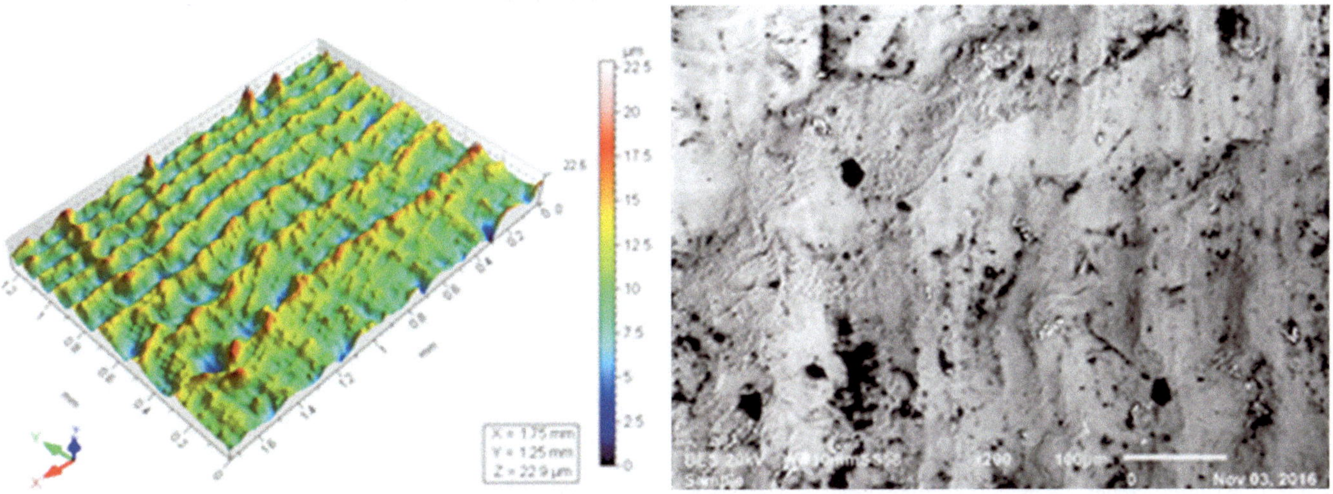

b

ISO25178	
Parameters for Height	
Sa	2.33 μm
Sq	3.00 μm
Space parameters	
Str	0.518

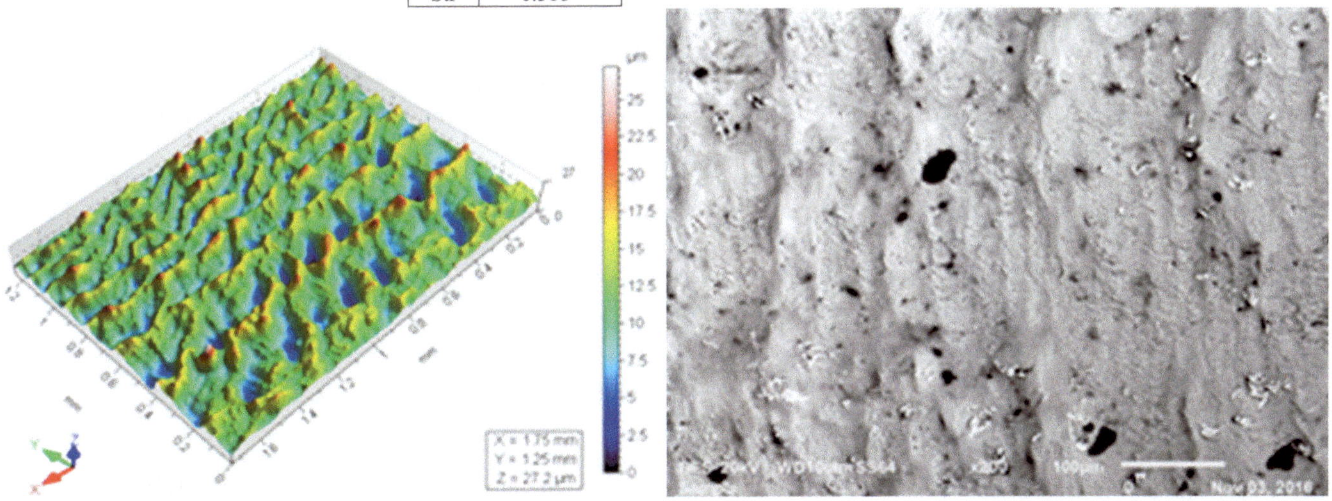

Fig. 3.6 (**a**) 3D surface topographies for workpieces profile of surface cutting edge at a power of 1400 Watts and with a speed of 0.4 m/min, Sa = 1,98 μm. (**b**) 3D surface topographies for workpieces profile of surface cutting edge at a power of 2000 Watts and with a speed of 0.4 m/min, Sa = 2.33 μm

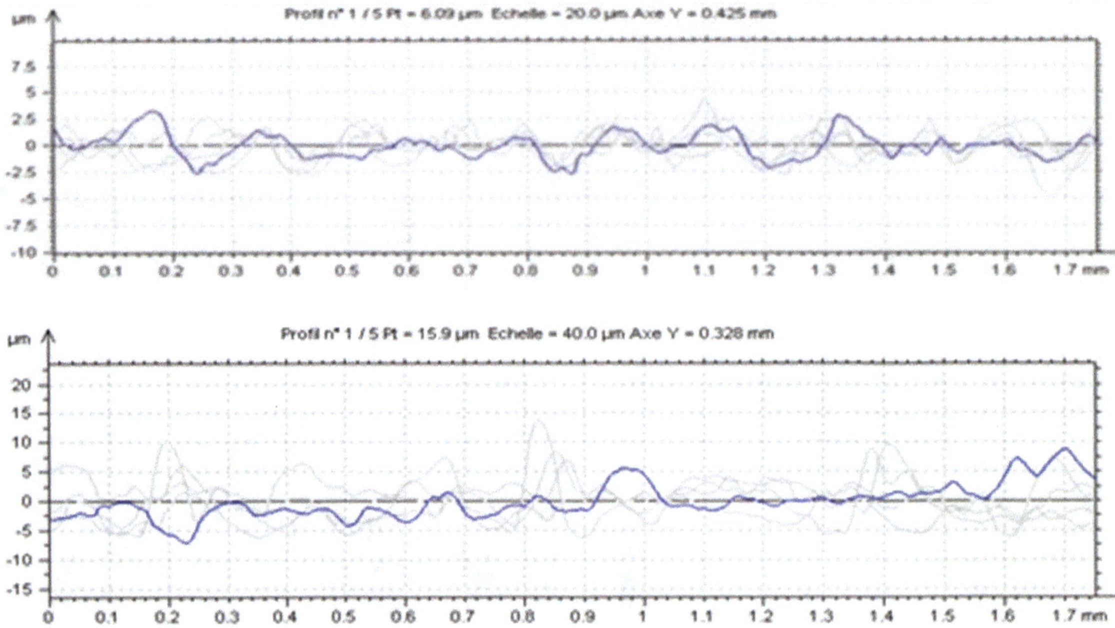

Fig. 3.7 (**a**) Surface roughness profile of cutting surface at a laser power of 2000 Watts with a cutting speed of 0.35 m/min. (**b**) Surface roughness profile of cutting surface at a laser power of 2000 Watts with a cutting speed of 0.50 m/min

In the work, TiAl intermetallic composite shows wide striation depth at the cutting edge surfaces. This is associated with the high oxidation potential of titanium, which results in sideways burning and deep molten layer formation at the kerf surface. It should be noted that high temperature exothermic reactions results in excessive energy generation contributing to the laser beam energy in the cutting section [9, 11, 14].

For evaluation of the microstructure after the laser cutting, a lot of amount microhardness measurements have been taken from cutting edge zone (melted zone) through heat affected zone up to base materials (centre of the specimen). These values can give significantly change of microstructure due to nonequilibrium heat treatment during the laser cutting process and can help the size of brittle zone.

Figure 3.9 summarizes evolution of microstructure with microhardness measurements that attached with a picture taken from microindentation device showing measurement method. As can be seen from these date explained in Fig. 3.9, all of the microhardness values show the highest value in the melted zone et the lowest value is found the value of the composite. These results are very similar each other for all of the cutting conditions.

3.3.4 Integrity of Surface After Laser Cutting

Optimization of laser cutting conditions was made on the two basic parameters power and cutting speed. The parameter used in this study to characterize surface roughness is the arithmetic mean heights of surface, (Sa), that is identified in the Eq. 3.1. Furthermore, the output process parameters were obtained using a 3D Scanning surface and treated by software program Mountains-Map.

$$S_a = \iint_a |Z\ (x, y)|\ dxdy \tag{3.1}$$

where "a" is the surface taken for observation and "z" the ordinate of the surface roughness.

Very simple optimization was made by this method and the first results are shown in Fig. 3.10. Evolution of the surface roughness values was identified as a function of two basic parameters, power and cutting speed. The first preliminary results are shown here and it should be noted that the effect of cutting speed on the surface roughness values much more effective regarding to power. These relations are given in the Eqs. 3.2 and 3.3.

a

ISO25178	
Parameters for Height	
Sa	1.09 µm
Sq	1.38 µm
Space parameters	
Str	0.527

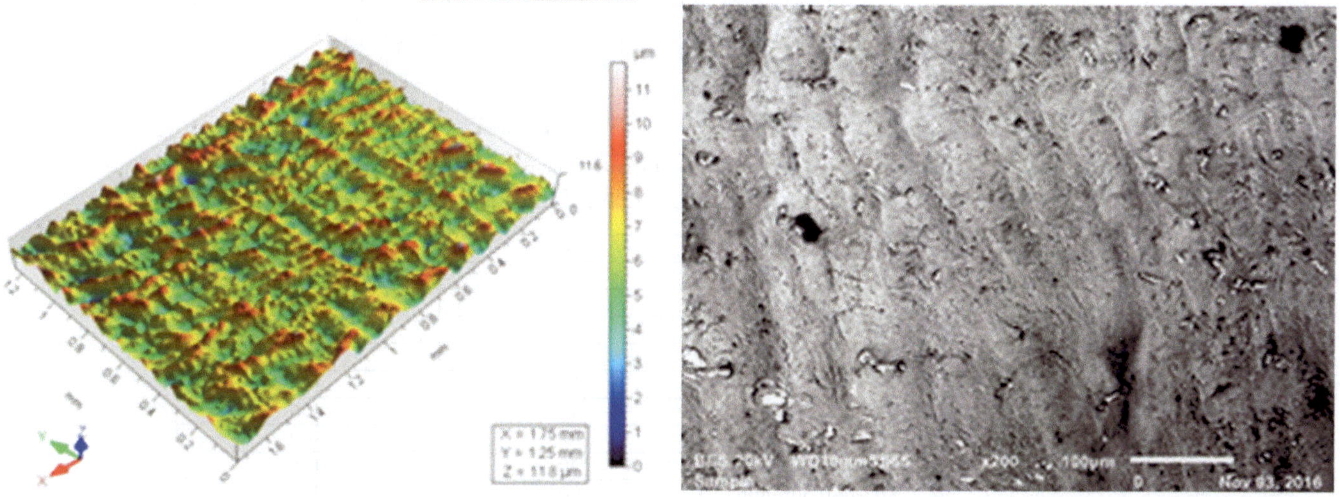

b

ISO25178	
Parameters for Height	
Sa	2.45 µm
Sq	3.15 µm
Space parameters	
Str	0.628

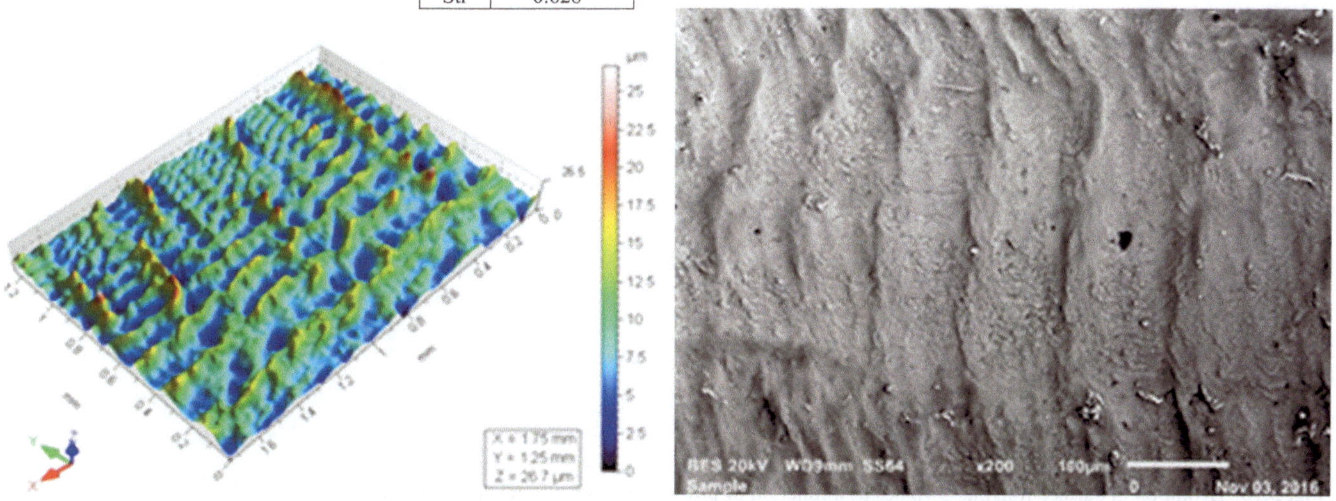

Fig. 3.8 (**a**) 3D surface topographies for workpieces profile of surface cutting edge at a power of 2000 Watts and with a speed of 0.35 m/min, Sa = 1.09 µm. (**b**) 3D surface topographies for workpieces profile of surface cutting edge at a power of 2000 Watts and with a speed of 0.5 m/min, Sa = 2.45 µm

$$\text{Ra}\ (\mu m) = -3.58 + 0.0068\ Power\ (W) - 0.000002\ Power\ (W)^2 \tag{3.2}$$

$$\text{Ra}\ (\mu m) = -3.01 + 13.94\ Speed\ (m/\text{min}) - 8.72\ Speed\ (m/\text{min})^2 \tag{3.3}$$

In the same way, an optimization of laser cutting process was made by software program and output parameters were obtained. These values were presented on the interaction plot for Ra values (Fig. 3.11) to see the effect of two basic parameters (power and cutting speed) at the same time on the graphic.

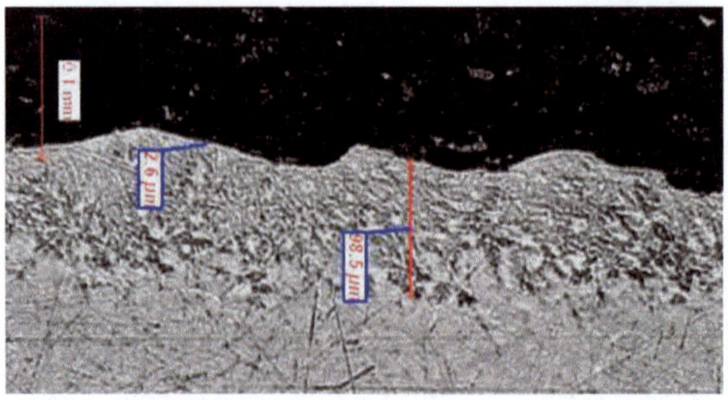

Laser Cutting conditions		Distance from cutting edge (µm)	Microhardness (HV)
Speed=0.4 m/min	Power= 1400W	20	470
		80	452
		300	353
	Power= 2000W	20	429
		80	418
		300	342
Power= 2000W	Speed=0.35 m/min	20	510
		80	476
		300	349
	Speed=0.5 m/min	20	446
		80	402
		300	355

Fig. 3.9 Evolution of microstructure with microhardness measurements that attached with a picture taken from microindentation device showing measurement method

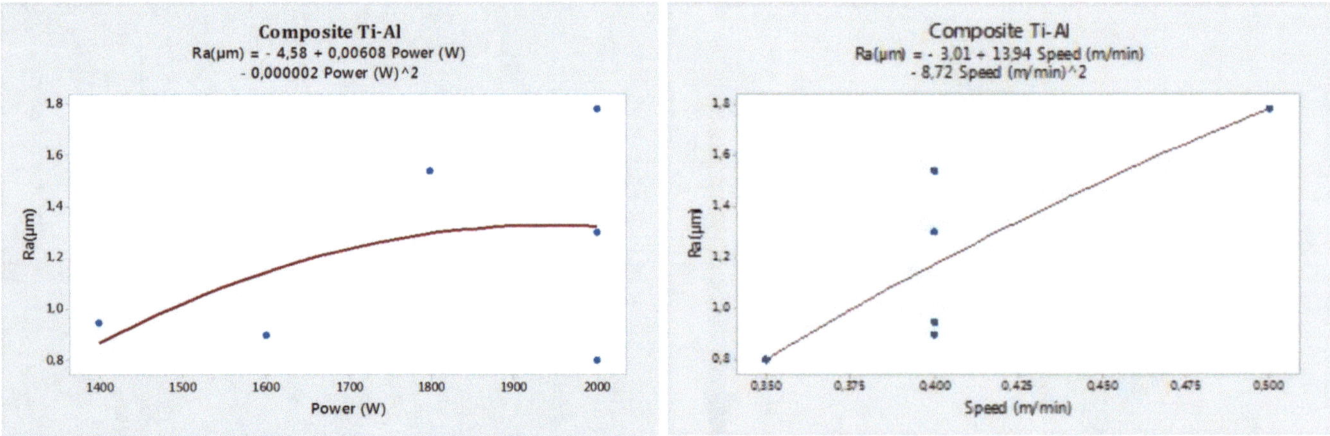

Fig. 3.10 Evolution of surface roughness values as a function of power (*left*) and cutting speed (*right*)

3.4 Conclusions

In this study, Titanium Aluminium (Ti-Al) based intermetallic composites were designed through combined method of powder metallurgy and thixoforming. This method is simple economic and safety for the intermetallic based composites. Laser cutting process has been carried out for showing the manufacturing facilities of these composites for industrial applications. It is a successful cutting process using certain parameters to obtain optimal cutting surface conditions depending on two basic parameters; power and cutting speed. They influence considerably surface roughness values, Ra. Laser cutting edges show in regular way increment of width size of striations – undulation at the cutting edge surfaces for high laser powers and cutting speeds. Micro-cracks were not observed at cutting surface under the experimental conditions that were used in this study. These values obtained here are preliminary results. Much more experimental tests are needed for a safety optimization multiplying laser cutting parameters with a reliable experimental design.

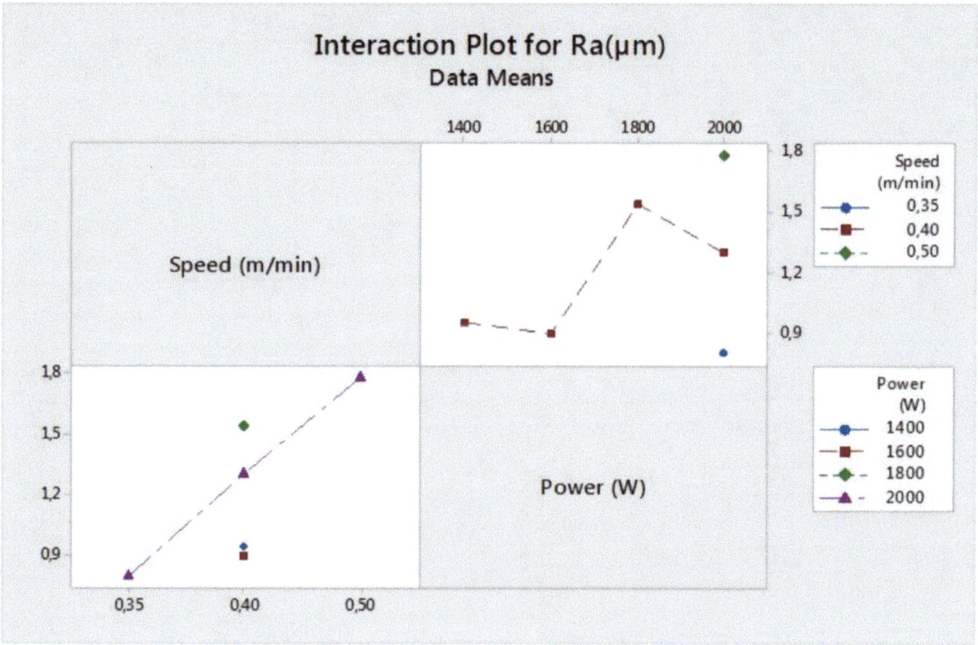

Fig. 3.11 Interaction effects of two parameters (power & speed) on the surface roughness values

References

1. Kim, Y.-W.: Intermetallic alloys based on gamma titanium aluminide. JOM. **31**(7), 24–30 (1989)
2. Stoloff, N.S., Liu, C.T., Deevi, S.C.: Emerging applications of intermetallics. Intermetallics. **8**, 1315–1320 (2000)
3. Krishnan, M., Natarajan, B., Vasudevan, V.K., Dimiduk, D.M.: Microstructure evolution in gamma titanium aluminides containing beta-phase stabilizers and boron addition. In: Proceeding of 2nd Symposium on Structural Intermetallics, Seven Springs, PA (USA), September 21–26, pp. 235–244, (1997)
4. Bayraktar, E., Bathias, C., Xue, H., Hao, T.: On the giga cycle fatigue behaviour of two-phase ($\alpha_2 + \gamma$) Ti-Al alloy. Int. J. Fatigue. **26**, 1263–1275 (2004)
5. Chen, Y., Chen, Y., Kong, F., Xiao, S.: Fabrication and processing of gamma titanium aluminides. Mater. Sci. Forum. **638-642**, 1281–1287 (2010)
6. Couret, A., Molenat, G., Galy, J., Thomas, M.: Microstructures and mechanical properties of TiAl alloys consolidated by spark plasma sintering. Intermetallics. **16 n9**, 1134–1141 (2008)
7. M. Thomas and M.-P. Bacos, Processing and characterization of TiAl-based alloys: Towards an industrial scale, J. Aerosp. Lab. 3, ALO3–5, 1–11 (2011)
8. Bhambri, Y., Sikka, V.K., Porter, W.D., Loria, E.A., Carneiro, T.: Effect of composition and cooling rate on the transformation of a to g phase in Ti-al alloys. Mater. Sci. Eng. A. **324**(1–2), 361–365 (2006)
9. Yilbas, B.S., Akhtar, S.S., Bayraktar, E., Gasem, Z.: Laser cutting of thin aluminum and silicon alloy: Influence of laser power on kerf width. Adv. Mater. Res. **345**, 442–447 (2012)
10. Alhas, H.A., Bayraktar, E.: New design of intermetallic composites. Annual research report. AMPT International Conference, Dubai, 17–20 November 2014
11. Yilbas, B.S., Akhtar, S.S., Karatas, C.: Laser hole cutting into Ti-6Al-4V alloy and thermal stress analysis. The Int. J. of Adv. Manuf. Technol. **59**(9), 997–1008 (2012)
12. Dubey, A.K., Yadava, V.: Multi-objective optimization of laser beam cutting process. Opt. Laser Technol. **30**(3), 562–570 (2008)
13. Wen, C.E., Yasue, K., Lin, J.G., Zhang, Y.G., Chen, C.Q.: The effect of lamellar spacing on the creep behavior of a fully lamellar TiAl alloy. Intermetallics. **8**, 525–529 (2000)
14. Melhem, O.A., Yilbas, B.S., Shuja, S.Z.: Jet impingement onto kerf: Effect of kerf wedge angle on heat transfer rates and skin friction. Opt. Laser Technol. **56**, 76–87 (2014)

Chapter 4
Mechanical Characterization of Epoxy: Scrap Rubber Based Composites Reinforced with Nanoparticles

A.B. Irez, I. Miskioglu, and E. Bayraktar

Abstract Epoxy rubber based structural composites (ERCs) are used in engineering applications mainly in the aeronautical area because they can meet the necessary requirements in new multifunctional systems. These composites exhibit good overall mechanical and thermal performance and they can potentially offer a large variety of functional properties. In order to realize an efficient engineering development process, the data for basic material parameters of these composites is essential. The present paper discusses the mechanical characterization of these composites. A combination of structural and energetic functions can be achieved by using different nanoparticle reinforcements in epoxy-rubber composites This type of material design gives an exigent task to the designers looking to integrate more functionality into the base material of their structure to achieve overall improved system performance. This paper is focused on the design of ERCs reinforced with nano powders in different ratios in a matrix of epoxy – fresh scrap rubber. It is expected this material would be attractive for industrial applications because of the readily available recycled materials that are utilized. The mechanical and some physical properties of these composite systems are studied in this research. Mechanical properties are evaluated by means of three-point bending tests, nano-indentation, drop weight tests and dynamic mechanical analysis. Scanning electron microscope is used to observe fracture surfaces and the microstructure.

Keywords Recycled composites • Epoxy-rubber • Creep-wear nanoindentation • Ceramic reinforcements

4.1 Introduction

Epoxy-rubber based structural composite materials are utilized more and more in the last decades thanks to their large spectrum of functional properties and performances. In order to realize an efficient engineering development process, the material property data needs to be obtained by different characterization steps and it is an exigence for material suppliers and engineers. The materials properties such as ease in manufacturing, flexibility and polymer-based dielectrics facilitate the usage of them in different industries [1]. To improve the fracture toughness of epoxies, adding a rubber phase such as commercially available styrene butadiene rubber with different functional terminal groups has been studied by different researchers [2]. The volume fraction of the toughening rubber phase generally varies from 5% to 20%. Nevertheless, the incorporation of rubber phase increases the viscosity of the epoxy resin mixture whereas the crosslink density, elasticity modulus and tensile strength diminish. Besides, the glass transition temperatures of epoxies dramatically reduce by the addition of rubber phase [3–4].

Another important point is that, the functionalities of these composites are increasing significantly in advanced electronics, electromagnetics and electric power systems such as actuators, AC motors, embedded capacitors, piezoelectric and high energy density pulsed-power devices [6]. By the way, microwave absorbing property also attracts certain attention for many researchers. In detail, there are certain parameters needed to be taken into consideration in manufacturing of microwave absorbing materials, such as the weight, thickness, types of fillers, filler content, environmental resistance and

A.B. Irez and I. Miskioglu contributed equally to this work.

A.B. Irez (✉) • E. Bayraktar
Supmeca-Paris, School of Mechanical and Manufacturing Engineering, Saint-Ouen, France
e-mail: irez.burak@gmail.com

I. Miskioglu (✉)
Michigan Technological University ME-EM Department, Houghton, MI, USA
e-mail: imiski@mtu.edu

P.R. Thakre et al. (eds.), *Mechanics of Composite and Multi-functional Materials, Volume 6*,
Conference Proceedings of the Society for Experimental Mechanics Series, DOI 10.1007/978-3-319-63408-1_4

mechanical strength [7]. In this respect, ferrites are considered to be the best magnetic material for electromagnetic wave absorbers thanks to their outstanding magnetic and dielectric properties. However, they are heavy and expensive.

In other respects, polymers are used to protect the electronic devices from electromagnetic interference (EMI) due to the flexibility, lightweight and cost effectiveness. Nevertheless, polymers are electrically insulating and transparent to electromagnetic wave. Therefore, ferrite materials are incorporated into polymer matrices to effectively eliminate EMI [8]. Many works have been done on polymer-based composites filled with magnetic materials in micrometer – size, such as Fe_3O_4/YIG [9]. However, conventional magnetic particles filled polymer-based composites have some drawbacks in corresponding the criterion in thin and lightweight microwave absorber because of high filler content [10].

Fe_3O_4 nanoparticle, was selected in this study mainly due to its genuine and novel physiochemical properties which can be attained according to their particle size (quantum size effect), morphology, shape and engineering form [11]. Also, it is found attractive to enhance the magnetic permeability of epoxy based composites, accordingly obtaining a good synchronization between the electrical and thermal conductivities [12]. Moreover, addition of Ni and Fe_3O_4 thermite powder mixtures to the matrix has been shown to have significantly affect the energetic of the reaction between the powders [13]. A balanced mixture of these characteristics can result in superior structural and energetic properties.

In this paper, preliminary results have been presented for the design procedure of an epoxy-scrap rubber matrix composite with the exothermic reinforcements mentioned above. During this experimental study, nano wear and creep behaviour were examined by means of nano-indentation technics. Compression tests were performed for mechanical characterization. In addition, surface hardness was determined by means of Shore D hardness measurement. Furthermore, impact behaviour was investigated by means of dynamic compression tests (drop weight testing). After all the tests and analyses, scanning electron microscopy (SEM) was used to observe the fracture surfaces and microstructure to study the distribution of different reinforcements and damage characteristics.

4.2 Experimental Conditions

4.2.1 Materials Processing

As well-known, epoxy resin and scrap rubber powder (already vulcanized) cannot make a chemical bonding. For this reason, scrap rubber powder should be treated with a chemical solution to activate surface of rubber powder and try to make a good mixture with epoxy resin powder. In the frame of the present work, recycling of rubber is aimed for the manufacturing of low cost composites for aeronautical and automotive applications. Scrap rubber powder was sent by sporting good company (sport shoes, etc.). At the first stage, a chemical-Silanization treatment was used to make a strong bonding very fine dry epoxy powder and fine scrap rubber powder (Styrene–butadiene rubber (SBR). Meanwhile, using of recycled rubber gives an economic perspective to this study. Epoxy and rubber were mixed by using blending and mortar mixture then dried in an oven to entirely eliminate humidity and also traces of chemicals due to treatment. This mixture was milled for an hour to obtain a homogeneous compound and then it was heated at 80 °C for 24 h. The resulting compound was used as the matrix for the proposed composites. In the second stage, the reinforcements Ni and Fe_3O_4 were added to the matrix in pre-defined ratios.

All of the compounds were then mixed in a blender and milled again by Fritsch® Pulverisette2 for 4 h. Longer milling times result in over heating of the mixture and this phenomenon influences the material properties. After that, the specimens (called as ENRF I-II hereafter) were manufactured by hot compacting (double uniaxial action) under a pressure of 70 MPa at 180 °C. The dwell time for compacting process was 15 min. All of the specimens (30 mm & 50 diameters, 5–6 mm thickness) were cooled down slowly. The post curing was concluded under isothermal conditions at a temperature of 80 °C for 24 h. General compositions of all the composites manufactured here are given in the Table 4.1 with specified weight ratios for each constituent.

Table 4.1 Composition of the epoxy-rubber based composites

Epoxy-Rubber based composition Epoxy – SBR rubber (10 phr)	ENRF I	ENRF II
Reinforcements (wt %)	10 (Fe_3O_4) 5 nickel	20 (Fe_3O_4) 5 nickel

Table 4.2 General conditions for compression tests of produced composites

Initial speed	10	mm/min
Load endpoint	10,000	N
Outer loop rate	100	Hz
Secondary speed	5	Mm/min
Strain endpoint	80	%

Table 4.3 Hardness values of ENRF specimens

Hardness Measurement	
Specimen	Shore D
ENRF I	92
ENRF II	93

4.2.2 Mechanical Behaviours and Microstructural Analyses

After manufacturing of the specimens, two specimens were submitted to macro-indentation compression tests by means of a 5 mm steel ball. Tests were carried out in a servo-hydraulic MTS Universal test device (model 5500R, equipped with a load cell of 25,000 kgf) with a quasi-static low speed (initial rate: 10 mm/min and second rate: 5 mm/min rate). All of the testing parameters were given in the Table 4.2. All of the static compression tests have been carried out under the same conditions to evaluate damage behaviour of these composites.

Microstructure and surface damage evaluations were made by means of optical (OM) and scanning electron microscopy (SEM). SEM observation was realized on fracture surface of the tested specimens with Scope/JSM-6010LA Jeol® electron microscope which has a magnification capacity of $\times$300,000 and with accelerating voltage from 500 V to 20 kV. In addition, hardness measurements were carried out with Shore test device in Shore D hardness scale. Ten measures are taken and then given as a mean value for each type of specimen. Results are given in Table 4.3.

Dynamic compression (drop weight tests) was carried out using a universal drop weight test device. A standard hemispherical punch with a nose diameter of 12.5 mm was used and all the drop weight tests have been carried out from a distance of 900 mm. Impact behavior of the manufactured composites were observed over the test specimens by using the force-time diagrams.

4.2.3 Creep and Wear Analyses Through Nanoindentation Tests

Creep tests using a nanoindenter were performed on the two compositions manufactured. On each sample 20 indents were performed on a 5 $\times$ 4 grid with a Berkovich indenter. The indents were spaced 50 μm along the 5 indent side and 75 μm along the 4 indent side. The load was increased at a rate of 1 mN/s to the max load and kept at the maximum load for 500 s then unloaded. A typical load-on-sample vs displacement-into-sample graph is shown in Fig. 4.1. Two maximum loads of 20 mN and 50 mN were used in these tests.

Scratch testing capability of a nanoindenter is utilized to perform relatively fast wear tests to compare the wear behavior of the different samples. In the wear tests conducted conical tip with a 90° cone angle was used. Wear tests were run under a normal load of 20 mN applied over a linear track of 500 nm for 50 cycles.

4.2.4 Damage Analysis by Means of Macro Scratch Test

Scratch tests results give a basic idea on the tribological behaviour of the epoxy and recycled elastomer based composites designed in the current research. After the scratch test, damage zone was investigated by a 3D optical surface scanner to assess damage in terms of scratch depth, average scratch roughness.

In the scratch test the contact between the sliding zirconium indenter (d = 2 mm) and the surface of the composite material during scratch test is analyzed. The normal and tangential forces on indenter are recorded and the tangential stress on the surface and the interfacial stresses can be obtained.

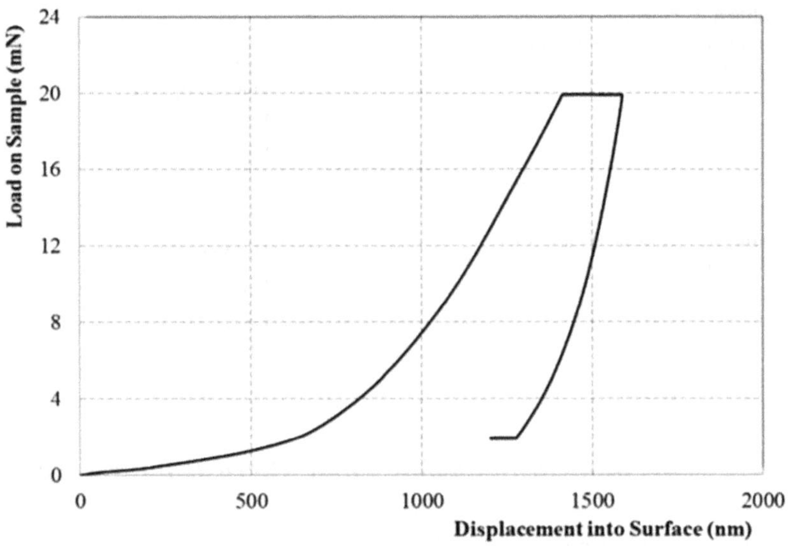

Fig. 4.1 Typical Load-Displacement curve during the creep test

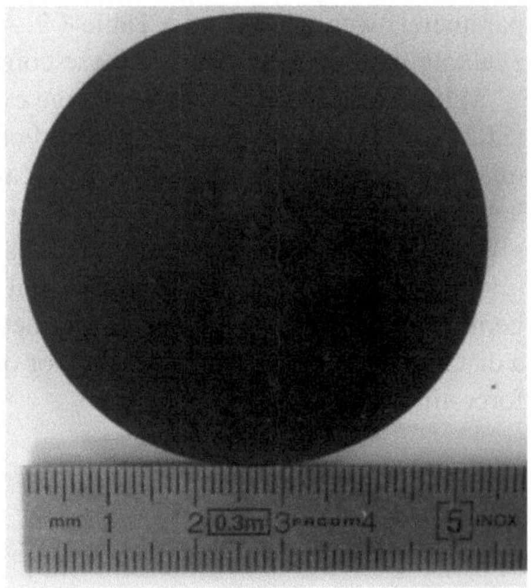

Fig. 4.2 Picture of the sample used for metallographic and mechanical characterizations

In the frame of the present work wear resistance was evaluated only under dry test conditions and 50.10^3 and 100.10^3 numbers of cycles for the composites presented here. After that surface wear resistance to scratch damage was evaluated in terms of the scratch depth, surface and volume lost during the subsequent to scratching.

4.3 Results and Discussions

4.3.1 Microstructure Analyses and Macro Indentation Compression Tests

In Fig. 4.2, a macrograph of the specimen was shown after the manufacturing. This type of specimen geometry was used in the scratch tests and also dynamic compression tests.

General microstructures in the transversal direction of the two compositions are shown in the Fig. 4.3. All of the compositions have shown quite homogenous distribution of the reinforcements in the structure. In the microstructure, some of the big rubber particles were observed. They are coming from the mixtures that could not be decreased the size of big rubber powders. Essentially, all of the microstructures show that a good cohesion of the rubber to the epoxy matrix and well bonding was successfully carried out after the simple chemical treatment and efficient milling process.

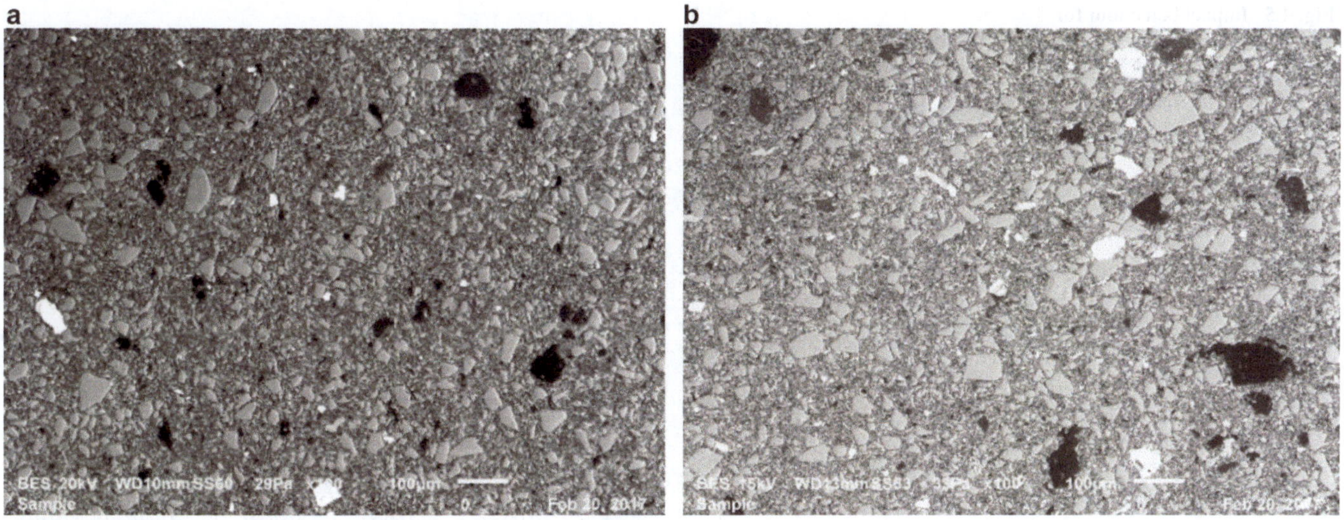

Fig. 4.3 Microstructure of ENRF I-II specimens transversal sectioning (**a**, **b**) respectively (ENRF I – II)

Fig. 4.4 Macro indentation compression test results of two types of ENRF composites

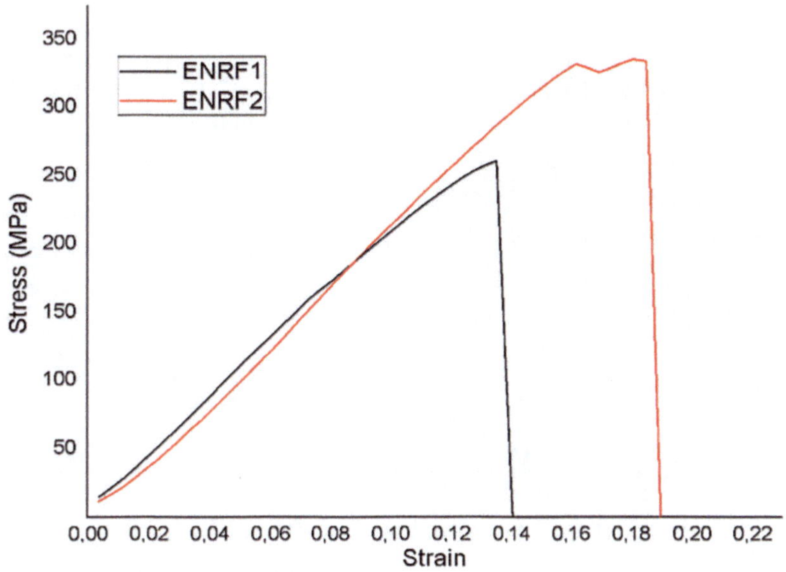

There is a good chemical bonding at interface between matrix and reinforcement elements. It means that activated surface of rubber powder and bonding effect give together a sound microstructure. However, more homogenous distribution of reinforcements and elimination of the big rubber powders need planetary ball milling to arrive at perfect conditions in the process.

Typical macroindentation compression tests have been carried out to eavluate the damage level of the composite strcuture. At the first stage, stress-strain curves obtained from macro indentation compression tests were evaluated in Fig. 4.4. Macro indentation strength values of two composites increase comparably with the iron oxide content and also Nickel added in the stucturevery good high stenght other mechanical tests also confirmed this improvement.

4.3.2 *Dynamic Compression (Impact) Testing*

As it is well known, epoxy has very brittle characteristics and to improve it rubber particles are incorporated to the matrix. Toughness of these composites was developed and impact behaviours of the specimens were studied. In Fig. 4.5 impact force

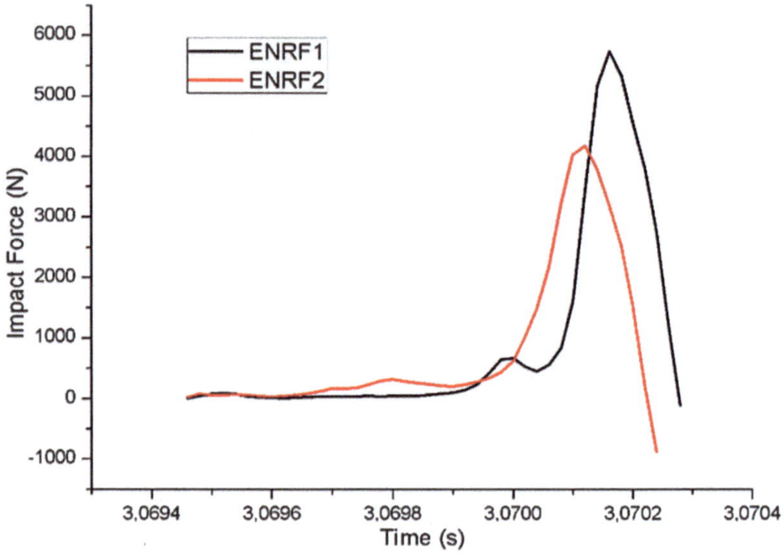

Fig. 4.5 Impact behaviour for two ENRF composites obtained by dynamic compression test

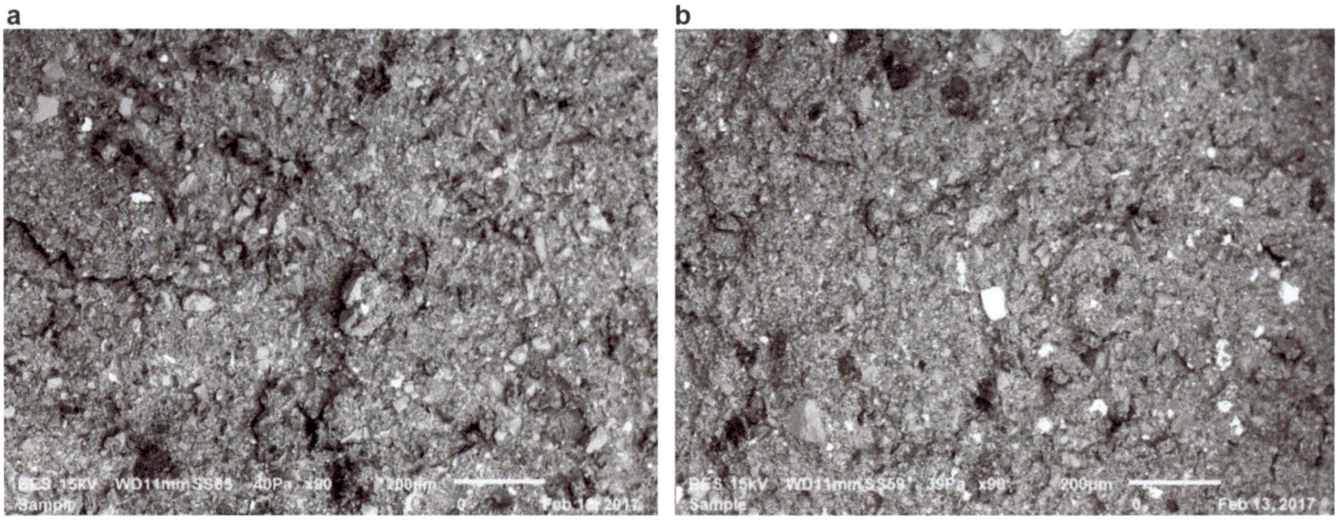

Fig. 4.6 (**a**) Fracture surfaces after impact testing ENRF I and (**b**) ENRF II respectively

was plotted as a function of time in this curve. From this curve one may mention that the effect of Fe_3O_4 in different percentage in the structure does not play a negative role on the toughness values of the composites. It seems that total absorbed energy for two composites are similar even if first group (ENRF-I) with 10% iron oxide content gives higher damping capacity regarding to that of ENRF-II with 20% iron oxide content. These values should be accepted as indicative values and number of test with other parameters are needed for high precision results.

Fracture surfaces of the 2 compositions are given in the Fig. 4.6a, b. All of the fracture surfaces of the specimens taken from 2 compositions indicate the brittle type fracture behaviour. Even if some local agglomerations of certain reinforcements in the structure, there is no decohesion during the impact choc, but small internal cracks were observed with very nice fine magnetic iron oxide in the structure.

All of the SEM observations on the fracture surfaces of the 2 compositions indicate that there is not any debonding of the reinforcement particles detected on the specimens.

4.3.3 Creep Testing by Nanoindentation

During each test, data collected was used to calculate the creep compliance and the stress exponent defined in Eq. 4.1 [14–18]:

$$\varepsilon(t) = \sigma_0 J(t) \tag{4.1}$$

where σ_0 is the constant stress applied and *J(t)* is calculated using Eq. 4.2

$$J(t) = A(t)/(1-\nu)P_0 \tan\theta \tag{4.2}$$

*In Eq. 4.2 *A(t)* is the contact area, P_0 constant applied load, θ is the effective cone angle which is 70.3° for a Berkovich indenter and the Poisson's ratio ν is assumed to be 0.3.

The strain versus time behavior during creep is characterized by a high strain rate $\dot{\varepsilon} = d\varepsilon/dt$ in the primary stage of creep and then in the secondary, steady state stage of creep, the strain rate is given in Eq. 4.3 can be written as

$$\dot{\varepsilon} = K\sigma^n \tag{4.3}$$

where K is a constant and n is the stress exponent. The strain rate is calculated in the software and in turn n is obtained from the log-log plot of train rate versus stress in the secondary stage of creep.

The materials under consideration are heterogeneous in nature and the fact that the nanoindenter testing is carried out over a small area/volume, a large scatter in the data is observed.

According to Fig. 4.7 it seems that the creep compliance values are not quite dependent to the iron oxide content of the compositions. It shows an alteration depending on the loading values. Therefore, effect of the iron oxide content to the creep compliance cannot be directly deducted from this type of curves.

The average indentation modulus and stress exponent values were obtained from the 20 indentations performed under 20 mN and 50 mN constant loads and they are presented in Fig. 4.8a, b with error bars showing ± one standard deviation.

From Fig. 4.6b it is seen that, increasing Fe_3O_4 amount results in a rise in the stress exponent. However, for the modulus it cannot be done the same straightforward inference. These results also demonstrate that these materials cannot conform to the idealized linear viscoelastic behaviour under the contact creep conditions used here. It means that there is no linear path during loading and unloading.

4.3.4 Wear Testing by Nanoindentation

During each nano wear test, one cycle is defined as a pass and return of the indenter over the track, so the total distance covered for one wear test was 0.050 m. Speed of the tip during wear tests was 50 μm/s.

Total of 10 wear tests for each sample were performed under the two normal loads. The wear in a track is characterized as the area between the initial profile and the residual profile of the wear track.

In the same way, the averages of the wear track deformation are shown in Fig. 4.9.

As expected the applied load has a significant effect on the wear track deformation values. However, wear track deformation values seem insensitive to the iron oxide content inside the composites. This situation can be explained by the anisotropy over the matrix.

4.3.5 Damage Analysis by Means of Scratch Test and 3D Optical Roughness Meter

After realizing two different group of macro scratch tests, three dimensional damages were obtained by 3D optical surface scanner. The results are presented in the Figs. 4.10 and 4.11.

Additionally, the filler particle size at a constant weight fraction has a significant influence on the scratch resistance of the composites. This is also confirmed with nano indentation tests. It means that certain fillers play an important role for the scratch resistance that they were used as reinforcement elements here.

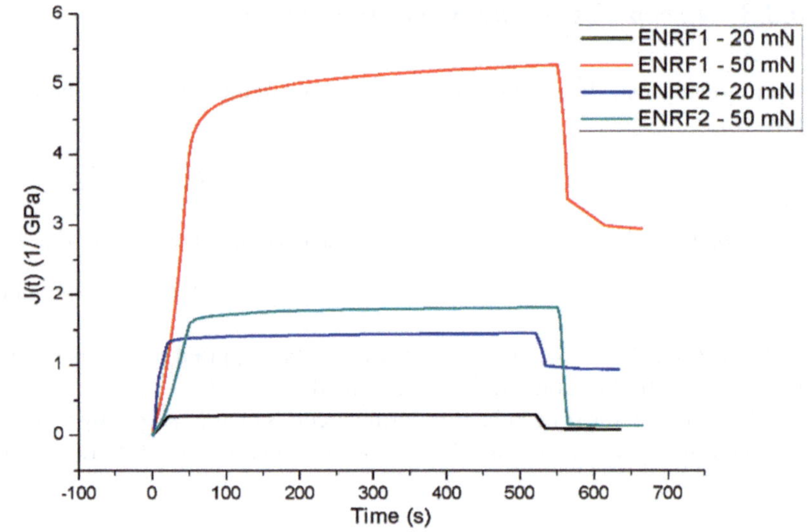

Fig. 4.7 Creep Compliance curves for sample ENRF I-II, under 20 mN and 50 mN load

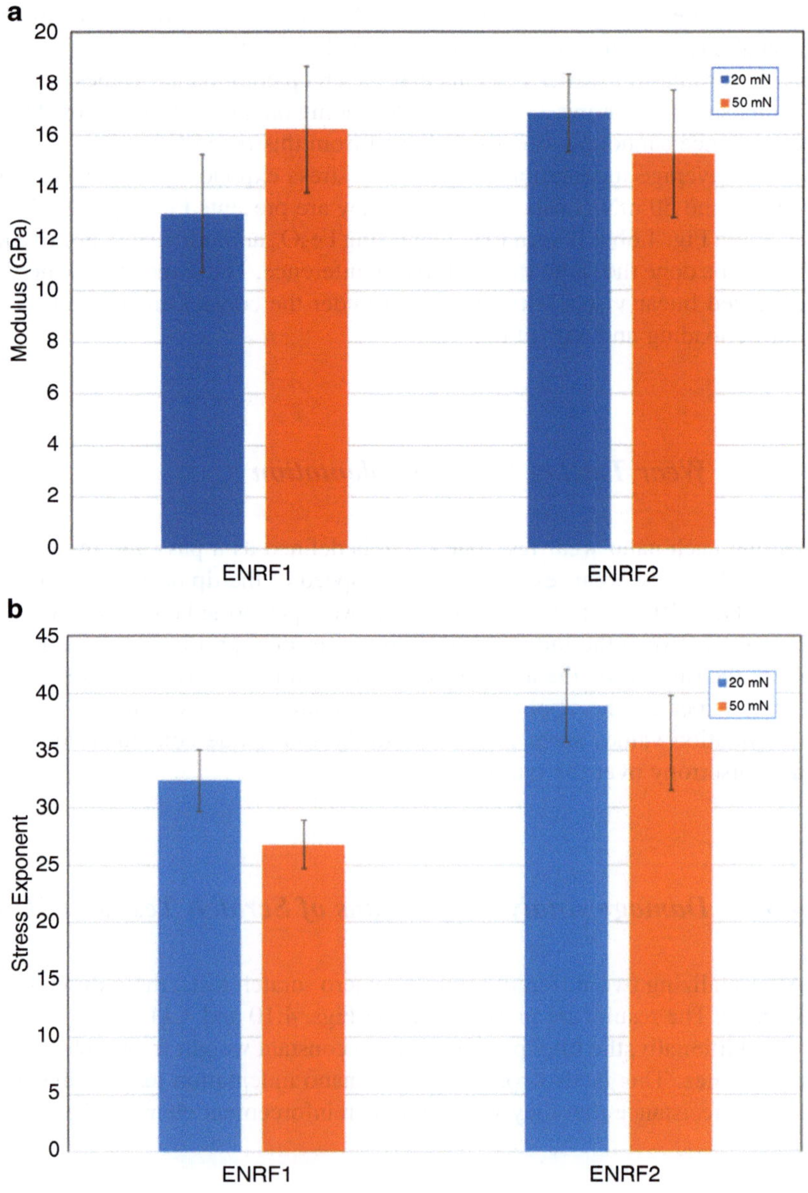

Fig. 4.8 (**a**) Indentation modulus (**b**) Stress exponent for the two composites under 20 mN and 50 mN

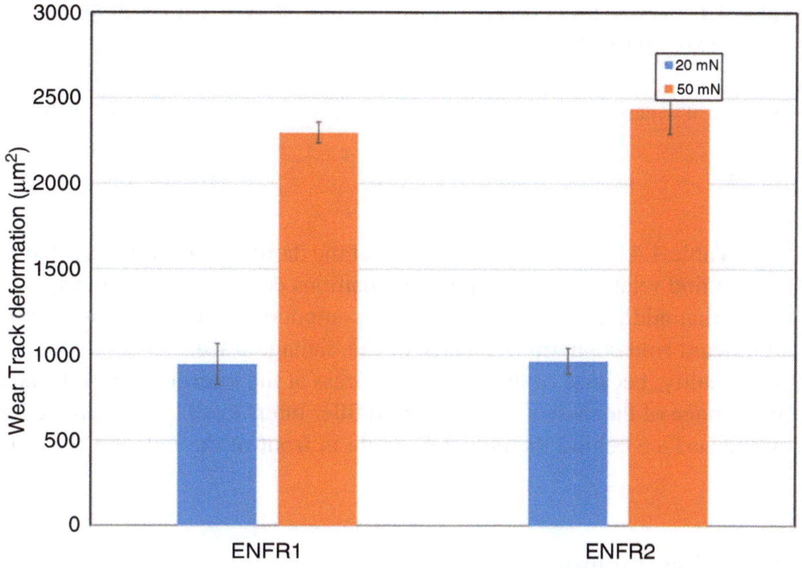

Fig. 4.9 Comparison of wear track deformation for ENRF composites

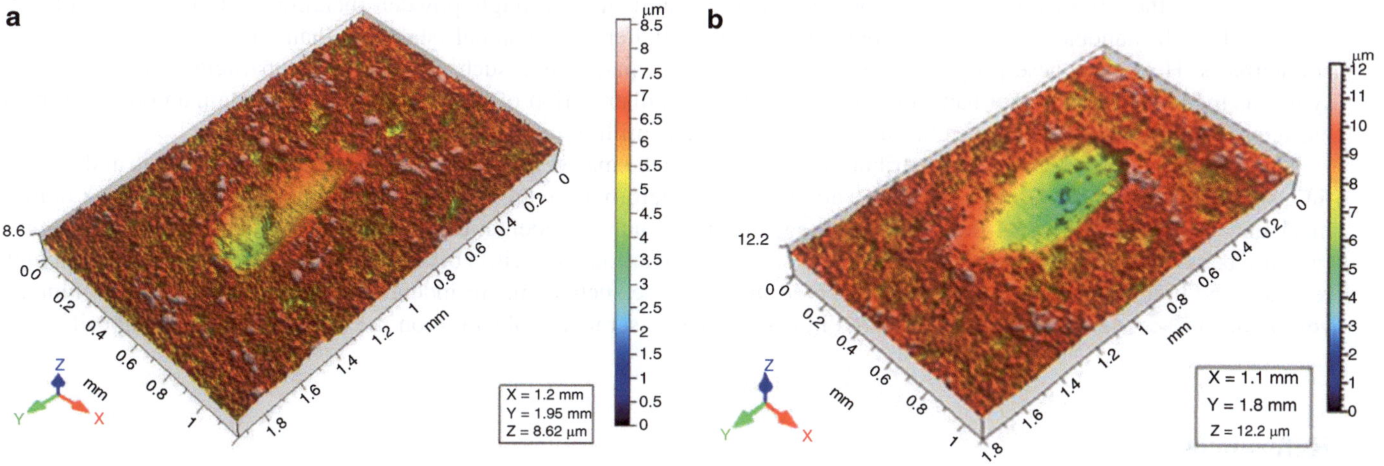

Fig. 4.10 (**a**) Damage traces obtained in the direction of width and length for the specimen ENRF I for 50 k cycles, and (**b**) for 100 k cycles

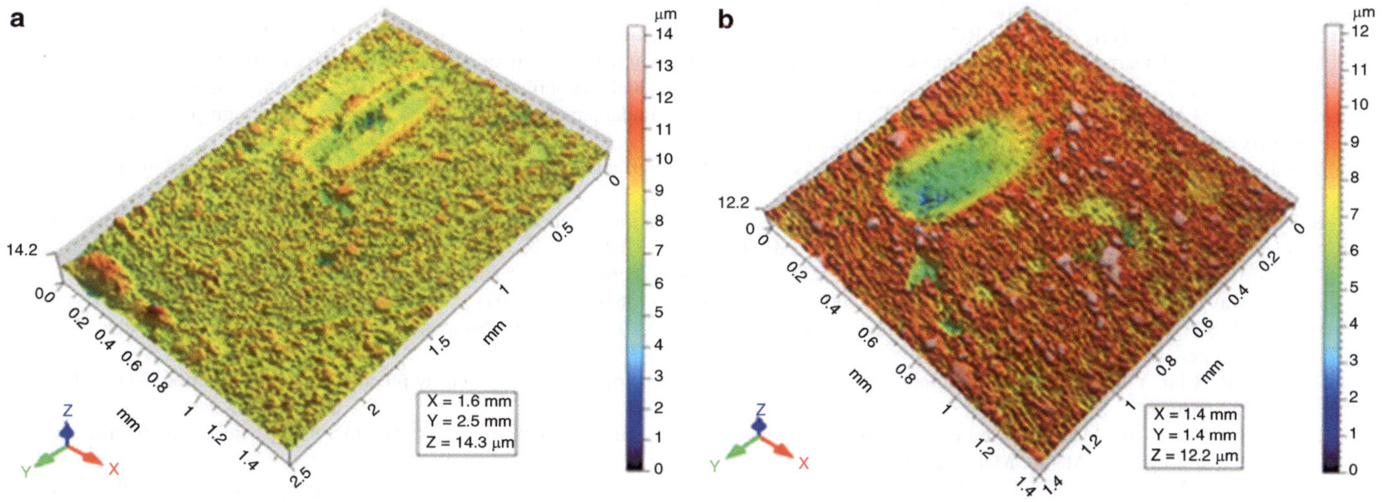

Fig. 4.11 (**a**) Damage traces obtained in the direction of width and length for the specimen ENRF II for 50 k cycles, and (**b**) for 100 k cycles

Table 4.4 Damage traces obtained in the direction of width and length for the specimens ENRF I-II for the number of cycle at 50.10^3 and 100.10^3 cycles respectively

	ENRF I		ENRF II	
Number of cycles	Worn surface (mm^2)	Worn volume (μm^3)	Worn surface (mm^2)	Worn volume (μm^3)
50.10^3	0.171	121,867	0.210	206,015
100.10^3	0.270	461,488	0.172	320,608

In Table 4.4 the volume and surface of the damage trace after scratch calculated from roughness test results are given. It is hard to find relation between the compositions concerning the surface of the damage trace. However, based on Table 4.4, it seems that addition of iron oxide to the compound resulted in less damage. In addition, the number of cycles doesn't play a significant role concerning surface of the damage trace.

In reality, because of the high shear stress at the interfaces the interfacial shear stress should probably be the main reason for damage of the matrix and reinforced filler interfaces [19]. When the indenter is slipping, tangential tensile stress is caused on the surface behind the indenter, while in front of the indenter the tangential stress is compressive.

4.4 Conclusions

In this study, the epoxy-rubber based composites were fabricated through powder metallurgy. Usually, composites reinforced with nanoparticles exhibit light weight and higher dimensional stability than conventional composite counterparts. However, these properties depend on the certain parameters such as type of nanoparticle, polymer matrix, synthesis methods, and polymer nanocomposites morphology, distribution of these particles, etc. Certain amount of rubber particles is found that could not be eliminated due to the short milling time.

This case should be improved the distribution of particles in the matrix homogeneously. It means that mixture and milling time should be optimized. Creep compliance and wear track deformation are very useful parameters that measure a unique mechanical characterization of the time-dependent materials such as indentation hardness and indentation modulus, etc. Fracture behaviour of the composites studied here show certain amount of brittle behaviour in case of impact chock but high resistant and high toughness beahviour is observed in other case such as macro indentation compression test, etc. Finally, very good surface wear resistance is observed in case of macro scratch application on the surface of these composites.

References

1. Ortiz RoP, A., Facchetti, T.J.: Marks, High-k organic, inorganic, and hybrid dielectrics for low-voltage organic field-effect transistors. Chem. Rev. **110**(1), 205–239 (2009)
2. Zaimova, D., Bayraktar, E., Katundi, D., Dishovsky, N.: Design of new elastomeric composites used in manufacturing engineering. In: 14th International Materials Symposium–IMSP. 2012. pp. 10–12
3. Shaw, S.J.: In: Collyer, A.A. (ed.) Rubber Toughened Engineering Plastics, p. 165e209. Chapman & Hall, London (1994)
4. Sultan, J.N., McGarry, F.J.: Effect of rubber particle size on deformation mechanisms in glassy epoxy. Polym. Eng. Sci. **13**, 29e34 (1973)
5. Manzione, L.T., Gillham, J.K.: Rubber-modified epoxies. I. transitions and morphology. J. Appl. Polym. Sci. **26**, 889e905 (1981)
6. Dang, Z.-M., Yuan, J.-K., Zha, J.-W., Zhou, T., Li, S.-T., Hu, G.-H.: Fundamentals, processes and applications of high-permittivity polymer-matrix composites. Prog. Mater. Sci. **47**(4), 660e723 (2012)
7. Folgueras, L.D.C., Nohara, E.L., Faez, R., Rezende, M.C.: Dielectric microwave absorbing material processed by impregnation of carbon fiber fabric with polyaniline. Mater. Res. **10**, 95–99 (2007)
8. Das, N.C., Khastgir, D., Chaki, T.K., Chakraborty, A.: Electromagnetic interference shielding effectiveness of carbon black and carbon fibre filled EVA and NR based composites. Compos. Part A. **31**, 1069–1081 (2000)
9. Yusoff, A.N., Sani, J.M., Abdullah, M.H., Ahmad, S.H., Ahmad, N.: Electromagnetic and absorption properties of some TPNR/Fe3O4/YIG microwave absorbers and specular absorber method. Sains Malaysiana. **36**, 65–75 (2007)
10. Ahmad, S.H.J., et al.: Magnetic and microwave absorbing properties of magnetite–thermoplastic natural rubber nanocomposites. J. Magn. Magn. Mater. **322**(21), 3401–3409 (2010)
11. Lin, C.R., Chu, Y.M., Wang, S.C.: Magnetic properties of magnetite nanoparticles prepared by mechanochemical reaction. Mater. Lett. **60**, 447–450 (2006)
12. Bayraktar, E., et al.: Manufacturing of aluminum matrix composites reinforced with iron oxide (Fe_3O_4) nanoparticles: microstructural and mechanical properties. Metall. Mater. Trans. B. **45**(2), 352–362 (2014)
13. Martin, M., Hanagud, S., Thadhani, N.N.: Mechanical behavior of nickel+ aluminum powder-reinforced epoxy composites. Mater. Sci. Eng. A. **443**(1), 209–218 (2007)

14. Bayraktar, E., Zaimova, D., Tan, M.-J., Miskioglu, I.: A preliminary study on an epoxy-rubber matrix with different reinforcements. In: G.P. Tandon (eds.), Experimental Mechanics of Composite, Hybrid, and Multifunctional Materials, vol. 6, no. 7, pp. 1–13. Springer Publication Springer link (2014)
15. Asaka, K., Ishihara, C.: Technical trends in soft magnetic parts and materials. Hitachi Powdered Metals Technical Report 420053 (2005)
16. Zaimova, D., Bayraktar, E., Miskioglu, I., Katundi, D., Hamouche, Z.: Manufacturing and damage analysis of filler reinforced epoxy based composites. Int. J. Mat. Prod. Technol. **48**(1–4), 47–65., Inter Science (ed.), 0268-1900/1741-5209 (2014)
17. Zaimova, D., Bayraktar, E., Miskioglu, I., Dishovsky, N.: Wear resistance of elastomeric based composites by continuous multi-cycle indentation used in manufacturing engineering. Adv. Mat. Res. **939**(1), 106–113., Trans Tech Publications (ed.), Switzerland (2014)
18. Zaimova, D., Bayraktar, E., Tan, M.J., Miskioglu, I.: Processing of epoxy-nickel matrix composites reinforced with aluminum and waste elastomers. Adv. Mat. Res. **939**(1), 98–105., Trans Tech Publications (ed.), Switzerland (2014)
19. Zaimova, D., et al.: Manufacturing and damage analysis of epoxy resin-reinforced scrap rubber composites for aeronautical applications. In: Experimental Mechanics of Composite, Hybrid, and Multifunctional Materials, vol. 6, pp. 65–76. ISBN 978-3-319-06992-0 (e-Book) DOI: 10.1007/978-3-319-06992-0. Springer Cham Heidelberg New York Dordrecht London (2014)

Chapter 5
Mechanical Characterization of Epoxy – Scrap Rubber Based Composites Reinforced with Nano Graphene

A.B. Irez, I. Miskioglu, and E. Bayraktar

Abstract The use of lightweight materials has accelerated recently due to the environmental restrictions and economic concerns. For this reason, in automotive and aerospace industries, designers and material manufacturers are in the search of durable materials in terms of their structural and energetic properties. Polymer matrix composites (PMCs) usually meet these expectations when proper reinforcements are used. Epoxy – rubber based composites reinforced with different materials are studied by several researchers. However, addition of scrap rubber to the matrix and graphene as reinforcement introduces a novelty in this area. Since, using of scrap rubber has favorable outcomes in economical and structural perspectives such as revaluation of scraps and compensating for the brittleness of epoxy. Also, due to its superior structural, thermal and physical characteristics, addition of graphene promises desired mechanical properties in case of a homogeneous distribution. This paper is focused on the design of PMCs reinforced with nano graphene in different ratios in a matrix of epoxy – fresh scrap rubber. The mechanical and some physical properties of these composite systems were studied in this research. Mechanical properties were evaluated by means of three-point bending tests, impact tests and nano indentation technics. Scanning electron microscope was used to observe fracture surfaces and the microstructure.

Keywords Three-point bending • Nano indentation • Epoxy – rubber • Graphene • SEM

5.1 Introduction

Recently, as a consequence of weight reduction precautions in automotive and aeronautic industry some metallic parts are being replaced by plastics. For this to happen, the feasibility of using these plastics has to be examined properly in terms of their mechanical properties.

Epoxy polymers have an indispensable place as matrix material in cutting edge applications thanks to their beneficial properties like high specific strength and stiffness, chemical resistance, ease of process, environmental stability and relatively low cost [1]. Nevertheless, they also have some unenviable properties such as brittleness or in other words low fracture resistance because of their highly cross-linked network structure. Therefore, for a few decades engineers have been working on the improvement in fracture toughness of epoxies in order to extend their applications [2, 3]. To toughen epoxy resins, secondary phase particles in the epoxy matrix can be incorporated. Secondary phase particles can be dispersed in the matrix and they can either be soft fillers such as thermoplastic particles, rubber or rigid fillers as silica beads or Titania or alumina [4].

Soft rubber particles have a significant toughening effect of epoxies. In addition, rubber phase debonding/cavitation, localized shear banding of matrix as well as rubber particle bridging/tear are known as the major toughening mechanisms [5, 6]. In other respects, modification using rigid fillers improves the stiffness and glass transition temperature (T_g) of the epoxy composites but the toughening effect was generally not remarkable. More effective rigid fillers have been demanded to overcome these trade-off relationships [7–10].

Recently with the help of the advancements in nano technology, nano particles such as nano silica, carbon nanotubes and nano graphene are utilized as reinforcements in composites. More importantly, rather than micro sized particles, nanoparticles have high surface area which means a low content of these nanoparticles may enhance the materials properties [11–13]. In this regard, nano graphene has unique mechanical, thermal and electrical properties. It has a two-dimensional

A.B. Irez (✉) • E. Bayraktar
Supmeca-Paris, School of Mechanical and Manufacturing Engineering, Saint-Ouen, France
e-mail: irez.burak@gmail.com

I. Miskioglu (✉)
Michigan Technological University ME-EM Department, Houghton, MI, USA
e-mail: imiski@mtu.edu

P.R. Thakre et al. (eds.), *Mechanics of Composite and Multi-functional Materials, Volume 6*,
Conference Proceedings of the Society for Experimental Mechanics Series, DOI 10.1007/978-3-319-63408-1_5

honeycomb structure coming from a single-atom-thick sheet of sp^2 bonded carbon atoms and this makes it a novel promising nanoscale filler for polymer nanocomposites. Its outstanding properties can be given as specific elasticity modulus (1 TPa) and mechanical strength (130 GPa), thermal conductivity (5000 W/mK) and high electrical conductivity (up to 6000 S/cm) [14–16]. However, some prominent issues restrain widely use of graphene. The most important of them can be the agglomeration problem. Due to the synthesis procedure of graphene, interlaminar van der Waal forces facilitate the agglomeration of the graphene sheet. As a result, during its mixing with polymers it is quite difficult to provide a homogeneous dispersion [17].

In this study, graphene nanoplatelets (GNPS) are used. Stacked two-dimensional graphene sheets constitute graphene nanoplatelets. Simple process and abundance of inexpensive naturally existing graphene precursor (graphite) lead to economical production of graphene in large quantities. Thus, outstanding balance between properties and cost make GNPS an ideal reinforcing nano-filler for polymer matrices [18].

This study focuses on the design of PMCs reinforced with nano graphene in different ratios in a matrix of epoxy – fresh scrap rubber. The mechanical and tribological properties of these composite systems were studied in this research. Mechanical properties were evaluated by means of three-point bending tests, impact tests, nano indentation and compression tests. Scanning electron microscope (SEM) was used to observe fracture surfaces and the microstructure. Surface wear resistance was evaluated with the help of a scratch test device.

5.2 Experimental Conditions

5.2.1 *Materials Processing*

In the first stage, in order to combine very fine dry epoxy powder and fine scrap rubber powder (Styrene–butadiene rubber (SBR)) series of chemical treatments were applied. To begin with, toluene and acrylic acid are applied to SBR and epoxy blend with pre-defined ratios then it is dried after treatment in the oven to prepare it for better adhesion to epoxy. This mixture was milled for an hour to obtain a homogeneous compound and then heated at 80 °C for 24 h. The resulting compound was used as the matrix for the proposed composites. At the last step of chemical treatment, vinyltriethoxysilane used for surface modification of GNPS by placing them in a chemical bath of vinyltriethoxysilane together with sulphuric acid solution. This step optimizes the interfacial interactions between GNPS and it helps to prevent agglomerations [19]. In the second stage, the reinforcements such as nano Al and GNPS were added to the matrix in pre-defined ratios.

All of the compounds were then mixed in a blender and milled by Fritsch® Pulverisette2 for 4 h. Longer milling times result in over heating of the mixture and this phenomenon influences the material properties. After that, the specimens (called as ERG I-II-III hereafter) were manufactured by hot compacting (double uniaxial action) under a pressure of 70 MPa at 180 °C. The dwell time for compacting process was 15 min. All of the specimens (30 mm & 50 diameters, 5–6 mm thickness) were cooled down slowly. The post curing was concluded under isothermal conditions at a temperature of 80 °C for 24 h.

General compositions of all the composites manufactured are given in the Table 5.1 with specified weight ratios for each constituent.

5.2.2 *Mechanical Tests and Microstructural Analyses*

After manufacturing of the specimens, three specimens were submitted to macro-indentation compression tests by means of a 5 mm steel ball. Tests were carried out in a servo-hydraulic INSTRON Universal test device (model Instron 5500R, equipped with a load cell of 25.000 kgf) with a quasi-static low speed (initial rate: 10 mm/min and second rate: 5 mm/min rate).

Table 5.1 Composition of the epoxy-rubber based composites

Epoxy-rubber based composition Epoxy – SBR rubber (10 phr)	ERG I	ERG II	ERG III
Reinforcements (wt %)	0,3 GNPS 5 Nano Al	0,5 GNPS 5 Nano Al	1 GNPS 5 Nano Al

Table 5.2 General conditions for compression tests of produced composites

Initial speed	10	mm/min
Load endpoint	10,000	N
Outer loop rate	100	Hz
Secondary speed	5	mm/min
Strain endpoint	80	%

Table 5.3 Hardness values of ERG specimens

Hardness measurement	
Specimen	Shore D
ERG I	95
ERG II	92
ERG III	91

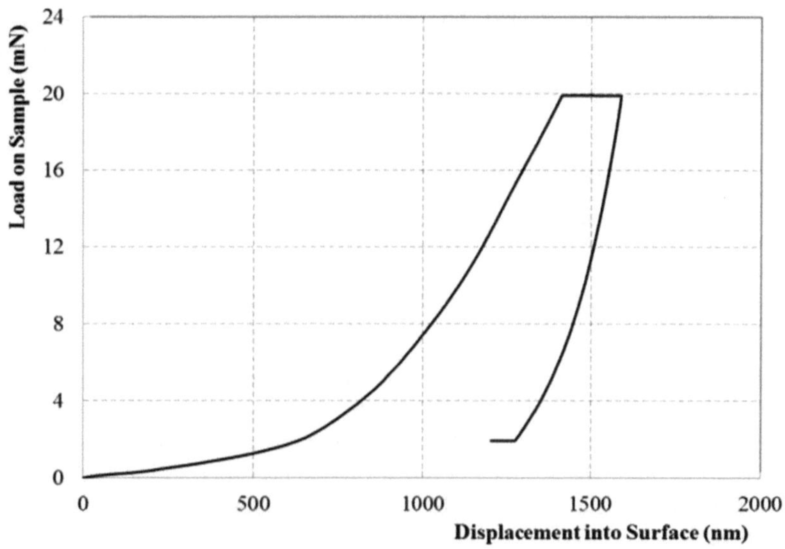

Fig. 5.1 Typical Load-Displacement curve during the creep test

All testing parameters were given in the Table 5.1. All of the static macro-indentation tests have been carried out under the same conditions to evaluate damage behaviour of this composite (Table 5.2).

After post curing, hardness and density measurements on the specimens were performed. Shore D hardness test measurements on the polished flat surfaces of the specimens were carried out according to ASTM D 2240 using durometer Shore test device, (type HBD-100-0). Hardness results were given in Table 5.3.

The bending specimens are prepared according to the ASTM D790 standard. The most common bending test for composite materials is thought as the three-point bending test. Deflection of the specimen is measured by the crosshead position. Bending strength, displacement and strain, are obtained from the test results. The test specimen is placed to the universal testing machine and force is applied until it fractured.

Dynamic compression (drop weight) tests were carried out using a universal drop weight test device. A standard conical punch was released from a height of 900 mm. Impact behavior of the manufactured composites were observed over the test specimens with the help of force-time curves.

Microstructure and surface damage evaluations were made by means of scanning electron microscopy.

5.2.3 Nanoindentation: Creep and Wear Tests

Creep tests using a nanoindenter were performed on the two compositions manufactured. On each sample 20 indents were performed on a 5x4 grid with a Berkovich indenter. The indents were spaced 50 μm along the 5 indent side and 75 μm along the 4 indent side. The load was increased at a rate of 1 mN/s to the max load and kept at the maximum load for 500 s then unloaded. A typical load-on-sample vs displacement-into-sample graph is shown in Fig. 5.1. Two maximum loads of 20 and 50 mN were used in these tests.

Scratch testing capability of the nanoindenter is utilized to perform relatively fast wear tests to compare the wear behavior of the different samples. In the wear tests conducted a conical tip with a 90° cone angle was used. Wear tests were run under a normal loads of 20 and 50 mN applied over a linear track of 500 nm for 50 cycles.

5.2.4 *Damage Analysis by Means of Scratch Test and 3D Optical Roughness Meter*

Scratch tests results give a basic idea on the tribological behaviour of the epoxy and recycled elastomer based composites designed in the current research. After the scratch test, damage zone was investigated by a 3D optical surface scanner to assess damage in terms of, scratch depth, average scratch roughness.

In the scratch test the contact between the sliding zirconium indenter and the surface of the composite material during scratch test is analyzed. The normal and tangential forces on indenter are recorded and the tangential stress on the surface and the interfacial stresses can be obtained. In the frame of this present work wear resistance was evaluated only under two test conditions 50,000 and 100,000 number of cycles for the composites presented here. Additionally, surface wear resistance was evaluated by using macro scratch test in terms of scratch depth, surface and volume lost subsequently analyzed.

5.3 Results and Discussion

5.3.1 *Microstructure Evaluation and Macro Indentation Compression Tests*

General microstructures in the transversal direction of the three compositions are shown in the Fig. 5.2. All of the compositions have shown a considerably homogenous distribution of the reinforcements with certain agglomerations in the structure. Essentially, all of the microstructures show that the adhesion of the rubber to the epoxy matrix is successfully carried out after the simple chemical treatment. Distribution and cohesion of the all other reinforcements with epoxy are similar. Some small local agglomerations are observed in the structure due to the mixture process; more homogenous distributions of the reinforcements need extended mixture times.

From the microstructures, the agglomerations of rubber particles were seen and they may influence the material properties.

In Fig. 5.3 stress-strain values obtained from macro indentation tests are given. From this figure it can be commented that increase in GNPS content results in decreasing stress values. However, this result is considered is due to the non-homogeneous distribution of GNPS, since bending tests showed different behavior for with respect to GNPS content.

5.3.2 *Dynamic Compression (Impact) Testing*

As it is well known, epoxy has very brittle characteristics and to improve it rubber particles are incorporated to the matrix. Toughness of these composites is developed and impact behaviour of the specimens was studied and it is given in Fig. 5.4. Impact force is plotted as a function of time in this curve. From this curve it is commented that increasing GNPs amount makes the composite weaker against to impacts and damping property of the composites improves with the increasing amount of GNPS.

In Fig. 5.5, SEM images on the fracture surface after drop weight tests show certain internal structure defects such as micro cracks because of the impact. After that, because of the opening of the micro cracks, matrix cannot maintain the integrity of the composite.

All of the fracture surfaces of the specimens taken from 3 compositions indicate the regular fracture deviations due to the effect of added reinforcements. Even if some local agglomerations observed on the structure; they play strong barriers against the propagation of the fracture. All of the SEM observations on the fracture surfaces of the 3 compositions indicate that there is not any debonding of the reinforcement particles detected on the specimens.

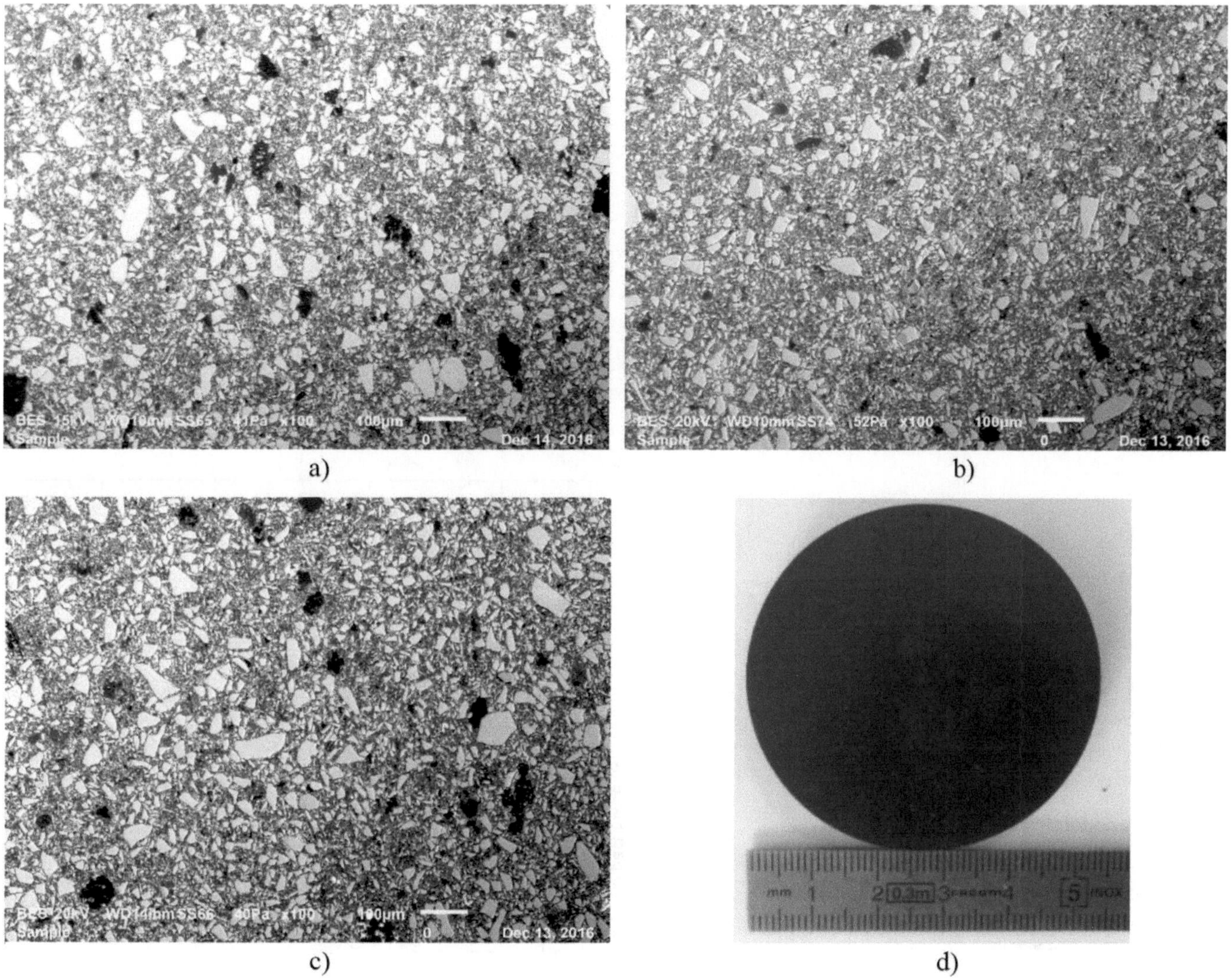

Fig. 5.2 Microstructure of ERG specimens transversal sectioning (**a–c** respectively ERG I – II – III) (**d**) Macrograph of the sample after compacting and post curing

5.3.3 Creep Testing by Nanoindentation

During each test, data collected was used to calculate the creep compliance and the stress exponent defined in Eqs. 5.1, 5.2, and 5.3 [20–24]:

$$\varepsilon(t) = \sigma_0 J(t) \tag{5.1}$$

where σ_0 is the constant stress applied and *J(t)* is calculated using Eq. (5.2)

$$J(t) = A(t)/(1 - \nu)P_0 \tan\theta \tag{5.2}$$

In Eq. (5.2) *A(t)* is the contact area, P_0 constant applied load, θ is the effective cone angle which is 70.3° for a Berkovich indenter and the Poisson's ratio ν is assumed to be 0.3.

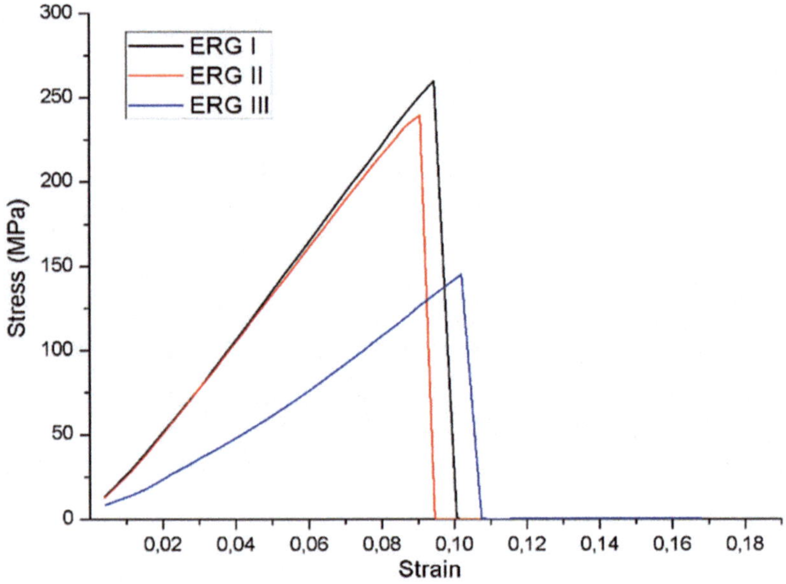

Fig. 5.3 Macro indentation compression test results of ERG composites

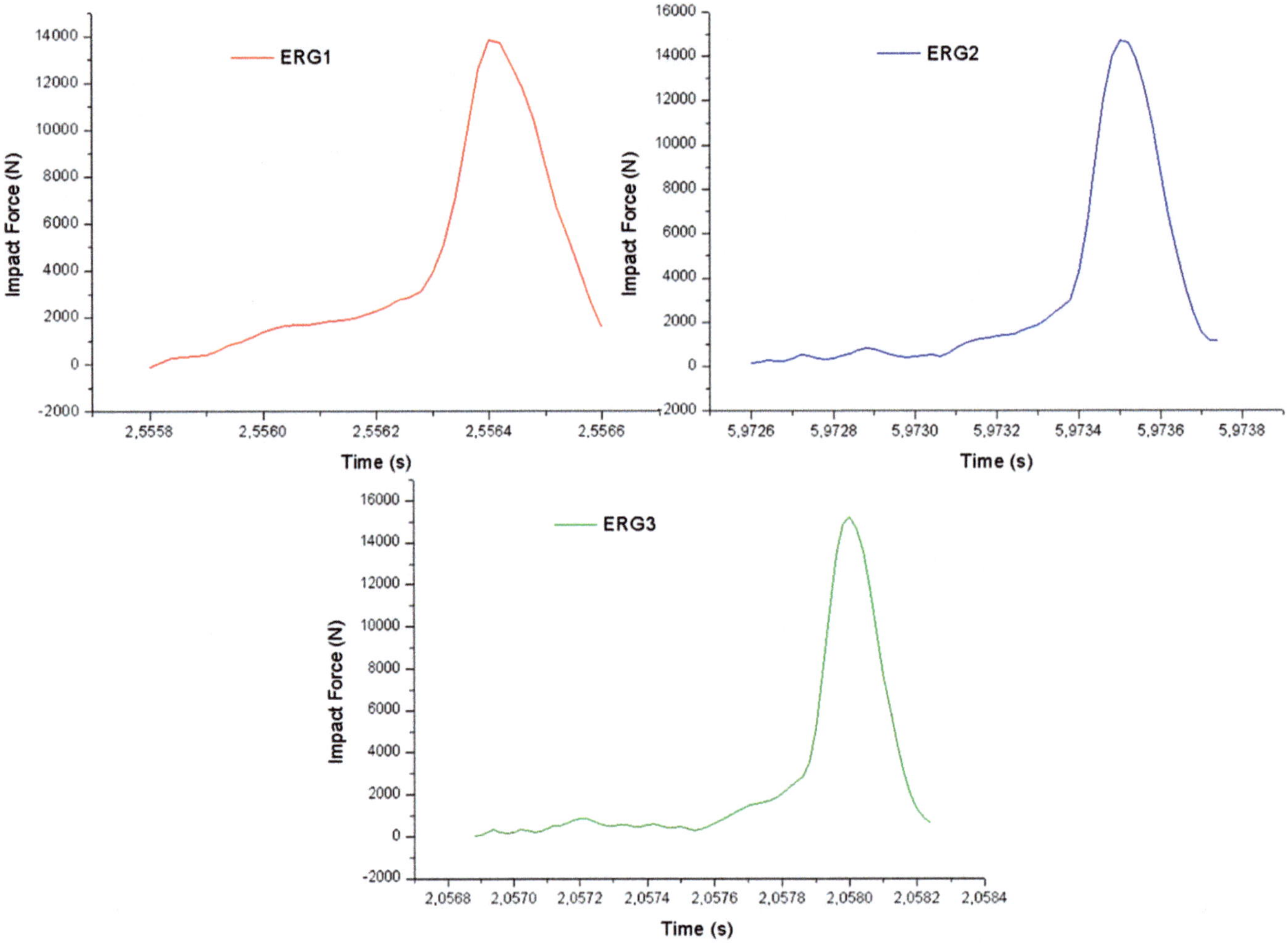

Fig. 5.4 Impact behavior of ERG composites by dynamic compression test

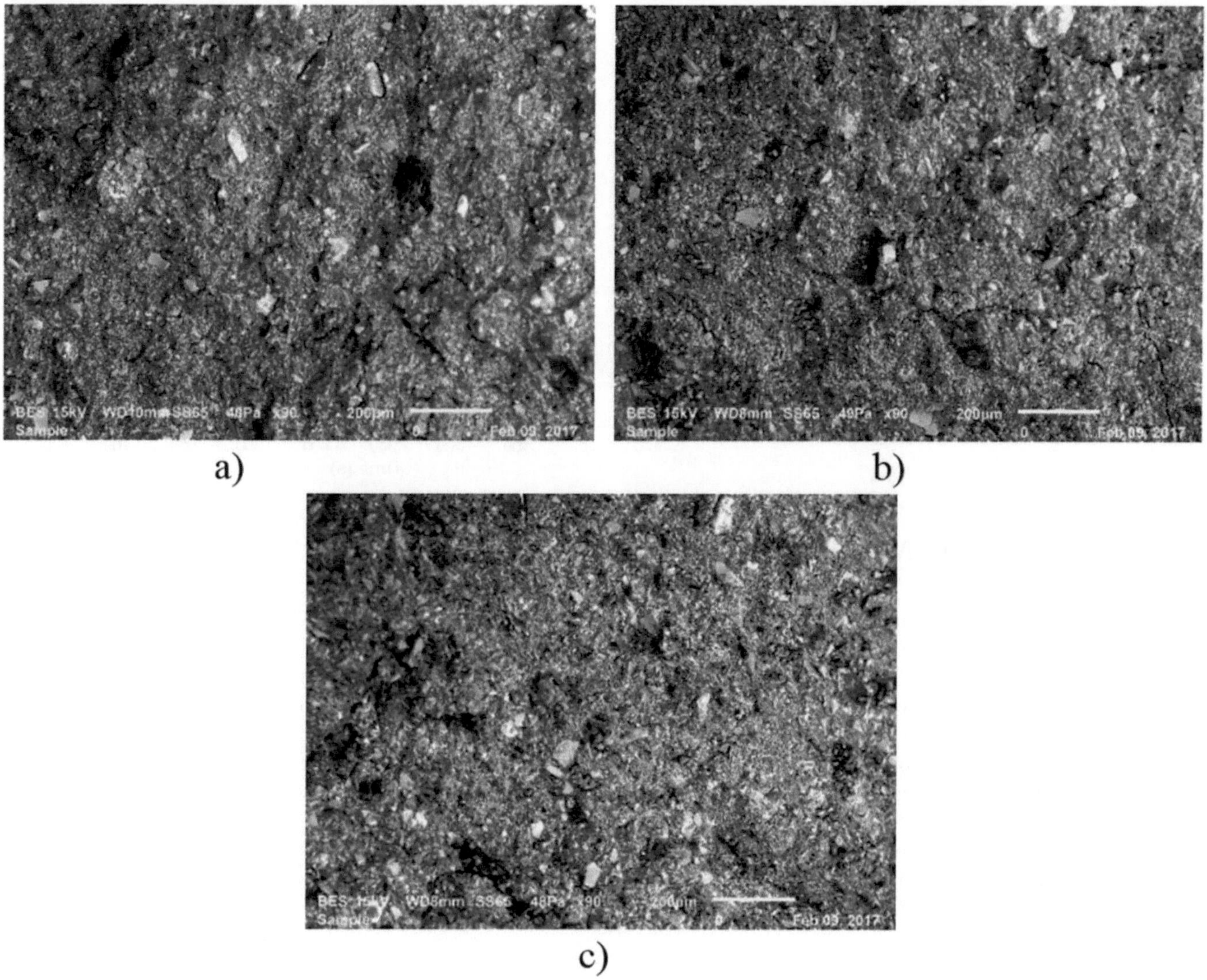

Fig. 5.5 Fracture surfaces after impact testing (**a**) ERG I (**b**) ERG II (**c**) ERG III

The strain versus time behavior during creep is characterized by a high strain rate $\dot{\varepsilon} = d\varepsilon/dt$ in the primary stage of creep and then in the secondary, steady state stage of creep, the strain rate is given in Eq. (5.3) can be written as

$$\dot{\varepsilon} = K\sigma^n \tag{5.3}$$

where K is a constant and n is the stress exponent. The strain rate is calculated in the software and in turn n is obtained from the log-log plot of strain rate versus stress in the secondary stage of creep.

The materials under consideration are heterogeneous in nature and the fact that the nanoindenter testing is carried out over a small area/volume, a large scatter in the data is observed.

According to Fig. 5.6 it is seen that creep compliance values are not quite dependent on the graphene content of the compositions. Also from this curve, it is quite difficult to generalize the creep compliance behavior as a function of force or graphene content.

The average indentation modulus and stress exponent values were obtained from the 20 indentations performed under 20 and 50 mN constant loads and they are presented in Fig. 5.7a–b with error bars showing ± one standard deviation.

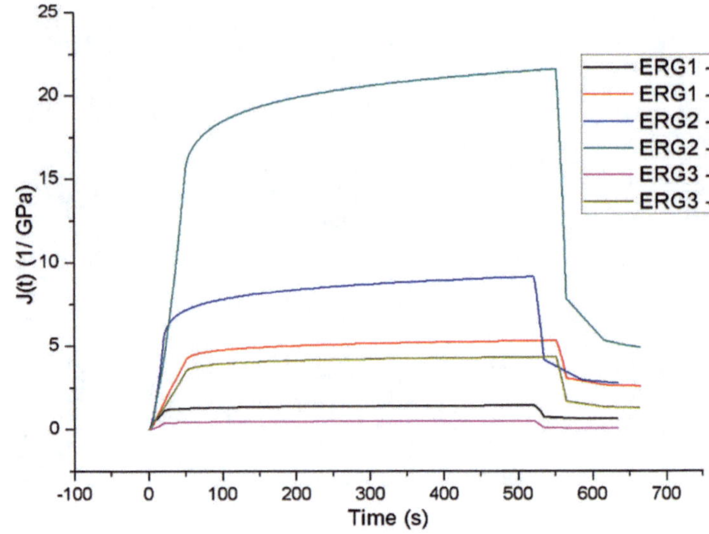

Fig. 5.6 Creep Compliance curves for specimens ERG I-II-III under 20 and 50 mN load

5.3.4 Wear Testing by Nanoindentation

During each nano wear test, one cycle is defined as a pass and return of the indenter over the track, so the total distance covered for one wear test was 0.050 m. Speed of the tip during wear tests was 50 μm/s.

Total of 10 wear tests for each sample were performed under the two normal loads of 20 and 50 mN. The wear in a track is characterized as the area between the initial profile and the residual profile of the wear track. In the same way, the averages of the wear track deformation are shown in Fig. 5.8.

As expected the applied load has a significant effect on the wear track deformation values. However, wear track deformation values seem insensitive to the graphene content inside the composites. This situation can be explained by the anisotropy over the matrix.

At the beginning nano indenter sweeps the surface to determine the initial profile of the wear track. Then, wear is done under different loadings. In Fig. 5.9, the difference in the depth is given as a comparison under a loading of 50 mN. Fluctuations on the graph are commented as the different hardness values of different elements of the composition.

5.3.5 Damage Analysis by Means of Scratch Test and 3D Optical Roughness Meter

After realizing two different group of macro scratch tests, three dimensional damages were obtained by 3D optical surface scanner. The results are presented in the Figs. 5.10, 5.11, and 5.12.

In Table 5.4 the volume and surface of the damage trace after scratch calculated from roughness test results are given. It is difficult to find a relation of worn volume and surface regarding the graphene contents of the compositions.

In reality, because of the high shear stress at the interfaces the interfacial shear stress should probably be the main reason for damage of the matrix and reinforced filler interfaces [25]. When the indenter is slipping, tangential tensile stress is caused on the surface behind the indenter, while in front of the indenter the tangential stress is compressive.

5.3.6 Bending Testing by Means of three Point Bending

Three-point bending test were conducted for each different type of composites. In tensile test, failed areas do not stay included in the specimen gauge length. Therefore, the bend test is more appropriate with our composites. Also, non-linear evolution of the curves is a consequence of the anisotropy of the inner structure.

Fig. 5.7 (**a**) Indentation modulus (**b**) Stress exponent diagrams for specimens ERG I-II-III under 20 and 50 mN

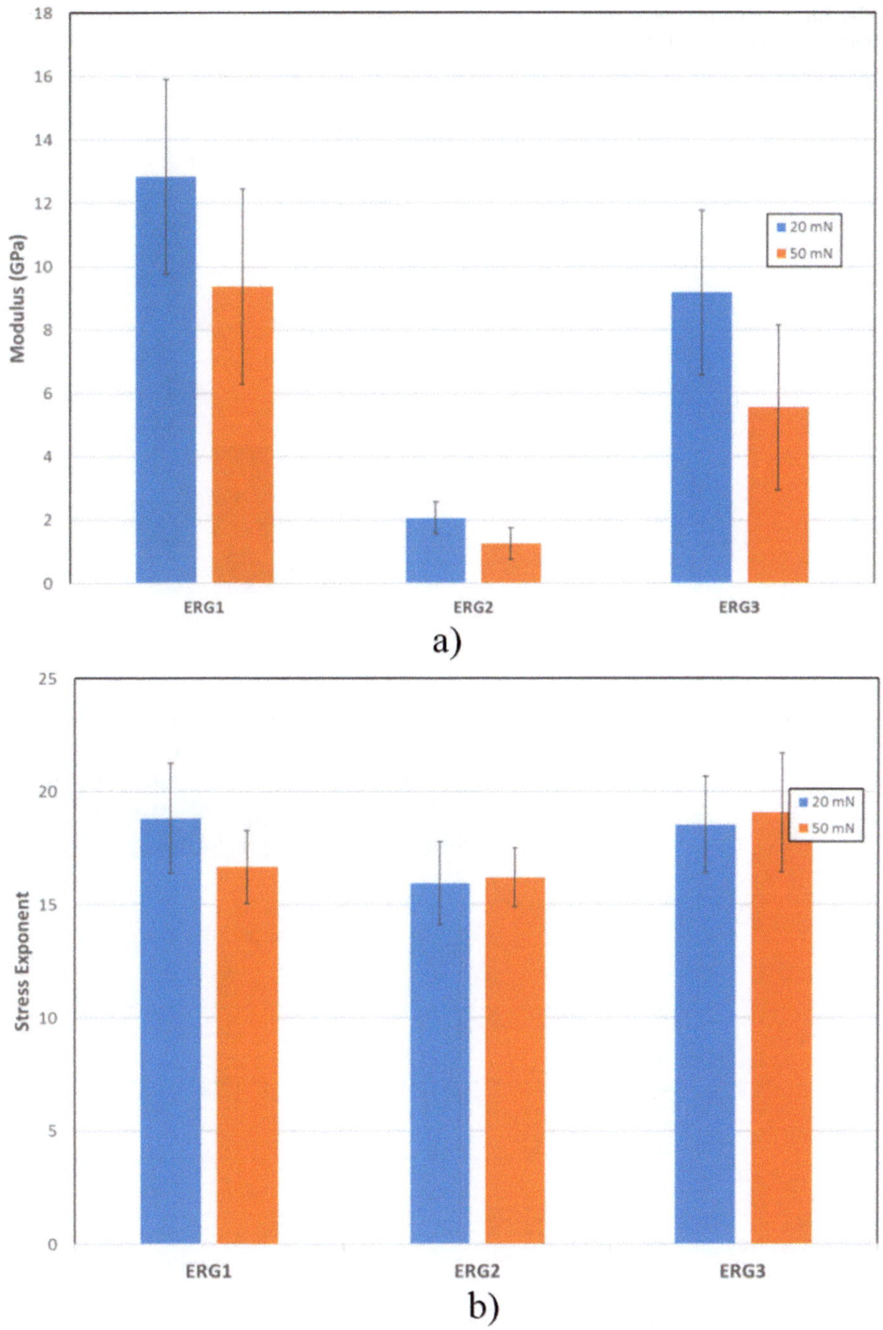

Bending stress is calculated of the composites tested according to the Eq. (5.4):

$$\sigma = \frac{3Pl}{2bh^2} \tag{5.4}$$

In this formula, l is the span length, P is the maximal bending load, b and h are the sample width and thickness, respectively.

An alteration was observed in the bending strength of specimens by the change of the mass rate of graphene amount which can be seen from Fig. 5.13. However, from the values this situation cannot be generalized. The most probable reason for this is thought as non-homogeneous distribution of GNPS.

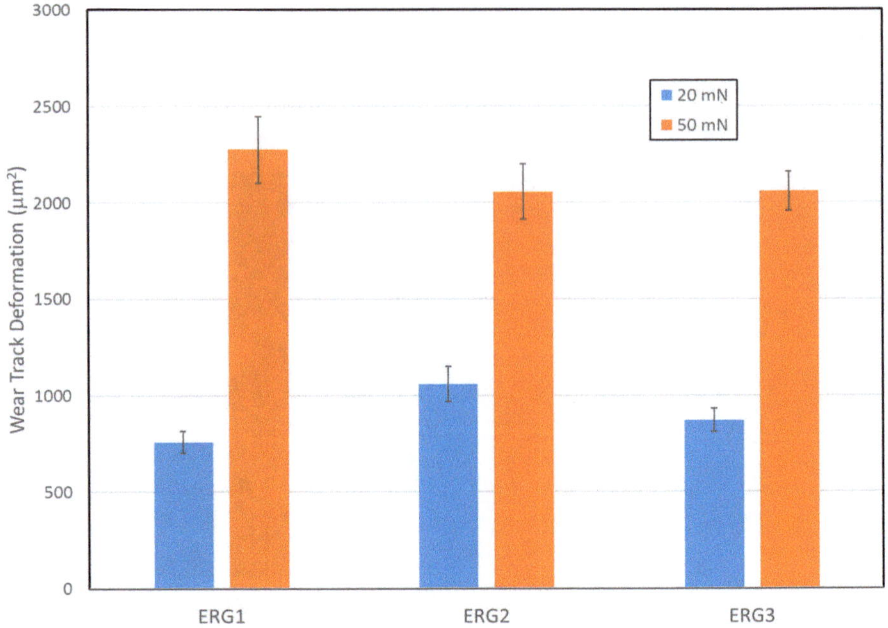

Fig. 5.8 Comparison of wear track deformation for ERG composites

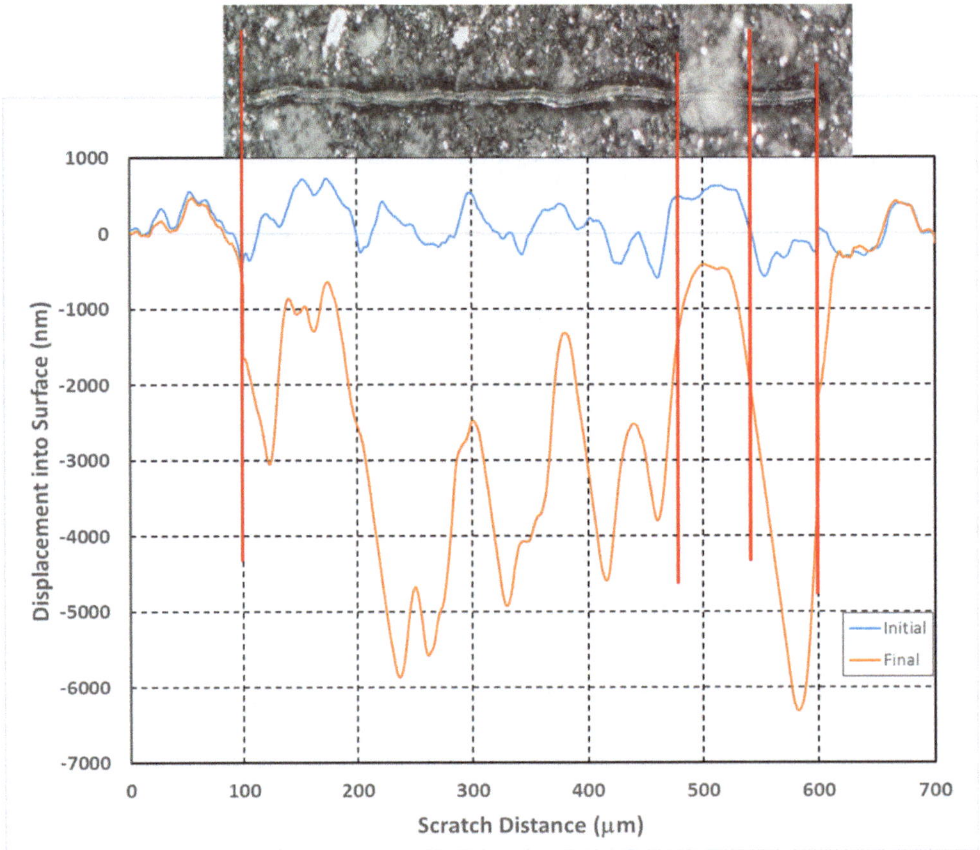

Fig. 5.9 Initial profile and the residual profile obtained at 50 mN on ERG1 for one of the wear tracks

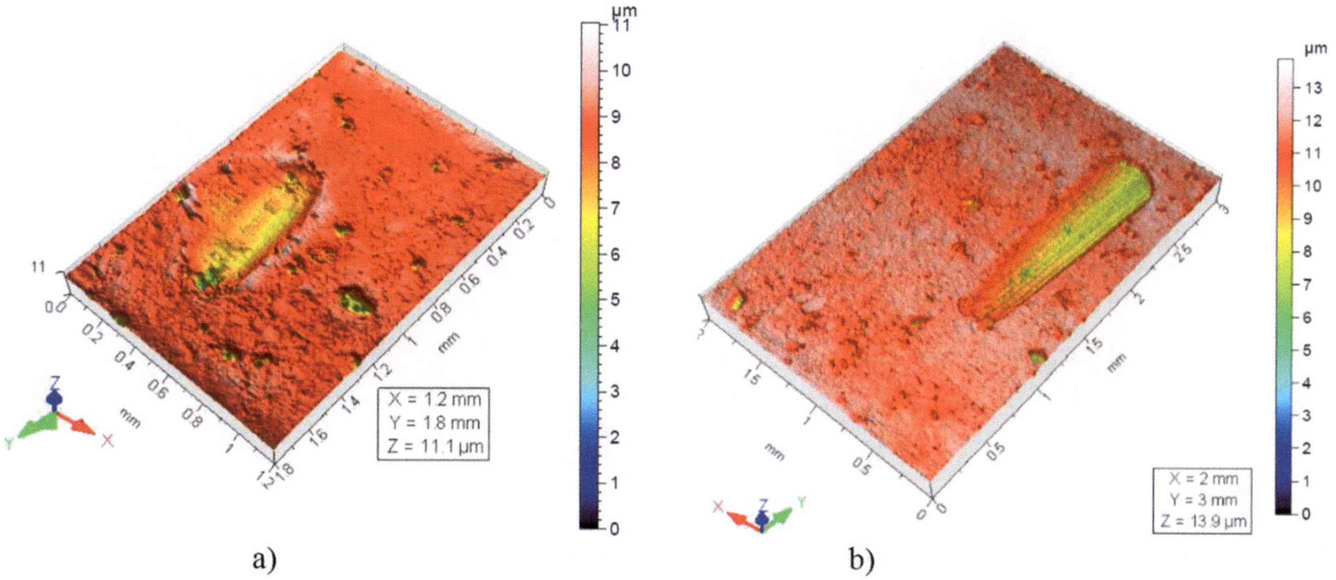

Fig. 5.10 (**a**) Damage traces obtained in the direction of width and length for the specimen ERG I for 50.10^3 cycles, and (**b**) for 100.10^3 cycles

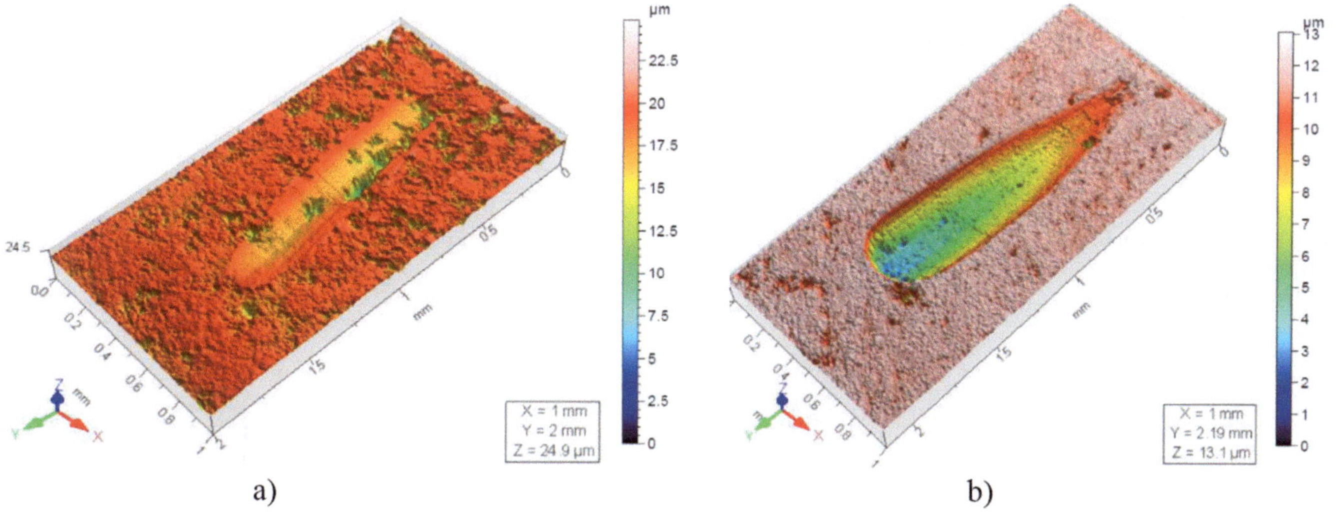

Fig. 5.11 (**a**) Damage traces obtained in the direction of width and length for the specimen ERG II for 50.10^3 cycles, and (**b**) for 100.10^3 cycles

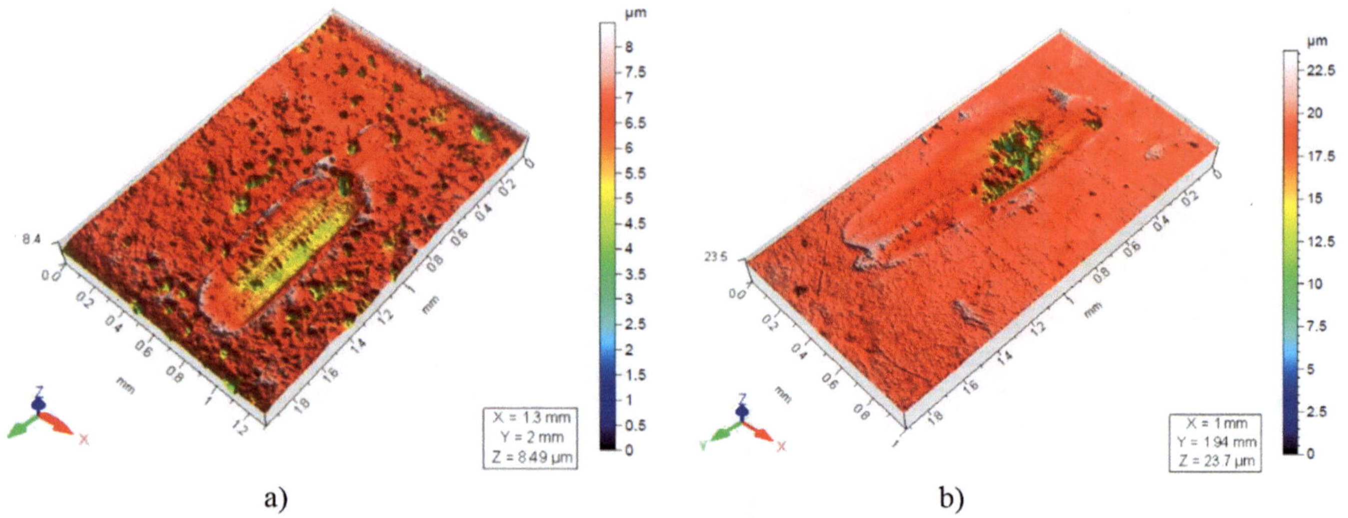

Fig. 5.12 (**a**) Damage traces obtained in the direction of width and length for the specimen ERG III for 50.10^3 cycles, and (**b**) for 100.10^3 cycles

Table 5.4 Damage traces in direction of width and length for the specimens: ERG I-II-III for 50.10^3 and 100.10^3 cycles

Numbers of cycles	ERG I		ERG II		ERG III	
	Worn surface (mm^2)	Volume lost (µm^3)	Worn surface (mm^2)	Volume lost (µm^3)	Worn surface (mm^2)	Volume lost (µm^3)
50.10^3	0.179	188,838	0.240	540,620	0.297	363,837
10.10^3	0.639	1,224,979	0.453	1,524,242	0.424	808,988

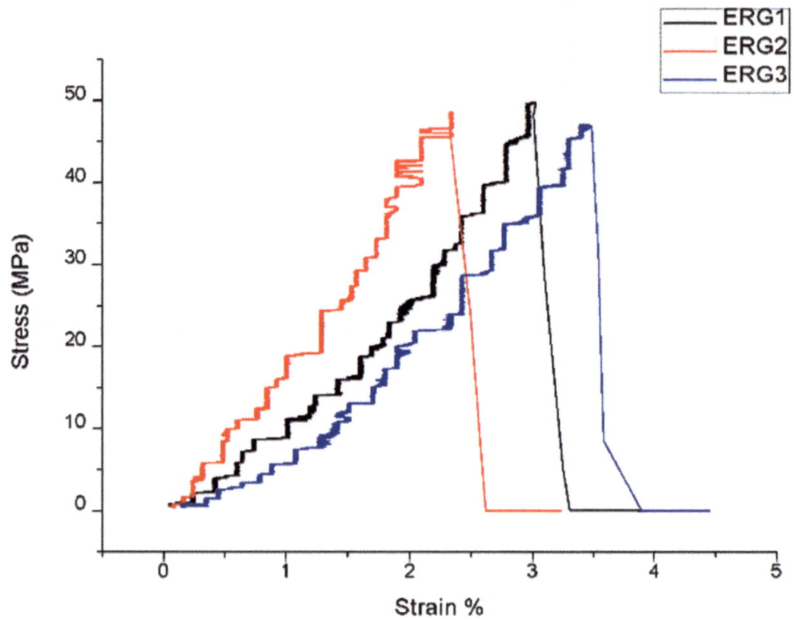

Fig. 5.13 Comparison of stress-strain behaviour for ERG composites

5.4 Conclusions

In this study, the epoxy-rubber based composites were fabricated by powder metallurgy. Usually, composites reinforced with nanoparticles exhibit light weight and higher dimensional stability than their conventional composite counterparts, however, these properties depend on the certain parameters such as type of nanoparticle, polymer matrix, synthesis methods, and polymer nanocomposites morphology, etc.

Very small amount of agglomerations of rubber particles have been observed due to the short milling time; increase in the milling time should improve the distribution of particles in the matrix homogeneously. However, long processing time may have resulted in over heating the matrix and the binding characteristics of the matrix are damaged.

Creep compliance and wear track deformation are very useful parameters that measure a unique mechanical characterization of the time-dependent materials. Finally, SEM results verify the brittle fracture and also give some signs about the root cause of damage such as debonding of reinforcements.

References

1. May, C.A.: Epoxy Resins-Chemistry and Technology, 2nd edn. Marcel Dekker, New York (1988)
2. Wetzel, B., Rosso, P., Haupert, F., Friedrich, K.: Epoxy nanocomposites–fracture and toughening mechanisms. Eng. Fract. Mech. **73**(16), 2375–2398 (2006)
3. Garg, A.C., Mai, Y.: Failure mechanisms in toughened epoxy resins-a review. Compos. Sci. Technol. **31**(3), 179–223 (1988)
4. Lee, J., Yee, A.: Inorganic particle toughening I: Micro-mechanical deformations in the fracture of glass bead filled epoxies. Polymer. **42**(2), 577–588 (2001)
5. Tripathi, G., Srivastava, D.: Studies on the physico-mechanical and thermal characteristics of blends of DGEBA epoxy, 3,4 epoxy cyclohexylmethyl, 30,40- epoxycylohexane carboxylate and carboxyl terminated butadiene coacrylonitrile (CTBN). Mater. Sci. Eng. A. **496**, 483–493 (2008)
6. Douglass, S.K., Beaumont, P.W.R., Ashby, M.F.: A model for the toughness of epoxy rubber particulate composites. J. Mater. Sci. **15**(5), 1109–1123 (1980)
7. Kinloch, A., Shaw, S., Tod, D., Hunston, D.: Deformation and fracture behaviour of a rubber-toughened epoxy: 1. Microstructure and fracture studies. Polymer. **24**(10), 1341–1354 (1983)
8. Yee, A., Pearson, R.: Toughening mechanisms in elastomer-modified 1. J. Mater. Sci. **21**, 2462–2474 (1986)
9. Pearson, R., Yee, A.: Toughening mechanisms in elastomer-modified 2. J. Mater. Sci. **21**, 2475–2488 (1986)
10. Zhao, Q., Hoa, S.V.: Toughening mechanism of epoxy resins with micro/nano particles. J. Compos. Mater. **41**(2), 201–219 (2006)
11. Wichmann, M.H., Schulte, K., Wagner, H.D.: On nanocomposite toughness. Compos. Sci. Technol. **58**(1), 329–331 (2008)
12. Liang, Y., Pearson, R.: Toughening mechanisms in epoxy–silica nanocomposites (ESNs). Polymer. **50**(20), 4895–4905 (2009)
13. Srivastava, I., Koratkar, N.: Fatigue and fracture toughness of epoxy nanocomposites. J. Mater. **52**(2), 50–57 (2010)
14. Geim, A.K., Novoselov, K.S.: The rise of graphene. Nat. Mater. **5**, 183–191 (2007)
15. Balandin, A.A., Ghosh, S., Bao, W., Calizo, I., Teweldebrhan, D., Miao, F., et al.: Superior thermal conductivity of single-layer graphene. Nano Lett. **8**, 902–907 (2008)
16. Scarpa, F., Adhikari, S., Phani, A.S.: Effective elastic mechanical properties of single layer graphene sheets. Nanotechnology. **20**, 1–11 (2009)
17. Pathak, A.K., et al.: Improved mechanical properties of carbon fiber/graphene oxide-epoxy hybrid composites. Compos. Sci. Technol. **135**, 28–38 (2016)
18. Wang, F., et al.: Enhancement of fracture toughness, mechanical and thermal properties of rubber/epoxy composites by incorporation of graphene nanoplatelets. Compos. A: Appl. Sci. Manuf. **87**, 10–22 (2016)
19. Zhang, G., et al.: Effect of functionalization of graphene nanoplatelets on the mechanical and thermal properties of silicone rubber composites. Materials. **9.2**, 92 (2016)
20. Bayraktar, E., Zaimova, D., Tan, M.-J., Miskioglu, I.: A preliminary study on an epoxy-rubber matrix with different reinforcements. In: Tandon, G.P. (ed.) Experimental Mechanics of Composite, Hybrid, and Multifunctional Materials, vol. 6, no. 7, pp. 1–13. Springer Publication Springer Link, USA (2014)
21. Asaka, K., Ishihara, C.: Technical trends in soft magnetic parts and materials, Hitachi Powdered Metals Technical Report 420053, (2005)
22. Zaimova, D., Bayraktar, E., Miskioglu, I., Katundi, D., Hamouche, Z.: Manufacturing and damage analysis of filler reinforced epoxy based composites. Int. J. Mater. Prod. Technol. **48**(1–4), 47–65., Inter Science (ed.), 0268-1900/1741-5209 (2014)
23. Zaimova, D., Bayraktar, E., Miskioglu, I., Dishovsky, N.: Wear resistance of elastomeric based composites by continuous multi-cycle indentation used in manufacturing engineering. Adv. Mater. Res. **939**(1), 106–113., Trans Tech Publications (ed.), Switzerland (2014)
24. Zaimova, D., Bayraktar, E., Tan, M.J., Miskioglu, I.: Processing of epoxy-nickel matrix composites reinforced with aluminum and waste elastomers. Adv. Mater. Res. **939**(1), 98–105., Trans Tech Publications (ed.), Switzerland (2014)
25. Zaimova, D., et al.: Manufacturing and damage analysis of epoxy resin-reinforced scrap rubber composites for aeronautical applications. In: Experimental Mechanics of Composite, Hybrid, and Multifunctional Materials, vol 6, pp. 65–76. Springer International Publishing, USA (2014)

Chapter 6
Mechanical Characterization of Epoxy – Scrap Rubber Based Composites Reinforced with Alumina Fibers

A.B. Irez, E. Bayraktar, and I. Miskioglu

Abstract Recently, encouraged by the idea of sustainable development, the emissions of CO_2 have become a very prominent issue to be solved. As a consequence, in the automotive and aeronautic industries, the reduction of weight has become a major issue and needs to be addressed for all possible structures in the vehicle. Therefore, some metallic parts are being replaced by polymer based composites. For this to happen, the feasibility of using these composites has to be examined properly in terms of their mechanical and energetic properties. Epoxy – rubber based composites reinforced with ceramics are studied by several researchers. However, addition of scrap rubber to the matrix and alumina fibers (Al_2O_3) as a reinforcement introduces a novelty in this area. Since, using of scrap rubber has positive outcomes in economical and structural perspectives such as revaluation of scraps and compensating the brittleness of epoxy. Also, due to its favorable structural characteristics such as interlocking properties of fibers, addition of alumina fibers ensures desired mechanical properties in case of a homogeneous distribution. This paper is focused on the design of epoxy – fresh scrap rubber matrix reinforced with different ratios of alumina fibers. The mechanical and some physical properties of these composite systems were studied in this research. Mechanical properties were evaluated by means of drop weight tests, three-point bending tests and nanoindentation technics. Scanning electron microscope (SEM) was used to observe fracture surfaces and the microstructure. After all, UV degradation of the material was observed by electron microscopy.

Keywords: Three-point bending • Nanoindentation • Epoxy – rubber • Alumina fibers • SEM

6.1 Introduction

In our day, breakthrough developments in the technology in the last decades bring novelties in the materials world. Out dated manufacturing methods and materials give their place to innovative ones. In this regard, novel, robust, light weight composite materials gradually substitute steels, aluminium alloys and so on.

Advances in the organic chemistry enable to produce durable polymers and certain types of these polymers are used as the matrix of the polymer based composites. For instance, epoxy is a thermoset polymer which is widely used in many different applications due to its favourable properties such as decent chemical resistance, outstanding mechanical properties, and ease in modification versatility as well as its inherent low viscosity and volatility [1]. However, the lack of ductility restricts its usage in certain applications. Therefore, in order to improve the fracture toughness of epoxies, adding a rubber phase such as commercially available styrene butadiene rubber with different functional terminal groups has been implemented by different researchers [2]. Meanwhile, using of recycled rubber gives an economic perspective to this study. The volume fraction of the toughening rubber phase generally varies from 5% to 20%. Nevertheless, the incorporation of rubber phase increases the viscosity of the epoxy resin mixture whereas the crosslink density, elasticity modulus and tensile strength diminish. Besides, the glass transition temperatures of epoxies dramatically reduce by the addition of rubber phase [3–5]. Another important point is that the modification of epoxies also done by the incorporation of some inorganic – particulate fillers such as silica [6, 6], alumina [8–11]. They generally improve the materials properties such as fracture toughness [12, 13], resin stiffness [14] and wear resistance [8]. Besides, they can reduce the coefficient of thermal expansion (CTE) [15]. The composites modified by these compounds utilized in different industries such as electronic packaging/under fill for circuit cards [16–18] dental restoration [19, 20] and automobile parts [21].

A.B. Irez (✉) • E. Bayraktar
Supmeca-Paris, School of Mechanical and Manufacturing Engineering, Paris, France
e-mail: alaeddin-burak.irez@supmeca.fr

I. Miskioglu (✉)
Michigan Technological University ME-EM Department, Houghton, MI, USA
e-mail: imiski@mtu.edu

P.R. Thakre et al. (eds.), *Mechanics of Composite and Multi-functional Materials, Volume 6*,
Conference Proceedings of the Society for Experimental Mechanics Series, DOI 10.1007/978-3-319-63408-1_6

In addition to previously mentioned properties of the fillers, alumina has good thermal conductivity, inertness to most acids and alkalis, high adsorption capacity, thermal stability and electrical insulation, and so on. Also, it is inexpensive, non-toxic and highly abrasive [22–24]. More specifically, in this study alumina has been used in fiber form and it has advantages compared to particle form. For instance, the diameter of alumina fiber is about two times larger than that of carbon fiber. Also, another beneficial parameters as interfacial strength and elastic modulus in a radial direction (Young's modulus of alumina fiber is much larger than that of carbon fiber in a radial direction) may also contribute to the difference in strength and fracture mechanism [25].

The present work presents processing of scrap rubbers treated with epoxy resin to create novel composites in an economic way. Main objective of this research was to determine the mechanical and tribological properties of these composites by using scrap rubber powders as a matrix reinforced with aluminium and alumina fibre. During this experimental study, nano wear and creep behaviour were examined by means of nanoindentation technics. Compression tests were performed for mechanical characterization. In addition, surface hardness was determined by means of Shore D hardness measurement. Furthermore, impact behaviour was investigated by means of drop weight testing. Scanning electron microscopy (SEM) was used to study the microstructure and fracture surfaces of these composites. After all, these polymer based composites were exposed to UV lights to see their degradation under these circumstances.

6.2 Experimental Conditions

6.2.1 Materials Processing

In the first stage, in order to combine very fine dry epoxy powder and fine scrap rubber powder (Styrene–butadiene rubber (SBR)) series of chemical treatments were applied. To begin with, toluene and acrylic acid are applied to SBR and then it is dried after treatment in the oven to prepare it for a better adhesion between epoxy. Also, vinyltriethoxysilane used as a surfactant to provide a better adhesion between this epoxy – rubber blend and reinforcements. Then this blend dried in an oven to eliminate liquid phase drawbacks. This mixture was milled for an hour to obtain a homogeneous compound and then heated at 80 °C for 24 h. The resulting compound was used as the matrix for the proposed composites. In the second stage, the reinforcements such as nano Al and Al_2O_3 fiber were added to the matrix in pre-defined ratios. All of the compounds were then mixed in a blender and milled by Fritsch® Pulverisette2 for 4 h. Longer milling times result in over heating of the mixture and this phenomenon influences the material properties. After that, the specimens (called as ERAL I-II hereafter) were manufactured by hot compacting (double uniaxial action) under a pressure of 70 MPa at 180 °C. The dwell time for compacting process was 15 min. All of the specimens (30 mm & 50 diameters, 5–6 mm thickness) were cooled down slowly. The post curing was concluded under isothermal conditions at a temperature of 80 °C for 24 h.

General compositions of all the composites manufactured are given in the Table 6.1 with specified weight ratios for each constituent.

6.2.2 Mechanical Tests and Microstructural Analyses

After manufacturing of the specimens, two specimens were submitted to macro-indentation compression tests by means of a 5 mm steel ball. Tests were carried out in a servo-hydraulic MTS Universal test device (model 5500R, equipped with a load cell of 25.000 kgf) with a quasi-static low speed (initial rate: 10 mm/min and second rate: 5 mm/min rate).

All testing parameters were given in the Table 6.2. All of the static compression tests have been carried out under the same conditions to evaluate damage behaviour of this composite.

Table 6.1 Composition of the epoxy-rubber based composites

Epoxy-Rubber based composition Epoxy – SBR rubber (10 phr)	ERAL I	ERAL II
Reinforcements (wt %)	10 alumina fiber (Al_2O_3) 5 Nano Aluminium	15 alumina fiber (Al_2O_3) 5 Nano Aluminium

Table 6.2 General conditions for compression tests of produced composites

Initial speed	10	mm/min
Load endpoint	10,000	N
Outer loop rate	100	Hz
Secondary speed	5	mm/min
Strain endpoint	80	%

Table 6.3 Hardness values of ERAL specimens

Hardness Measurement	
Specimen	Shore D
ERAL I	91
ERAL II	91

Microstructure and surface damage evaluations were made by means of optical (OM) and scanning electron microscopy (SEM). SEM observation was realized on fracture surface of the tested specimens with Scope/JSM-6010LA Jeol® electron microscope which has a magnification capacity of x300,000 and with accelerating voltage from 500 V to 20 kV.

After the post curing, hardness and density measurements on the specimens were performed. Shore D hardness test measurements on the polished flat surfaces of the specimens were carried out according to ASTM D 2240 using durometer Shore test device, (type HBD-100-0). Hardness results were given in Table 6.3.

The bending specimens were prepared according to the ASTM D790 standards. The most common bending test for composite materials is thought as the three-point bending test. Deflection of the specimen is measured by the crosshead position. Bending strength, displacement and strain, are obtained from the test results. The test specimen was placed in the universal testing machine and force was applied until it fractured.

Dynamic compression tests (drop weight tests) were carried out using a universal drop weight test device. A standard conical punch was released from a height of 900 mm. Impact behavior of the manufactured composites were observed over the test specimens with the help of force-time curves.

6.2.3 Nanoindentation: Creep and Wear Tests

Creep tests using a nanoindenter were performed on the two compositions manufactured. On each sample 20 indents were performed on a 5×4 grid with a Berkovich indenter. The indents were spaced 50 μm along the 5 indent side and 75 μm along the 4 indent side. The load was increased at a rate of 1 mN/s to the max load and kept at the maximum load for 500 s then unloaded. A typical load-on-sample vs displacement-into-sample graph is shown in Fig. 6.1. Two maximum loads of 20 mN and 50 mN were used in these tests.

Scratch testing capability of a nanoindenter is utilized to perform relatively fast wear tests to compare the wear behavior of the different samples. Wear tests were conducted by a conical tip with a 90° cone angle. Tests were run under a normal load of 20 mN and 50 mN applied over a linear track of 500 nm for 50 cycles.

6.2.4 Damage Analysis by Means of Scratch Test

Scratch tests results give a basic idea on the tribological behaviour of the epoxy and recycled elastomer based composites designed in the current research. After the scratch test, damage zone was investigated by a 3D optical surface scanner to assess damage in terms of scratch depth and average scratch roughness. In the scratch test the contact between the sliding zirconium indenter and the surface of the composite material during scratch test is analyzed. The normal and tangential forces on indenter are recorded and the tangential stress on the surface and the interfacial stresses can be obtained. In the frame of this present work wear resistance was evaluated only under dry conditions and 50×10^3 and 100×10^3 numbers of cycles of wear for the composites presented here.

In the frame of the current research, the resistance to scratch deformation is evaluated in terms of scratch depth, surface and worn volume subsequent to scratching.

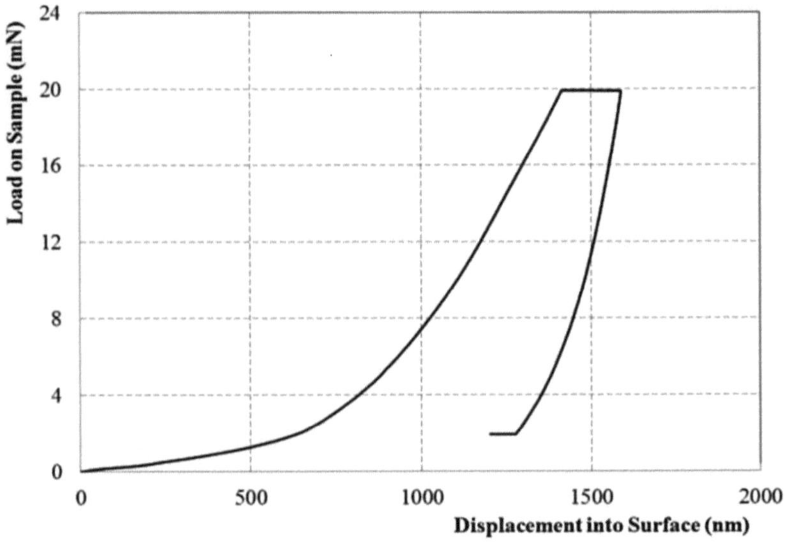

Fig. 6.1 Typical Load-Displacement curve during the creep test

6.3 Results and Discussions

6.3.1 Microstructure of the Composites and Macro Indentation Tests

In Fig. 6.2 two different specimens were shown in different diameters for different characterizations. They were manufactured by means of hot compaction techniques.

General microstructures in the transversal direction of two compositions are shown in the Fig. 6.3. All of the compositions have shown a considerably homogenous distribution of the reinforcements in the structure. Essentially, all of the microstructures show that the adhesion of the rubber to the epoxy matrix is very successfully carried out after the simple chemical treatment (Silanization). In addition, white circular particles in Fig. 6.3 show the alumina fibers in different sizes. Some small local agglomerations are observed in the structure due to the short time mixture process. Since, homogenous distributions of the reinforcements need more mixture process with distribution small size particle reinforcements.

In Fig. 6.4, stress-strain values were shown obtained from macro indentation compression tests. Macro indentation strength of the composites increases slightly with the iron oxide content. However, maximum strength values are quite close to each other for these composites that show high toughness behaviour.

6.3.2 Dynamic Compression (Impact) Testing

As it is well known, epoxy has a very brittle characteristic. To improve it rubber particles are incorporated to the matrix. Toughness of these composites is developed and impact behaviour of the specimens was studied and it is given in Fig. 6.5. Impact forces were evaluated depending on the time for two composites and time in this curve. From this curve, one may note that the main reinforcement, Al_2O_3 fiber does not affect seriously toughness level of these composites Total absorbed energy during impact is considerable for two composites. May be an optimization of the reinforcement level can be kept minimum for damping properties of the composites. In fact, these results should be accepted as indicative results obtained under certain experimental conditions in laboratory scale that give a practical idea about the basic roles of the alumina fibers as reinforcement particles.

Again, fracture surfaces of the two compositions are given in the Fig. 6.6 obtained after impact tests. All of the fracture surfaces of the specimens indicate the regular brittle type of fracture influenced by brittle effect of the epoxy resin. Even if recycled scrap rubber powder are used for creating a ductile matrix by using different binding elements to create a good chemical bonding between the matrix and reinforcements, the matrix keep the brittle behaviour. It should be change the ratio epoxy: rubber by increasing the rubber percentage. Another microstructural defect coming from the big size of recycled rubber powder should be eliminated. During the milling, it is difficult to decrease the big rubber particles. These particles can

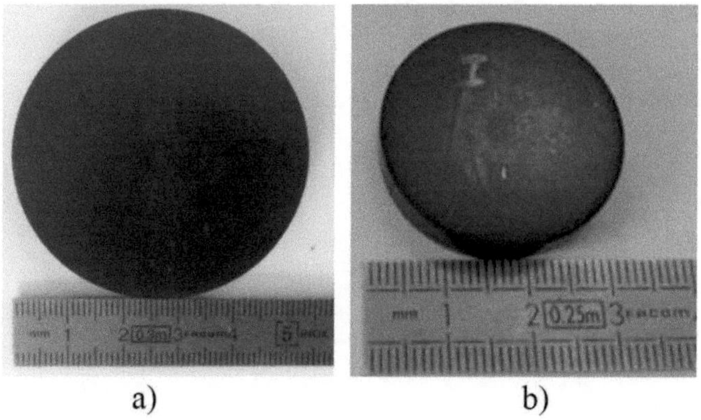

Fig. 6.2 (**a**) Macrograph of the specimens after compacting and post curing (Al + Al_2O_3+ epoxy-rubber) d = 50 mm (**b**) d = 30 mm

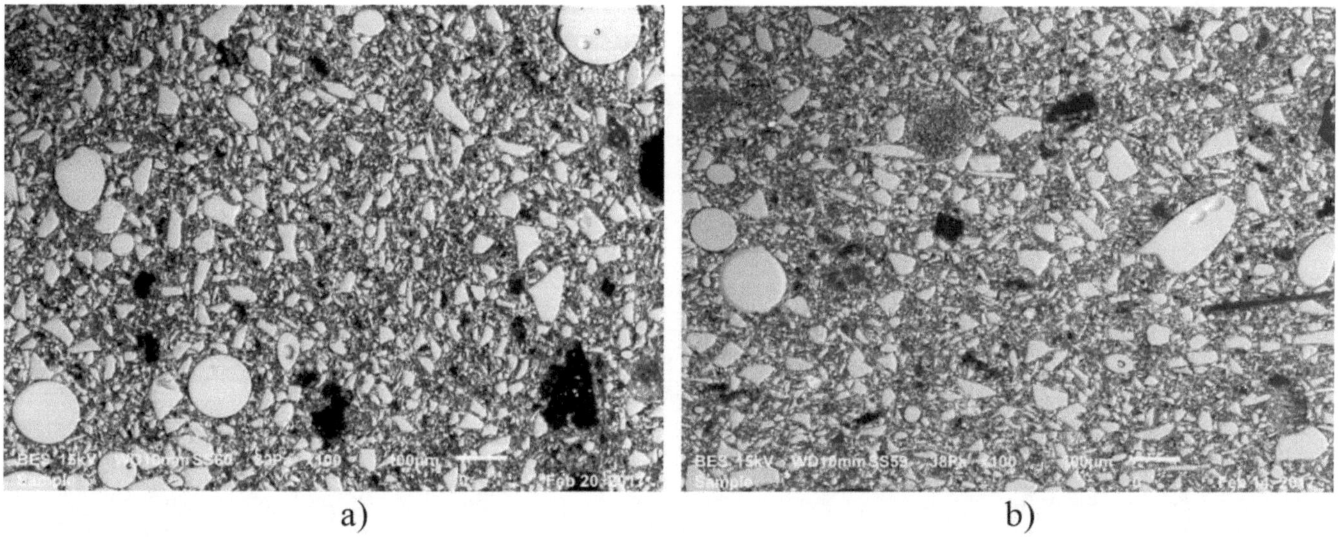

Fig. 6.3 Microstructure of ERAL specimens transversal sectioning (**a**) ERAL-I, (**b**) ERAL-II, respectively

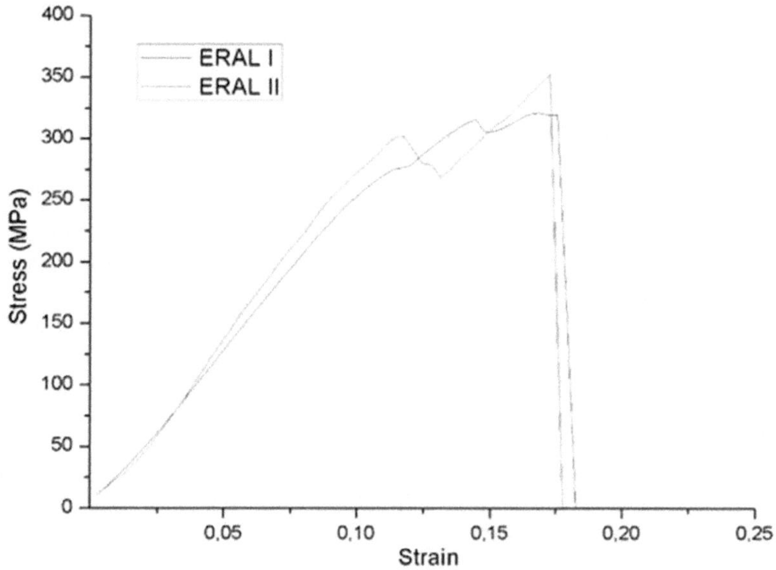

Fig. 6.4 Macro indentation test results of ERAL composites

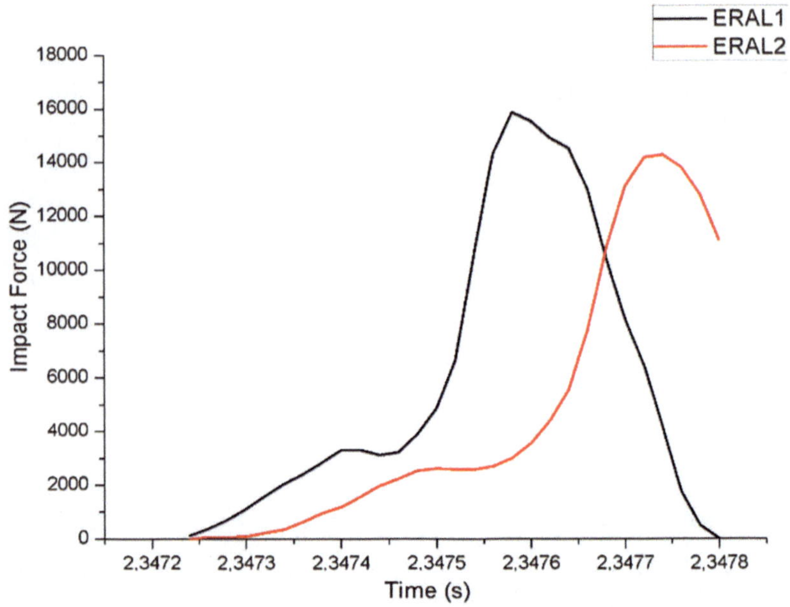

Fig. 6.5 Impact behaviors of two types of ERAL composites by dynamic compression test

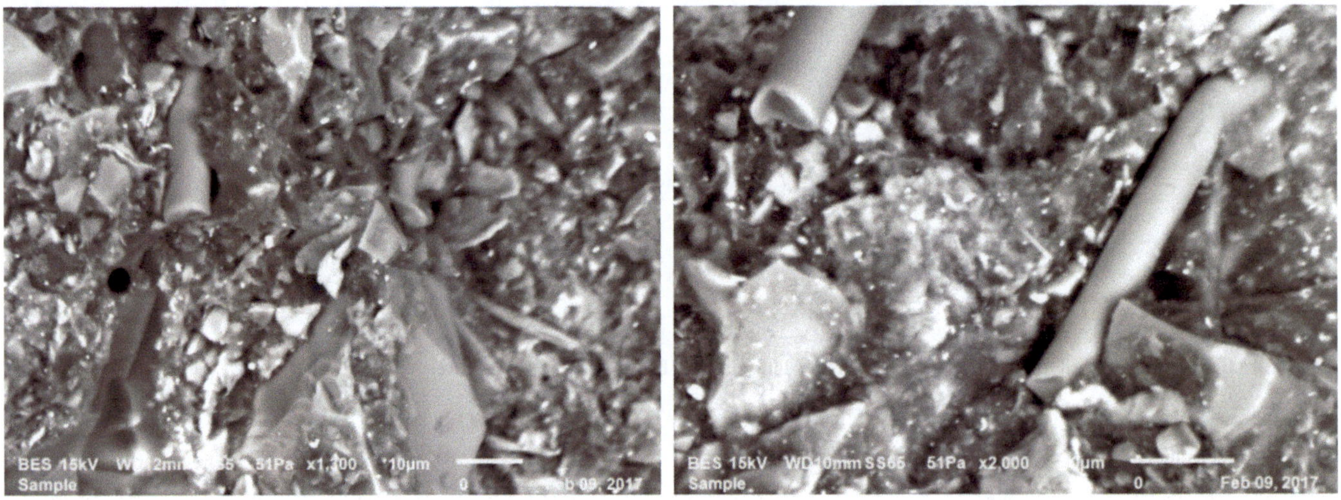

a) b)

Fig. 6.6 (**a**) Fracture surfaces observed after impact testing ERAL I and (**b**) ERAL II

cause in decohesion between matrix and reinforcement particles that may be decrease of impact beahviour. Finally, rubber powders in the matrix can improve ductility and also reinforcements play strong barriers against the propagation of the fracture. SEM images on the fracture surface observed after drop weight tests showed that certain fracture of the individual particles and also the fibers play a role of bridge during the choc. Even if the brittleness effect of epoxy resin in matrix, these composites are very suitable for many engineering applications where not so high impact forces are not required.

6.3.3 Creep Testing by Nanoindentation

During each test, data collected was used to calculate the creep compliance and the stress exponent defined in Eq. (6.1) [26–30]:

$$\varepsilon(t) = \sigma_0 J(t) \tag{6.1}$$

where σ_0 is the constant stress applied and *J(t)* is calculated using Eq. (6.2)

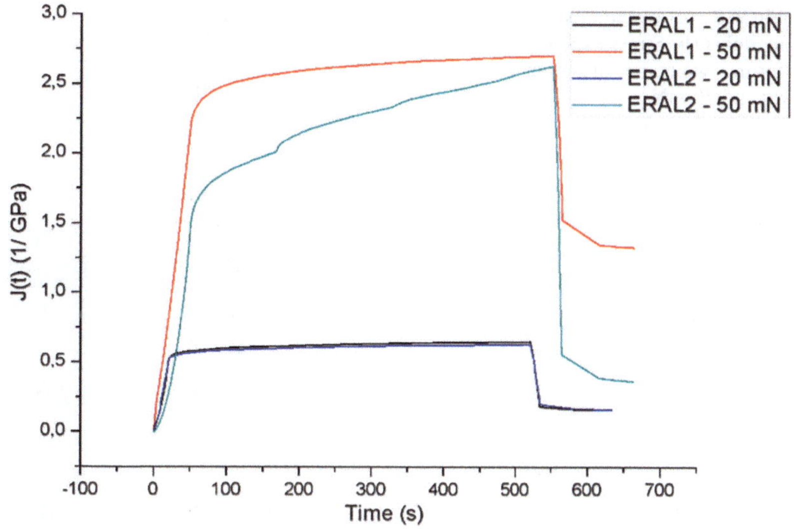

Fig. 6.7 Creep Compliance curves for two composites ERAL I-II, under 20 and 50 mN load

$$J(t) = A(t)/(1 - \nu)P_0 \tan\theta \tag{6.2}$$

In Eq. (6.2) $A(t)$ is the contact area, P_0 constant applied load, θ is the effective cone angle which is 70.3° for a Berkovich indenter and the Poisson's ratio ν is assumed to be 0.3.

The strain versus time behavior during creep is characterized by a high strain rate $\dot{\varepsilon} = d\varepsilon/dt$ in the primary stage of creep and then in the secondary, steady state stage of creep, the strain rate is given in Eq. (6.3) can be written as

$$\dot{\varepsilon} = K\sigma^n \tag{6.3}$$

where K is a constant and n is the stress exponent. The strain rate is calculated in the software and in turn n is obtained from the log-log plot of strain rate versus stress in the secondary stage of creep.

The materials under consideration are heterogeneous in nature and the fact that the nanoindenter testing is carried out over a small area/volume, a large scatter in the data is observed.

According to Fig. 6.7 it is seen that creep compliance values are not quite dependent to the alumina fiber content of the compositions. For instance, creep compliance is not very sensitive to fiber content at 20 mN of loading.

The average indentation modulus and stress exponent values were obtained from the 20 indentations performed under 20 and 50 mN constant loads and they are presented in Fig. 6.8a, b with error bars showing ± one standard deviation.

From Fig. 6.8b, it is seen that, increasing Al_2O_3 fiber amount results in a rise in the stress exponent. On the other hand, for different forces, Young modulus does not show a constant tendency whether a rise or descent.

6.3.4 *Wear Testing by Nanoindentation*

Nano-indenter was used to obtain scratches over the specimen. One cycle is defined as a pass and return of the indenter over the track, so the total distance covered for one wear test was 0.050 m. Speed of the tip during wear tests was 50 μm/s.

Total of 10 wear tests for each specimen were performed under the two normal loads. The wear in a track is characterized as the area between the initial profile and the residual profile of the wear track.

In the same way, the averages of the wear track deformation are shown in Fig. 6.9.

There is not a significant change was observed in the wear track deformations for the 20 mN force value. However, increasing rate of alumina fibers resulted in a more resistant nano-wear behavior for more considerable loading values.

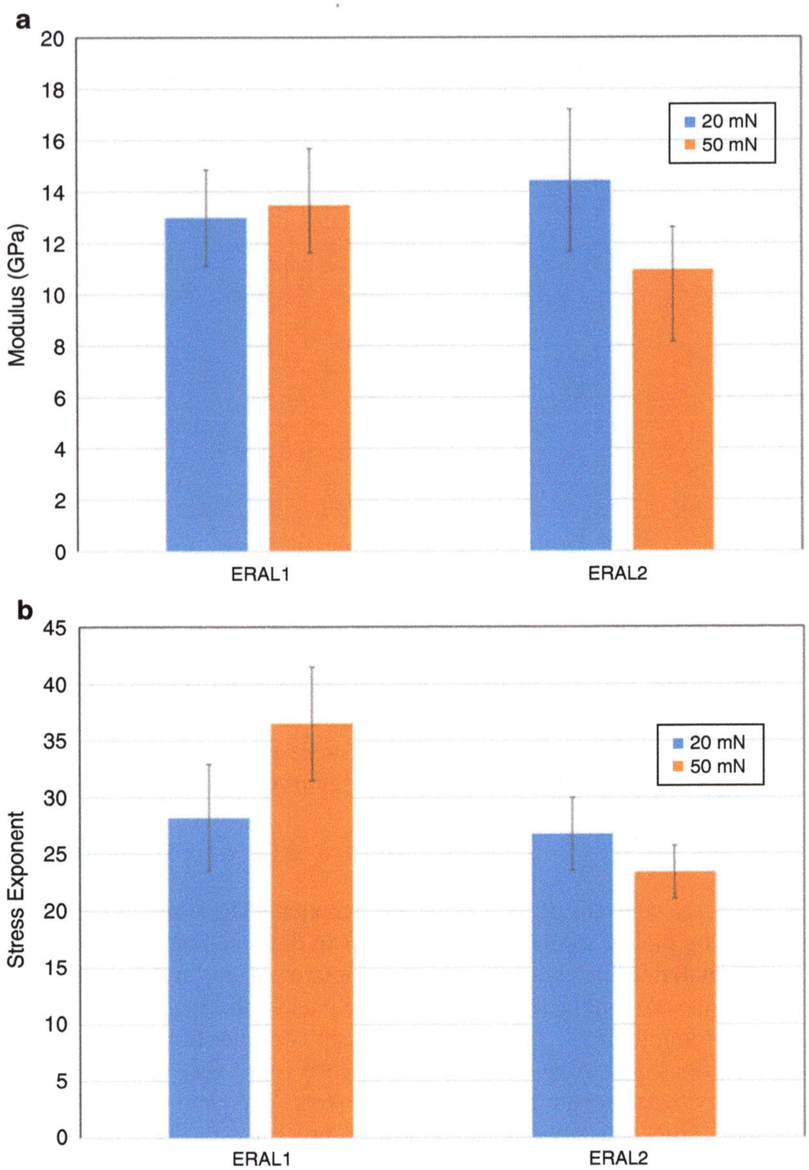

Fig. 6.8 (**a**) Indentation modulus (**b**) Stress exponent for the two composites under 20 and 50 mN

6.3.5 Damage Analysis by Means of Scratch Test and 3D Optical Roughness Meter

After realizing two different group of macro scratch tests, three dimensional damages were obtained by 3D optical surface scanner. The results are presented in the Figs. 6.10 and 6.11.

In Table 6.4 the volume and surface of the damage trace after scratch calculated from roughness test results are given. It is hard to find relation between two compositions concerning the surface of the damage trace. However, based on Table 6.4, it seems that addition of alumina to the compound resulted in less damage. In addition, the number of cycles doesn't play a significant role concerning surface of the damage trace.

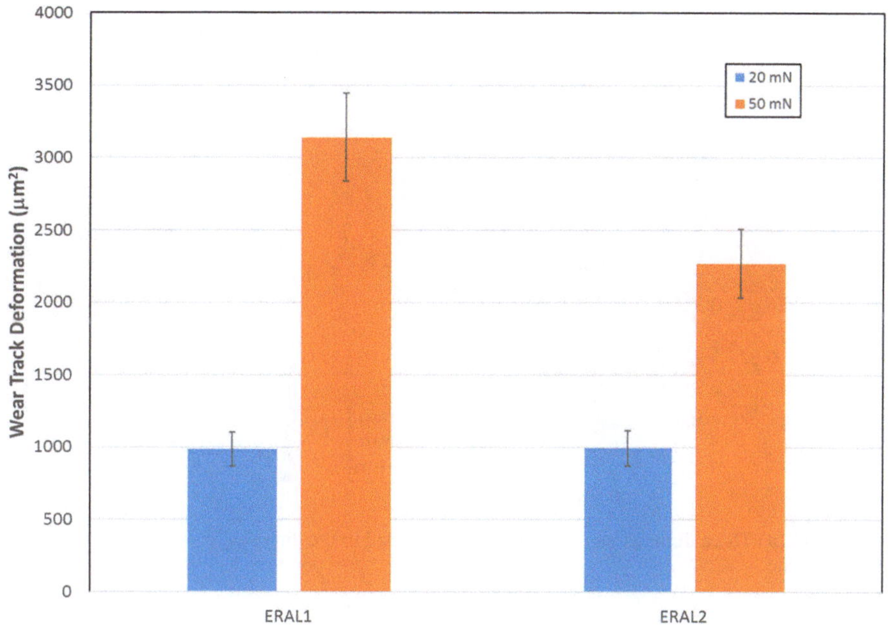

Fig. 6.9 Comparison of wear track deformation for ERAL composites

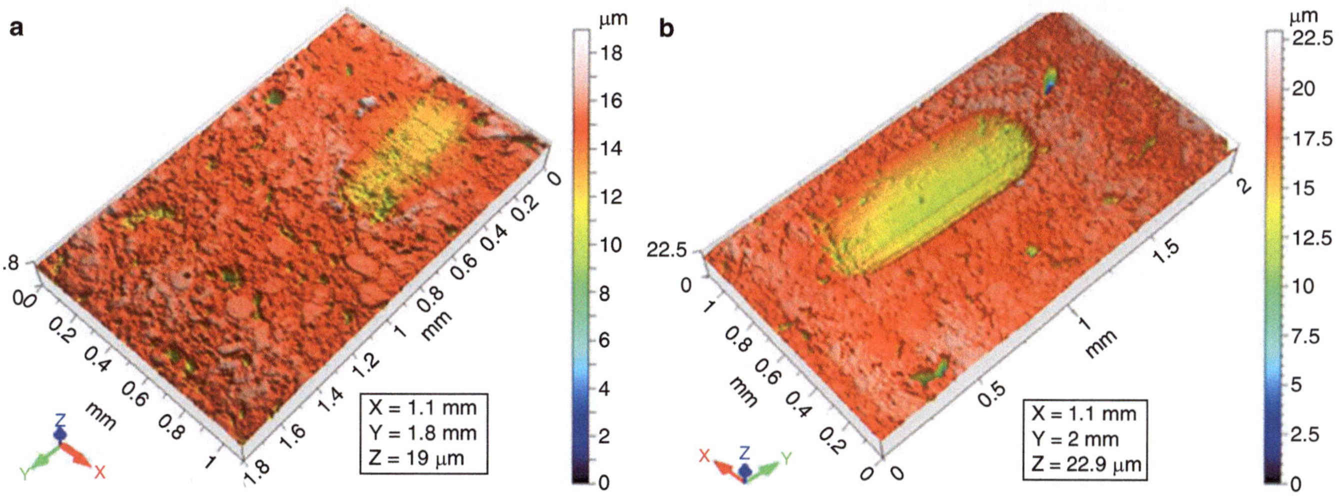

Fig. 6.10 (**a**) Three dimensional damage traces obtained in the direction of width and length for the specimen ERAL I for 50 k cycles, and (**b**) for 100 k cycles

In reality, because of the high shear stress at the interfaces the interfacial shear stress should probably be the main reason for damage of the matrix and reinforced filler interfaces [31]. When the indenter is slipping, tangential tensile stress is caused on the surface behind the indenter, while in front of the indenter the tangential stress is compressive.

6.3.6 Bending Testing by Means of three Point Bending

Three-point bending tests were conducted for each different type of composites and the preliminary results were given in Fig. 6.12. It is known that from literature, bending test is always necessary for the composites to see well the behaviour of the reinforcements. Additionally, non-linear evolution of the curves is a consequence of the anisotropic internal structure. In

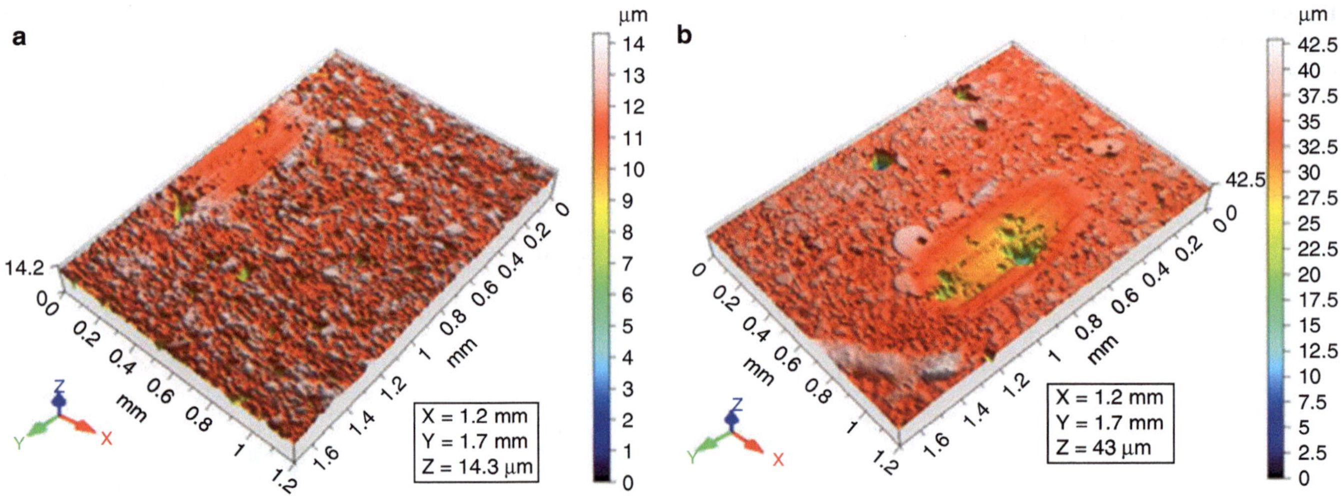

Fig. 6.11 (**a**) Three dimensional damage traces obtained in the direction of width and length for the specimen ERAL II for 50 k cycles, and (**b**) for 100 k cycles

Table 6.4 Damage traces obtained in the direction of width and length for two composites ERAL I and ERAL-II for different numbers of cycles under the same test conditions

	ERAL I		ERAL II	
Number of cycles	Worn surface (mm^2)	Worn volume (μm^3)	Worn surface (mm^2)	Worn volume (μm^3)
50.10^3	0.203	325,093	0.128	58,112
100.10^3	0.397	1,279,721	0.234	856,060

Fig. 6.12 Comparison of stress-strain behaviour for ERAL composites

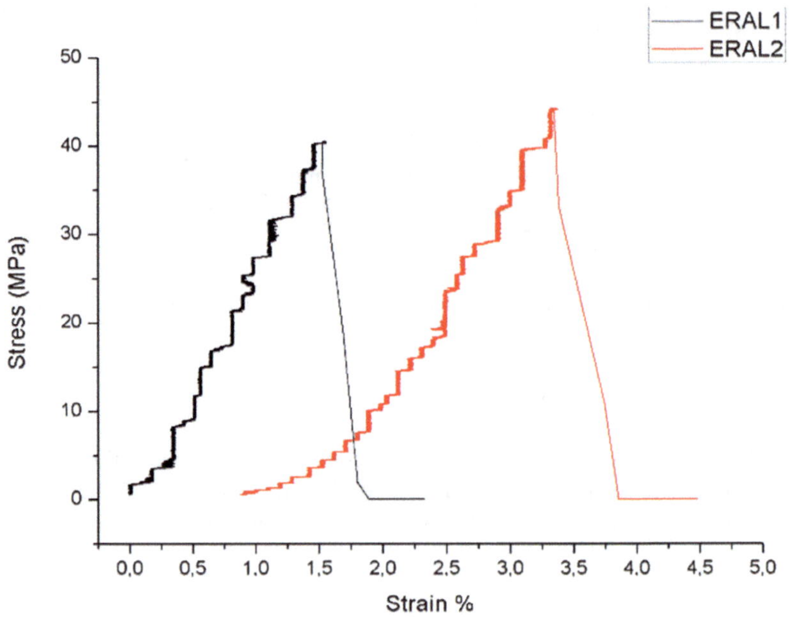

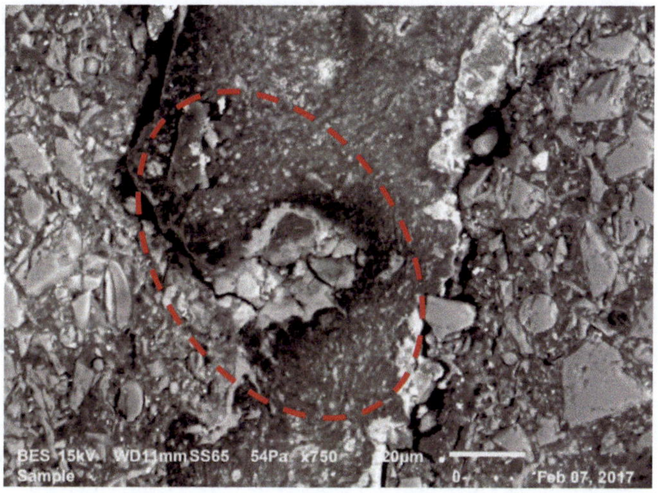

Fig. 6.13 Typical damage zone (in *red circle*) created by UV exposition for a long period (2 months here) at interface between matrix and reinforcements (Example: specimen taken from ERAL-I)

case of these composites, bending stress was calculated of the composite tested according to the Eq. (6.4) by considering the standard test conditions.

$$\sigma = \frac{3xPxl}{2xbxh^2} \tag{6.4}$$

In this formula, l is the span length, P is the maximal bending load, b and h are the sample width and thickness, respectively.

It seems that, a slight improvement is observed in the bending strength values for the specimens containing higher alumina fiber.

6.3.7 UV Degradation on Specimens

After manufacturing of the composites, certain amount of the specimens were exposed to UV light in a closed ambient for 2 months to see the damage effect of UV lights on these composites. Only as an example, a typical damaged zone was presented in the Fig. 6.13 taken by SEM microstructure. The same specimen was observed before UV exposition and under regular period was checked in SEM and finally this zone was detected as damage area found on the specimen ERAL-I after 2 months UV exposition. A simple decohesion of the rubber particles (around 40 µm) is observed from epoxy matrix due to drying and decreasing the bonding liaison of the rubber particles during the UV exposition. During this period, shrinkage in volume can create an internal stress and cause in small crack sources around this interface as observed in the Fig. 6.13.

6.4 Conclusions

In this study, the epoxy-rubber based composites were fabricated through powder metallurgy. Usually, composites reinforced with nanoparticles exhibit light weight and higher dimensional stability than conventional composite counterparts, however, these properties depend on the certain parameters such as type of nanoparticle, polymer matrix, synthesis methods, and polymer nanocomposites morphology, distribution of these particles, etc. Certain amount of rubber particles is found that could not be eliminated due to the short milling time. This case should be improved the distribution of particles in the matrix homogeneously. It means that mixture and milling time should be optimized. Creep compliance and wear track deformation are very useful parameters that measure a unique mechanical characterization of the time-dependent materials such as indentation hardness and indentation modulus, etc. Fracture behaviour of the composites studied here show certain amount of brittle behaviour in case of impact but high resistant and high toughness behaviour is observed in other case such as macro indentation compression test, bending test, etc. Finally, very good surface wear resistance is observed in case of macro scratch application on the surface of these composites.

References

1. McGrath, L.M., et al.: Investigation of the thermal, mechanical, and fracture properties of alumina–epoxy composites. Polymer. **49**(4), 999–1014 (2008)
2. Zaimova, D, Bayraktar, E, Katundi, D, Dishovsky, N.: Design of new elastomeric composites used in manufacturing engineering. In: 14th International Materials Symposium–IMSP, pp. 10–12. Denizli (2012)
3. Shaw, S.J.: Rubber modified epoxy resins. In: Collyer, A.A. (ed.) Rubber Toughened Engineering Plastics, pp. 165–209. Chapman & Hall, London (1994)
4. Sultan, J.N., McGarry, F.J.: Effect of rubber particle size on deformation mechanisms in glassy epoxy. Polym. Eng. Sci. **13**, 29–34 (1973)
5. Manzione, L.T., Gillham, J.K.: Rubber-modified epoxies. I. transitions and morphology. J. Appl. Polym. Sci. **26**, 889–905 (1981)
6. Nakamura, Y., Yamaguchi, M., Okubo, M., Matsumoto, T.: Effects of particle size on mechanical and impact properties of epoxy resin filled with spherical silica. J. Appl. Polym. Sci. **45**(7), 1281–1289 (1992)
7. Okazaki, M., Murota, M., Kawaguchi, Y., Tsubokawa, N.: Curing of epoxy resin by ultrafine silica modified by grafting of hyperbranched polyamidoamine using dendrimer synthesis methodology. J. Appl. Polym. Sci. **80**(4), 573–579 (2001)
8. Wetzel, B., Haupert, F., Zhang, M.Q.: Epoxy nanocomposites with high mechanical and tribological performance. Compos. Sci. Technol. **63** (14), 2055–2067 (2003)
9. Zee, R.H., Huang, Y.H., Chen, J.J., Jang, B.Z.: Properties and processing characteristics of dielectric-filled epoxy resins. Polym. Compos. **10** (4), 205–214 (1989)
10. Jin Kim, D., Hyun Kang, P., Chang Nho, Y.: Characterization of mechanical properties of γAl_2O_3 dispersed epoxy resin cured by γ-ray radiation. J. Appl. Polym. Sci. **91**, 1898–1903 (2004)
11. Arayasantiparb, D., McKnight, S., Libera, M.: Compositional variation within the epoxy/adherend interphase. J. Adhes. Sci. Technol. **15**(12), 1463–1484 (2001)
12. Nielsen, L.E., Landel, R.F.: Mechanical Properties of Polymers and Composites, 2nd edn. Marcel Dekker, Inc, New York (1994)
13. Rothon, R. (ed.): Particulate-Filled Polymer Composites. Longman Scientific and Technical, London (1995)
14. Rothon, R.N.: Mineral fillers in thermoplastics: filler manufacture and characterisation. Adv. Polym. Sci. **139**, 67–107 (1999)
15. Zhang, Z., Wong, C.P.: Double-layer no-flow underfill materials and process. IEEE Trans. Adv. Packag. **26**(2), 199–205 (2003)
16. Madenci, E., Shkarayev, S., Mahajan, R.: Potential failure sites in a flip-chip package with and without underfill. Trans ASME J Elect Pack. **120**(4), 336–341 (1998)
17. Wu, T.Y., Tsukada, Y., Chen, W.T.: Materials and mechanics issues in flip-chip organic packaging. Proc Electron. Comp. Technol. Conf. **46**, 524–534 (1996)
18. Vo, H.T., Todd, M., Shi, F.G., Shapiro, A.A., Edwards, M.: Towards model-based engineering of underfill materials: CTE modeling. Microelectron. J. **32**(4), 331–338 (2001)
19. Craig, R.G. (ed.): Restorative Dental Materials, 10th edn. Mosby-Year Book, Inc, St. Louis, Springer Cham Heidelberg New York Dordrecht London (1997)
20. Stannard, J.G.: Materials in Dentistry, 2nd edn. Denali Pub, Hanover (1988)
21. Babbington, D.A., Enos, J., Cox, J.M.: Barron J. Mod Plast. **64**(8), 72, 74, 76 (1987)
22. Mirjalili, F., Mohamad, H., Chuah, L.: Preparation of nano-scale α-Al2O3 powder by the sol–gel method. Ceram-Silikáty. **55**, 378–383 (2011)
23. Amirshaghaghi, A., Kokabi, M.: Tailoring size of α-Al2O3 nanopowders via polymeric gel–net method. Iran. Polym. J. **19**, 615–624 (2010)
24. Zhang, Z., Lei, H.: Preparation of α-alumina/polymethacrylic acid composite abrasive and its CMP performance on glass substrate. Microelectron. Eng. **85**, 714–720 (2008)
25. Williams, J.G.: Fracture Mechanics of Polymers. Halsted Press, New York (1984)
26. Bayraktar, E., Zaimova, D., Tan, M.-J., Miskioglu, I.: A preliminary study on an epoxy-rubber matrix with different reinforcements, In: G.P. Tandon (eds.) Experimental Mechanics of Composite, Hybrid, and Multifunctional Materials, vol. 6, no. 7, pp. 1–13. Springer Publication Springer link, (2014)
27. Asaka, K., Ishihara, C.: Technical trends in soft magnetic parts and materials. Hitachi Powdered Metals Technical Report 420053 (2005)
28. Zaimova, D., Bayraktar, E., Miskioglu, I., Katundi, D., Hamouche, Z.: Manufacturing and damage analysis of filler reinforced epoxy based composites. Int J Mat Prod Technol. **48**(1–4): 47–65. Inter Science (ed.), 0268-1900/1741-5209 (2014)
29. Zaimova, D., Bayraktar, E., Miskioglu, I., Dishovsky, N.: Wear resistance of elastomeric based composites by continuous multi-cycle indentation used in manufacturing engineering. Adv. Mat. Res. **939**(1):106–113. Trans Tech Publications (ed.), Switzerland (2014)
30. Zaimova, D., Bayraktar, E., Tan, M. J., Miskioglu, I.: Processing of epoxy-nickel matrix composites reinforced with aluminum and waste elastomers. Adv. Mat. Res. **939**(1): 98–105. Trans Tech Publications (ed.), Switzerland (2014)
31. Zaimova, D., et al.: Manufacturing and damage analysis of epoxy resin-reinforced scrap rubber composites for aeronautical applications. Experimental mechanics of composite, hybrid, and multifunctional materials, vol. 6, pp. 65–76. Springer International Publishing, (2014)

Chapter 7
Scaled Composite I-Beams for Subcomponent Testing of Wind Turbine Blades: An Experimental Study

Mohamad Eydani Asl, Christopher Niezrecki, James Sherwood, and Peter Avitabile

Abstract Full-scale testing of large composite structures is time-consuming and very expensive. Scaled-down models can facilitate time- and cost-effective experimental evaluations of large structures. Due to the special characteristics of laminated composite structures, the design of a scaled-down model of a large composite structure can be challenging. Moreover, the similarity and correlation between the reduced-scaled model and the full-scale structure need to be demonstrated before the experimental data from testing of a scaled-down model can be used for predicting the behavior of the full-scale structure. In this study, laminated composite I-beam reduced-scale models in three different size scales are designed from the spar caps and the shear web geometry of a utility-scale wind turbine blade. The designed composite I-beams are manufactured and tested in a quasi-static four-point bending configuration, and the strain distributions of the loaded small, medium and large beams are measured using the digital image correlation technique. The strain fields of the three beams are compared and similarity of the strain distribution in the three scales is demonstrated.

Keywords Wind turbine blade • Similitude • Subcomponent testing • Scaling • Prototype

7.1 Introduction

The introduction of new materials within a wind turbine blade starts with coupon testing of the materials and culminates with the full-scale testing of the blade. Coupon testing is not always representative of the structural durability of the blade and full-scale testing of the blade is time consuming and very expensive (i.e. in order of hundreds of thousands of dollars). If some small-scale tests can be designed to emulate the structural performance of the blade during the full-scale test, they can potentially expedite and facilitate the identification of viable new materials and the blade certification process.

Numerous studies have been carried out to investigate and design small-scale subcomponents emulating a wind turbine blade. Mandell et al. [1] fabricated and tested composite I-beams with shear webs and flanges representative of the cross section of a wind turbine blade. The root section of the blade was studied by Cairns et al. [2] in which the specimens emulating a single insert of a blade to hub joint were evaluated by using a pull out test. A significant amount of static and fatigue strength data were generated, however the primary focus of the study was manufacturing. Mandell et al. [3] conducted a study on sandwich panel terminations and skin-stiffener intersections. The objective was to estimate skin-stiffener fracture loads and evaluate the performance at the transition location between sandwich panel and normal laminate by using tension tests. Gharghabi et al. [4] developed a subcomponent test to investigate the electrical properties and change of structural parameters of carbon fiber composite materials used in wind turbine blades in response to current impulses electrical.

There are a few studies focusing on the adhesive joints and bond-line durability in wind turbine blades by means of the subcomponent testing concept. The tests were designed with the intent to simulate the complex stress field which bond-lines experience in the working condition of a utility-scale blade. An asymmetric three-point bending test was performed on what is called a Henkel beam by Sayer et al. [5]. The goal was to introduce a comparable amount of shear and normal stress to the bond-line to certify the adhesive performance. Zarouchas et al. [6] proposed a static four-point bending test on two symmetric and asymmetric composite I-beams made from materials incorporated in the spar caps and shear web of the blade. Although the proposed subcomponent tests for the wind turbine blades up to this date vary in size and details, none of them investigated whether there is correlation between the designed subcomponents and the parent components of the blade.

One goal of the research in this paper is to help understand how to design a scaled-down subcomponent to replicate the behavior its full scale component. It is important to verify that the designed subcomponent is able to predict the behavior of

M.E. Asl (✉) • C. Niezrecki • J. Sherwood • P. Avitabile
Department of Mechanical Engineering, University of Massachusetts Lowell, Lowell, MA, USA
e-mail: Mohamad_eydaniasl@student.uml.edu

P.R. Thakre et al. (eds.), *Mechanics of Composite and Multi-functional Materials, Volume 6*,
Conference Proceedings of the Society for Experimental Mechanics Series, DOI 10.1007/978-3-319-63408-1_7

its corresponding full-scale component and data from the subcomponent test can be used to predict or estimate the response of the full-scale component. As blades are made of composite materials, scaling is a challenge because thickness of the plies in a laminate cannot be scaled down to any desired ratio. Therefore a deliberate methodology should be considered to scale down laminated components in the process of subcomponent design. This possible connection between subcomponent (referred to as "model") and its corresponding full-scale component (referred to as "prototype") must be based on the existence of the same field equations and structural parameters that govern the response of both model and prototype when subjected to different load conditions. If such structural parameters could be found, a scaled down model (subcomponent) can be built that duplicates the behavior of its prototype but on a much smaller scale. The descriptor which describes this relationship is "structural similarity". A scaled model which is structurally similar to its full-scale prototype can be used for mechanical testing, verification and validation and system identification of the large structures [7, 8].

Structural similitude analysis can provide the necessary and sufficient similarity conditions between two systems. Simitses and Rezaeepazhand [9] derived the similarity conditions for buckling of the orthotropic and symmetric cross-ply laminated plates using the direct approach. They applied the same approach to extract the scaling laws for the vibration of scaled laminated rectangular plates [10] and to investigate the effect of axial and shear loads on stability of the scaled laminated rectangular plates [11]. Later, the same method was extensively used in their works regarding the prediction of vibration response of laminated shells [12]. According to their results, ply-level scaling yields models with perfect accuracy in predicting the response of the prototype. However, no systematic approach was suggested to design the scaled models that were partially similar to their prototype. A study on the vibration of normalized stepped annular functionally graded plates was performed by Hosseini-Hashemi et al. [13]. They showed that the smooth variation of material in a plate thickness direction can be scaled down without affecting its dynamic properties. Ilbeigi et al. [14] developed a method to design reduced scale models for an euler-bernoulli beam and later they extended their approach to study reduced scale models for systems with disparate spatial and temporal scales [15]. Xu et al. [16] derived the scaling laws for the shape recovery temperature for the shape memory polymers [17]. In the authors' previous work [18, 19], a deliberate methodology was proposed to design the partially similar models for a simply supported laminated plate. The same approach was improved in the later studies [20–22] to design scaled composite I-beams representative of the static response of the spar caps and shear web of a utility-scale wind turbine blade using partial similarity. The design of the composite I-beams that are representative of the dynamic characteristics of the shear web and spar caps of a utility-scale wind turbine blade was discussed in an analytical study by Asl et al. [23]. In the current paper which is an extension of authors' previous work [24], experimental validation of the proposed technique was investigated by manufacturing and testing scaled composite I-beams that were designed using the proposed approach and are partially similar to the I-beam geometry inside a utility-scale blade [25].

The spar caps and shear web are the backbone of a wind turbine blade and make an interesting case study for subcomponent test design. In this study, the objective is to design a scaled-down composite I-beam that has a strain distribution that is the same as its full-scale prototype under adjusted loads. Similarity of the strain fields between two scales can be used to characterize the structural performance of the new materials on a smaller scale. Structural similitude analysis is applied to the I-beam structure inside a utility scale wind turbine blade. The scaled-down models are designed using the Distorted Layup Scaling technique [23, 26, 27] which are partially similar to their full-scale prototype inside the blade. The designed models are manufactured and loaded in a quasi-static four-point flexural bending test and their strain fields are measured using the digital image correlation (DIC) technique [28] for different prescribed load steps. The measured strain values of the beams with different length scale and lamination schemes are compared to each other to demonstrate the static similarity between different scales.

7.2 Governing Equations

In this section, the governing equations for flexural bending of a shear deformable thin-walled composite I-beam are introduced and similarity conditions are derived for strain distribution and transverse displacement using the similitude analysis. The flexural bending equations in the absence of thermal effects are considered for a symmetric I-beam. Neglecting all the coupling effects due to the symmetric geometry, the equations governing the bending with respect to x-axis are given by [29]:

$$(GA_x)_{com}\left(V'' + \Psi'_x\right) = p \tag{7.1a}$$

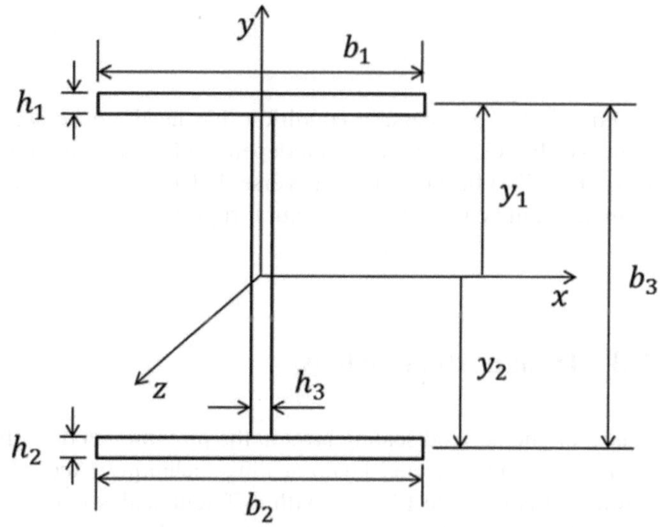

Fig. 7.1 Geometry and coordinate of the I-beam [20]

$$(EI_x)_{com}\Psi''_x - (GA_x)_{com}(V' + \Psi_x) = 0 \tag{7.1b}$$

where Ψ_x denotes the rotation of the cross section with respect to x-axis shown in Fig. 7.1, V the displacement in y direction, p the point load, the prime (′) is used to indicate differentiation with respect to z-axis, $(GA_x)_{com}$ and $(EI_x)_{com}$ are composite shear and flexural rigidities with respect to x-axis, respectively that are expressed as:

$$(EI_x)_{com} = \left[A_{11}^{\alpha}Y_{\alpha}^{(2)} - 2B_{11}^{\alpha}Y_{\alpha} + D_{11}^{\alpha}\right]b_{\alpha} + \frac{b_3^{(3)}}{12}A_{11}^{3} \tag{7.1c}$$

$$(GA_x)_{com} = A_{55}^{\alpha}b_{\alpha} + A_{66}^{3}b_3 \tag{7.1d}$$

where A_{11}, A_{66}, A_{55}, B_{11} and D_{11} are elements of extensional, coupling and bending stiffness matrices for a composite layup. The superscript in the parenthesis () denotes the power of the exponent, and repeated index denotes summation. Index α varies from 1 to 3 where the indices 1 and 2 represent the top and bottom flanges, and 3 is for the web, respectively as shown in Fig. 7.1 and b_α denotes width of the flanges and web. The closed-form solution for the normal strain ε_{zz} and transverse deflection υ_0 can be derived from for Eq. (8.1) for simply-supported boundary condition and point load as [29]:

$$\varepsilon_{zz} = y\frac{\partial\Psi}{\partial z} \tag{7.2a}$$

$$\upsilon_0(z) = \int\left(\frac{p}{(GA_x)_{com}} - \Psi\right)dz \tag{7.2b}$$

To derive the scaling laws, all variables of the governing equations of the model (x_m) are assumed to be connected to their corresponding values in prototype (x_p) by one to one mapping. Then the scale factor may be defined as $\lambda_x = x_p/x_m$ and standard similitude procedure [9] can be applied to the Eq. (8.2) that results in the following scaling laws:

$$\lambda_l{}^2 = \frac{\lambda_{(EI_x)_{com}}}{\lambda_{(GA_x)_{com}}} \tag{7.3}$$

$$\lambda_{\upsilon_0} = \frac{\lambda_p\lambda_l{}^2\lambda_z}{\lambda_{(EI_x)_{com}}} = \frac{\lambda_p\lambda_z{}^3}{\lambda_{(EI_x)_{com}}} = \frac{\lambda_p\lambda_l{}^2\lambda_z}{\lambda_{(EI_x)_{com}}} = \frac{\lambda_p\lambda_z}{\lambda_{(GA_x)_{com}}} \tag{7.4}$$

$$\lambda_{\varepsilon_{zz}} = \frac{\lambda_p \lambda_y \lambda_z}{\lambda_{(EI_x)_{com}}} \tag{7.5}$$

Eq. (7.3) is a necessary condition that needs to be satisfied to use Eq. (7.4) for predicting the transverse deflection. However Eq. (7.5) can be used independently to predict the strain field of the different scales. Eq. (7.3) is a prerequisite for using Eq. (7.4) because the transverse deflection of the beam depends on both shear and flexural rigidity terms. However, normal strain is a function of flexural rigidity.

7.3 Design Methodology

Based on the derived scaling laws in the previous section, the similarity conditions are discussed and the scaled models are designed using distorted layup scaling technique. According to Eq. (7.3) the necessary condition for similarity of two laminated composite I-beams with different scales is the equivalency between their flexural to shear stiffness ratios. The flexural to shear stiffness ratio must be equal to square of the length scale for complete similarity between the composite beams with different scales. Eq. (7.3) is considered as the basis for the design of the composite I-beams that are different in terms of length scale and layup schemes. However, the designed beams have the same specified flexural to shear stiffness ratios and therefore are expected to show similar responses in a static flexural bending test. The prototype geometry and the lay-up scheme considered in this study emulates the spar caps and the shear web of the Sandia 10 MW wind turbine blade [25] near the max chord. This area was chosen for analysis because many failures in blades occur near maximum chord, and the area of interest is responsible for carrying most of the aerodynamic loads. The prototype is assumed to be an I-beam consisting of two identical laminated plates having an overall dimension of 29.3 × 2.446 × 1.5 m^3 with layup $[\pm 45_2/0_{46}]_s$ and ply thickness of t = 1.36 *mm*. The shear web is a sandwich panel with the foam thickness of 160 *mm* and two face sheets with layup $[\pm 45]_{50}$ with an overall thickness of 17.5 *mm*. Because the blade is tapered throughout its length and the geometry of the structure changes significantly, the geometry of the full-scale component (i.e. first 1/3 of the Sandia blade from its root) was averaged to simplify analysis. All computations are implemented for the glass/epoxy materials with following material characteristics [25]: $E_1 = 41.8\ GPa$, $E_2 = E_3 = 14\ GPa$, $G_{12} = G_{13} = 2.63\ GPa$, $G_{23} = 1.83\ GPa$, $\upsilon_{12} = \upsilon_{13} = 0.28$, $\upsilon_{23} = 0.47$.

For the considered prototype geometry, nine scaled-down models were considered to be designed in three different length scales small, medium and large. The scale of the small beams were considered to be 1/50 of the prototype (i.e. $l_p/l_m = 50$), 1/25 for the medium beams and 1/8.33 for the large beams. To select the best layup for the flanges of the scaled models in order to satisfy Eq. (7.3) between models and the prototype, distorted layup scaling approach was applied [21] and the error criterion was defined as the following:

$$e = \left[\left(\lambda_l^2 - \frac{\lambda_{EI}}{\lambda_{GA}}\right)/\lambda_l^2\right]^* 100 \tag{7.6}$$

For small, medium and large beams, models with a total of 2, 4 and 12 plies in the flanges were searched respectively including 0, 90 and ±45 plies. The selected layups for each length scale including the overall geometry and error percentage in satisfying the design scaling law Eq. (7.6) are listed in Table 7.1. The possible symmetric and balanced layups for the small scale beams turned out to be $[0]_2$ and $[90]_2$ where $[0]_2$ was considered for all three small beams because of the less discrepancy in satisfying the design scaling law Eq. (7.3). For the medium and large beams different layups were considered as their total number of plies in their flanges allowed for more possible combinations of 0, 90 and ±45 plies. The 0, 90 and ±45 plies were selected intentionally to make the designed model practical to manufacture. The prescribed load steps for a quasi-static four-point flexural bending test were calculated based on Eq. (7.5) to induce the same level of strain in the beams with different scales and lamination schemes (i.e. $\lambda_{\varepsilon_{zz}} = 1$) as listed in Table 7.1. The designed beams were manufactured and tested based on the calculated loads and their strain fields were measured and are compared to each other as described in the next section.

Table 7.1 The geometry, layup scheme and prescribed load steps for the designed models

Beam scale	Overall dimensions (cm^3)	Flange Layup	Face sheet layup	e(%)	Prescribed load steps (N)
Small	58.6 × 4.93 × 3	$[0]_2$	[±45]	11.4	
		$[0]_2$	[±45]	11.4	
		$[0]_2$	[±45]	11.4	0–445–890-1335-1557 (0–100–200-300-350 *lbf*)
Medium	117.2 × 10.6 × 6	$[0]_4$	$[\pm45]_2$	10.2	0–1780–3560-5338-6227 (0–400–800-1200-1400 *lbf*)
		$[0/90]_s$	$[\pm45]_2$	19.1	0–1156–2402-3558-4181 (0–260–540-800-940 *lbf*)
		$[90]_4$	$[\pm45]_2$	10.7	0–578–1156-1734-2001 (0–130–260-390-450 *lbf*)
Large	351.6 × 31.8 × 18	$[0]_{12}$	$[\pm45]_6$	11.4	0–8000–16,000-24,000-28,000 (0–1800–3600-5400-6300 *lbf*)
		$[0_5/90]_s$	$[\pm45]_6$	2.9	7100–14,200–21,300-24,900 (0–1600–3200-4800-5600 *lbf*)
		$[0_4/90_2]_s$	$[\pm45]_6$	6.2	6200–12,450–18,700-21,800 (0–1400–2800-4200-4900 *lbf*)

Fig. 7.2 The design I-beams were manufactured in three scales: small (58.6 cm length), medium (117.2 cm length) and large (3.51 cm length)

7.3.1 Experimental Results and Discussion

Based on the structural similarity analysis, nine composite beams were manufactured which their layups and geometries specified in the previous section. The flanges and the face sheets were made from glass/epoxy materials. The sandwich shear webs were processed separately. Then, the three pieces (i.e. two flanges and a shear web) were carefully bonded together using a fixture to control the bond-line thickness throughout the length span of the beams and were processed for a final cure. The flanges of the beams were painted and prepared for DIC measurement as shown in Fig. 7.2. The small, medium and large beams were loaded in a quasi-static four-point bending configuration as shown in Fig. 7.3. The prescribed load for each beam was based on the predicted loads from similarity analysis as shown in Table 7.1. The DIC measurement was performed on the bottom flange of the I-beams in a region away from the from the stress concertation areas (i.e. support and loading tips). The noise floor for the normal strain ε_{zz} during the measurement was ±20 μm/m. The field of view in DIC measurement was 55 cm for the small beams and 130 cm for medium and large beams. The speckle size for the small, medium and large beams were 1 mm, 2 mm and 6 mm respectively. Because the primary purpose of the experiment was to investigate the accuracy of the derived scaling laws derived from a formulation valid for the linear elastic bending, the beams were loaded in their elastic bending region to keep both governing equations and the scaling laws valid for the entire study. The normal strain distribution on the midline of the bottom flanges of the beams along the length span were selected for different loading steps. The strain fields of the beams in each scale set (i.e. small, medium and large) were averaged for each load step. The strain values for small and medium beams were expanded in length direction using the length scale factor and plotted over those of the large beams. Figure 7.4 shows the average normal strain distribution along the specified path for small, medium and large

Fig. 7.3 Quasi-static flexural bending test setup for small beams (*left*), medium beams (*middle*) and large beams (*right*)

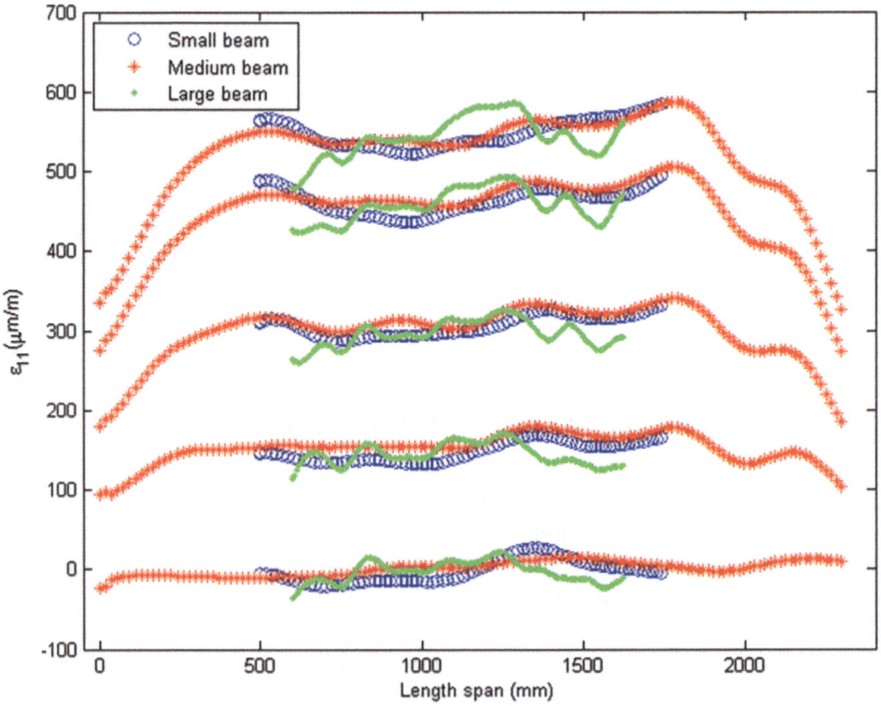

Fig. 7.4 Normal strain along the midline of the bottom flanges of the small, medium and large beams for different loading steps

beams for specified loads in Table 7.1. The strain fields for all the beams in each set (i.e. small, medium and large) were averaged and plotted in Fig. 7.4 with labels small, medium and large. For instance, the medium beams were loaded based on the loads specified in Table 7.1 and then the three strain fields were average and plotted in Fig. 7.4 as medium beam. As shown in Fig. 7.4, all three scales show a comparable level of normal strain for different loads regardless of their lamination schemes. This observation suggests that the designed beams using the structural similitude show a normal strain field similar to each other regardless of their scales and lamination scheme. The strain field of the small beams is representative of those of the medium and large beams and the prescribed loads based on the similarity analysis work accurately with different scales which induce the desired normal strain levels in the flanges of the beams. This observation is important as it suggests that a certain strain level for the composite beam geometry with specific scale and layup can be accurately replicated in a smaller scale with a different layup. Because large structures are tested for fatigue life must routinely withstand cyclic strain levels over a long period of time (i.e. on order of weeks), the presented results suggest that small-scale models designed using structural similitude can facilitate and expedite the fatigue testing process of the large composite structures. The scaled-down models can be manufactured with a reduced cost and time and be tested on higher frequencies and similar induced strain levels as their full-scale parent component for a fatigue test to achieve the desired number of cycles for certification purposes.

7.4 Conclusion

In this study, composite laminated I-beams with different length scales and layups were designed using the structural similitude and distorted layup technique. The designed models were manufactured and tested in a quasi-static four-point bending test under adjusted loads. The strain fields of the beams were measured along the midline of the bottom flange of the beams for increasing loading steps. The normal strain distributions of the beams for the load steps were compared to each other to demonstrate the static similarity. According to the results, all the designed beams with different layups and length scales showed a comparable normal strain field for the prescribed loads. The similarity of the strain fields between different scales and applicability of the distorted layup scaling in design of the scaled laminated beams were experimentally demonstrated.

Acknowledgments This material is based upon work supported by the National Science Foundation under Grant Number 1230884 (Achieving a Sustainable Energy Pathway for Wind Turbine Blade Manufacturing). Any opinions, findings, and conclusions or recommendations expressed in this material are those of the author(s) and do not necessarily reflect the views of the National Science Foundation.

References

1. Mandell, J.F., Creed Jr., R.J., Pan, Q., Combs, D.W., Shrinivas, M.: Fatigue of fiberglass generic materials and substructures. Wind Energy. **207**, (1994)
2. Cairns, D.S., Skramstad, J.D., Mandell, J.F.: Evaluation of Hand Lay-Up and Resin Transfer Molding in Composite Wind Turbine Blade Structures, pp. 967–80. International SAMPE Symposium and Exhibition: SAMPE (1999)
3. Mandell, J.F., Combs, D.E., Samborsky, D.D.: Fatigue of fiberglass beam substructures. Wind Energy. **16**, 99 (1995)
4. Gharghabi, P., Lee, J., Mazzola, M., Lacy, T.: Development of an experimental setup to analyze carbon/epoxy composite subjected to current impulses. In: Proceedings of the American Society for Composites: Thirty-First Technical Conference. https://scholar.google.com/scholar?q=Development+of+an+experimental+setup+to+analyze+carbon%2Fepoxy+composite+subjected+to+current+186+impulses&btnG=&hl=en&as_sdt=0%2C22&sciodt=0%2C22&cites=3034633313243656090&scipsc (2016)
5. Sayer, F., Antoniou, A., van Wingerde, A.: Investigation of structural bond lines in wind turbine blades by sub-component tests. Int. J. Adhes. Adhes. **37**, 129–135 (2012)
6. Zarouchas, D.S., Makris, A.A., Sayer, F., Van Hemelrijck, D., Van Wingerde, A.M.: Investigations on the mechanical behavior of a wind rotor blade subcomponent. Compos. Part B. **43**, 647–654 (2012)
7. Behmanesh, I., Yousefianmoghadam, S., Nozari, A., Moaveni, B., Stavridis, A.: Effects of prediction error bias on model calibration and response prediction of a 10-story building. In: Model Validation and Uncertainty Quantification, vol. 3, pp. 279–291. Springer International Publishing, Orlando (2016)
8. Yousefianmoghadam, S., Song, M., Stavridis, A., Moaveni, B.: System Identification of a Two-Story Infilled rc Building in Different Damage States. Improving the Seismic Performance of Existing Buildings and Other Structures 2015, pp. 607–618. American Society of Civil Engineers (ASCE), San Francisco (2015)
9. Simitses, G.J., Rezaeepazhand, J.: Structural similitude for laminated structures. Compos. Eng. **3**, 751–765 (1993)
10. Simitses, G.J., Rezaeepazhand, J.: Structural similitude and scaling laws for buckling of cross-ply laminated plates. J. Thermoplast. Compos. Mater. **8**, 240–251 (1995)
11. Rezaeepazhand, J., Simitses, G.J., Starnes Jr., J.H.: Design of scaled down models for stability of laminated plates. AIAA J. **33**, 515–519 (1995)
12. Rezaeepazhand, J., Simitses, G.J., Starnes Jr., J.H.: Design of scaled down models for predicting shell vibration response. J. Sound Vib. **195**, 301–311 (1996)
13. Hosseini-Hashemi, S., Derakhshani, M., Fadaee, M.: An accurate mathematical study on the free vibration of stepped thickness circular/annular Mindlin functionally graded plates. App. Math. Model. **37**(6), 4147–4164 (2013)
14. Ilbeigi, S., Chelidze, D.: Model order reduction of nonlinear Euler-Bernoulli beam. In: Nonlinear Dynamics, vol. 1, pp. 377–385. Springer International Publishing, Orlando (2016)

15. Ilbeigi, S., Chelidze, D.: Reduced order models for systems with disparate spatial and temporal scales. In: Rotating Machinery, Hybrid Test Methods, Vibro-Acoustics & Laser Vibrometry, vol. 8, pp. 447–455. Springer International Publishing, Garden Grove (2016)
16. Xu, H., Yu, C., Wang, S., Malyarchuk, V., Xie, T., Rogers, JA.: Deformable, programmable, and shape-memorizing micro-optics. Adv. Funct. Mater. **23**, 3299–3306 (2013)
17. Taherzadeh, M., Baghani, M., Baniassadi, M., Abrinia, K., Safdari, M.: Modeling and homogenization of shape memory polymer nanocomposites. Compos. Part B Eng. **91**, 36–43 (2016)
18. Asl, M.E., Niezrecki, C., Sherwood, J., Avitabile, P.: Predicting the vibration response in subcomponent testing of wind turbine blades. In: Special Topics in structural dynamics, vol. 6, pp. 115–123. Springer, Orlando (2015)
19. Asl, M.E., Niezrecki, C., Sherwood, J., Avitabile, P.: Application of structural similitude theory in subcomponent testing of wind turbine blades. In: Proceedings of the American Society for Composites, pp. 8–10. San Diego (2014)
20. Asl, M., Niezrecki, C., Sherwood, J., Avitabile, P.: Similitude analysis of the strain field for loaded composite i-beams emulating wind turbine blades. In: Proceedings of the American Society for Composites: Thirty-First Technical Conference. Williamsburg (2016)
21. Asl, M.E., Niezrecki, C., Sherwood, J., Avitabile, P.: Similitude analysis of composite i-beams with application to subcomponent testing of wind turbine blades. Exp. Appl Mech. **4**.: Springer, 115–126 (2016)
22. Asl, M.E., Niezrecki. C., Sherwood. J., Avitabile, P.: Design of scaled-down composite i-beams for dynamic characterization in subcomponent testing of a wind turbine blade. In: Shock & Vibration, Aircraft/Aerospace, Energy Harvesting, Acoustics & Optics, vol. 9, pp. 197–209. Springer International Publishing, (2016)
23. Eydani Asl, M., Niezrecki, C., Sherwood, J., Avitabile, P.: Vibration prediction of thin-walled composite I-beams using scaled models. Thin-Walled Struct. **113**, 151–161 (2017)
24. Asl, M.E., Niezrecki, C., Sherwood, J., Avitabile, P.: Similitude Analysis of the Frequency Response Function for Scaled Structures. Model Validation and Uncertainty Quantification, vol. 3, pp. 209–217. Springer, Cham
25. Griffith, D.T., Ashwill, T.D.. The Sandia 100-meter all-glass baseline wind turbine blade: SNL100–00. Sandia National Laboratories, Albuquerque, Report NoSAND2011–3779 (2011)
26. Asl, M.E., Niezrecki, C., Sherwood, J., Avitabile, P.: Experimental and theoretical similitude analysis for flexural bending of scaled-down laminated I-beams. Compos. Struct. **176**, 812–822 (2017)
27. Asl, M.E., Niezrecki, C., Sherwood, J., Avitabile, P.: Similitude analysis of thin-walled composite I-beams for subcomponent testing of wind turbine blades. Wind Eng. **0**, 0309524X17709924 (2017)
28. Baqersad, J., Poozesh, P., Niezrecki, C., Avitabile, P.: Photogrammetry and optical methods in structural dynamics–A review. Mech. Syst. Signal Process. **86**, 17–34 (2017)
29. Lee, J.: Flexural analysis of thin-walled composite beams using shear-deformable beam theory. Compos. Struct. **70**, 212–222 (2005)

Chapter 8
Development Analysis of a Stainless Steel Produced by High Energy Milling Using Chips and the Addition of Vanadium Carbide

C.S.P. Mendonça, F. Gatamorta, M.M. Junqueira, L.R. Silveira, J.H.F. Gomes, M.L.N.M. Melo, and G. Silva

Abstract Duplex stainless steels are preeminent materials that have been receiving special attention nowadays, due its considerable strength, toughness and exceptional corrosion resistance. This paper aims the production of a duplex stainless steel by powder metallurgy through cutting chips, X ray diffraction was used to identify the ferrite phase and austenite phase, also this paper compares the normal milling process to the addition of vanadium carbide to improve the milling action. To assure the significant changes by using the carbide and the change on time parameters, data was collected, and it was used statistical analysis by factorial design. Adding vanadium carbide at 0%–3%, the reduction in average particle size appeared to be significant, compared to the material without the carbide addition. The smallest particle size was obtained by the high energy milling that was performed in a planetary ball mill with ball to powder weight ratio 20:1, and mill speed of 350 rpm milled in argon atmosphere for 50 h, and adding 3% of vanadium carbide. Static data acknowledged that addition of carbides in the process is the most influential term, followed by the time of milling.

Keywords Stainless steel chips • High-energy milling • Particle size • Factorial design • Vanadium carbides

8.1 Introduction

Duplex stainless steels are becoming more and more traditional materials with increasing applications in the naval, industrial, construction and chemical processing industries due to their excellent combination of mechanical properties and corrosion resistance [1, 2]. An important class of duplex steels is the 2205 (UNS S31803), which are used in the oil and gas industry in flexible pipes for oil extraction and in the pumps and paper industries in product storage tanks, equipment and reagents Chemicals [3]. The conventional manufacturing processes of duplex stainless steels are complex, while the powder metallurgy process, which is a near net shaping process, reduces processing costs and is attractive for stainless steel (SS) production [4]. The automotive market introduces newly designed sintered parts in large amounts in produced cars. The new cars constructions are equipped with at least six powder metal flanges piece. Stainless steel is the preferred material for powder metal flanges because of its resistance to corrosion and oxidation [5]. High energy milling is one of the most widely used plastic deformation methods to achieve the extreme refinement of the structure [6]. The reuse of industrial waste has several environmental advantages, such as, reduction of energy consumption and emission of effluents. According to Padilha (2000) to produce metals from the recycling, only 15% of the energy needed to obtain the same quantity is used through the primary processing [7]. The recycling of materials avoids the disposal of materials in the environment, reduces the extraction of natural resources and reduces the energy input in the manufacture of numerous types of parts and the extraction of natural resources [8]. There has been increasing application of powder metallurgy (PM) for the production of corrosion resistant alloy parts [9]. Brytan et al. (2009) obtained duplex stainless steel from ferritic stainless steel powder (16.86% Cr, 1.15% Si, 0.18% Mn, 0.02% C) with controlled addition of powders, synthesized at 1250 ° C in a vacuum oven with an argon atmosphere [10]. Yonekubo (2010) obtained a super duplex stainless steel from machining chips of a plate of a UNS S32250 superdúplex steel, through high energy milling, performed with Spex, Attritor and Planetary mills [11]. The

C.S.P. Mendonça • M.M. Junqueira • L.R. Silveira • J.H.F. Gomes • M.L.N.M. Melo • G. Silva
University of Itajubá, Av. BPS, 1303, Bairro Pinheirinho. Itajubá, Minas Gerais, Brazil
e-mail: claudiney.sales@unifei.edu.br; mateusjunqueira@unifei.edu.br; lucasrafael09@yahoo.com.br; ze_henriquefg@yahoo.com.br; mirianmottamelo@unifei.edu.br; gilbert@unifei.edu.br

F. Gatamorta (✉)
University of Campinas, Av mendeleiev, 200 campinas, São Paulo, Brazil
e-mail: fabiog@fem.unicamp.br

P.R. Thakre et al. (eds.), *Mechanics of Composite and Multi-functional Materials, Volume 6*,
Conference Proceedings of the Society for Experimental Mechanics Series, DOI 10.1007/978-3-319-63408-1_8

high energy milling powder processing technique allows obtaining more homogeneous materials [12]. In the milling process, particle size is an important factor in stainless steel processed by powder metallurgy that affects the compressibility, thus increasing the densification and the properties of the sintered products [9]. As the studies related to high energy milling involves the variables time, rotation, mass / sphere ratio [12], a factorial experimental design was applied to different systems to obtain the smallest particle size. Factorial experiment is used to reduce the number of experiments, time, cost of the overall process, being more indicated in situations in which there are several factors and it is desired to know which are the most influential [13] being used in the initial phase of planning experiment, defining the most important factors and studying the effects on the response variable that was chosen. A major advantage of SS is its ability to be recyclable; in this respect, this work aims to process the chip of a duplex stainless steel by means of the powder metallurgy process (PM), with and without additions of Carbides as reinforcements, these elements acting as blockers of movement of dislocations raising mechanical strength and also inhibit high growth of grain size.

8.2 Experimental

8.2.1 Materials

In this work, UNS S31803 duplex stainless steel chips were used as the starting material for the milling process with and without the presence of vanadium carbide with initial size between 5 and 15 μm.

8.2.2 Experimental Statistical

Chip milling was performed using high energy milling in a planetary ball mill with an inert argon atmosphere to prevent oxidation of the powders. The parameters analyzed were; time of rotation, rotation, relation: mass/ball, and carbide percentage. The validation of the experimental mathematical model was made by analysis of variance (ANOVA) using the Minitab 16 program. The experimental data were generated by a complete factorial planning (24) with one repetition totaling 32 experiments. The values of rotation, milling time, mass / ball ratio and percentage of vanadium carbide were varied. The values of these levels are found in Table 8.1. These values were chosen based on the literature and, mainly, empirical tests performed in high energy milling.

8.2.3 Experimental Procedure

To characterize the morphology, size identification and particle distribution, was made by a scanning electron microscope of the Carl Zeiss brand and model EVO MA15. After milling process, the material was subjected to a stress relief process at 1050 ° C for 30 min, and then milled powder samples were compacted using hydraulic pressing machine with a load of 700 MPa, vacuum sintered with a heating ramp of 5 °C/min to a temperature of 1200 ° C for 1 h. In order to identify the phases present in the microstructure, the X-ray diffraction technique was used. The parameters used were: 40 kV voltage and 40 mA current, with a scanning range of 45 to 135 °, a step size of 0.02 ° per second, the cobalt tube (1.7890 Å) was used. Magnetic characterization was performed by a VSM (vibrating sample magnetometer). For the magnetization curves versus the applied magnetic field of the material, a maximum applied field of 14,000 G and a constant time of 15 min were used. The metallographic experiments were performed using Olympus optical microscopy, model BX41. The density was

Table 8.1 Real levels of the factors studied

		Levels	
Parameters	Notation	Low (−)	High (+)
Time of rotation (h)	A	10	50
Rotation (rpm)	B	250	350
Relation mass/ball	C	1/10	1/20
Carbide percentage (%)	D	0	3.0

evaluated using the Archimedes principle. The sample was immersed in distilled water for 24 h, according to NBR 6220: 2011 [14]. The porosity of the polished samples was studied through image analysis, in which the pores were identified as black pixels and it was possible to calculate the area of each pore by percentage of surface area.

8.3 Results and Discussion

8.3.1 Analysis Statistics of the Milling Parameters

Minitab 16 software was used to analyze the milling variables and the main effects were determined. The model presented a square correlation adjusted coefficient of R^2 (adjustment) of 90.37%. The mathematical model coded by 2^4 factorial can be given by Eq. (8.1). Where R is the mean particle size, X0 is the overall mean and A, B, C, D represents the variable rotation, milling time, mass / sphere ratio and percentage of vanadium carbide, respectively. Table 8.2 shows the average particle size values for each condition stipulated in the experiment design for the variable rotation, time, mass/ball relation and percentage of carbide.

$$R = 939.04 - 1.31A + 12,78B - 1800.81C + 72.29D - 0.054AB + 3.89AC - 0.36AD - 90.57BC \\ - 6.77BD - 884.04CD + 0.28ABC + 0.021ABD + 2.15ACD + 49.72BCD - 0.141ABCD \quad (8.1)$$

Analyzing the p_value (descriptive level) that is a statistic used to synthesize the result of a test of hypotheses, for a level of significance ($\alpha = 0.05$), considering a 95% confidence interval, it has the following results:

- The parameters of rotation, time, relation and carbide are significant factor, because showed a p_value = 0.000 ie, lower than the level of significance ($\alpha = 0.05$).
- The second interaction parameters, such as the relation (rotation * relation) and (rotation * relation) presented a p_value = 0.000 and the variable (relation * carbide) presented a p_value = 0.009, these values of p_value are smaller than Level of significance ($\alpha = 0.05$), making these factors significant.
- The second order parameters (rotation * time), (rotation * carbide) and (time * relation) values are not significant, since the values of p_value is greater than the level of significance ($\alpha = 0.05$).

According to the normal probability graphs (Fig. 8.1), the points that are close to a line in the middle group of points represent those estimated factors that do not demonstrate any significant effect on the response variables [13]. Points away from the line probably represent the effects of the "real" factor, so the percentages of carbide, time, rotation, mass / ball ratio,

Table 8.2 Milling Parameters: rotation, milling time, mass/ball relation and percentage of carbide and the experimental values for the variable medium particle size

Sample	Rotation (Rpm)	Time (h)	Relation mass/ball	Carbide (%)	Medium particle size (μm)	Medium particle size (μm) – Replication
1	−1	−1	−1	−1	498.69	489.83
2	+1	−1	−1	−1	378.10	375.10
3	−1	+1	−1	−1	360.60	415.90
4	+1	+1	−1	−1	186.79	151.80
5	−1	−1	+1	−1	386.46	427.60
6	+1	−1	+1	−1	347.58	292.46
7	−1	+1	+1	−1	243.95	168.30
8	+1	+1	+1	−1	143.49	137.90
9	−1	−1	−1	+1	299.59	386.61
10	+1	−1	−1	+1	230.36	174.86
11	−1	+1	−1	+1	209.19	255.34
12	+1	+1	−1	+1	98.40	77.91
13	−1	−1	+1	+1	214.61	206.79
14	+1	−1	+1	+1	142.17	108.36
15	−1	+1	+1	+1	85.33	118.10
16	+1	+1	+1	+1	40.60	52.50

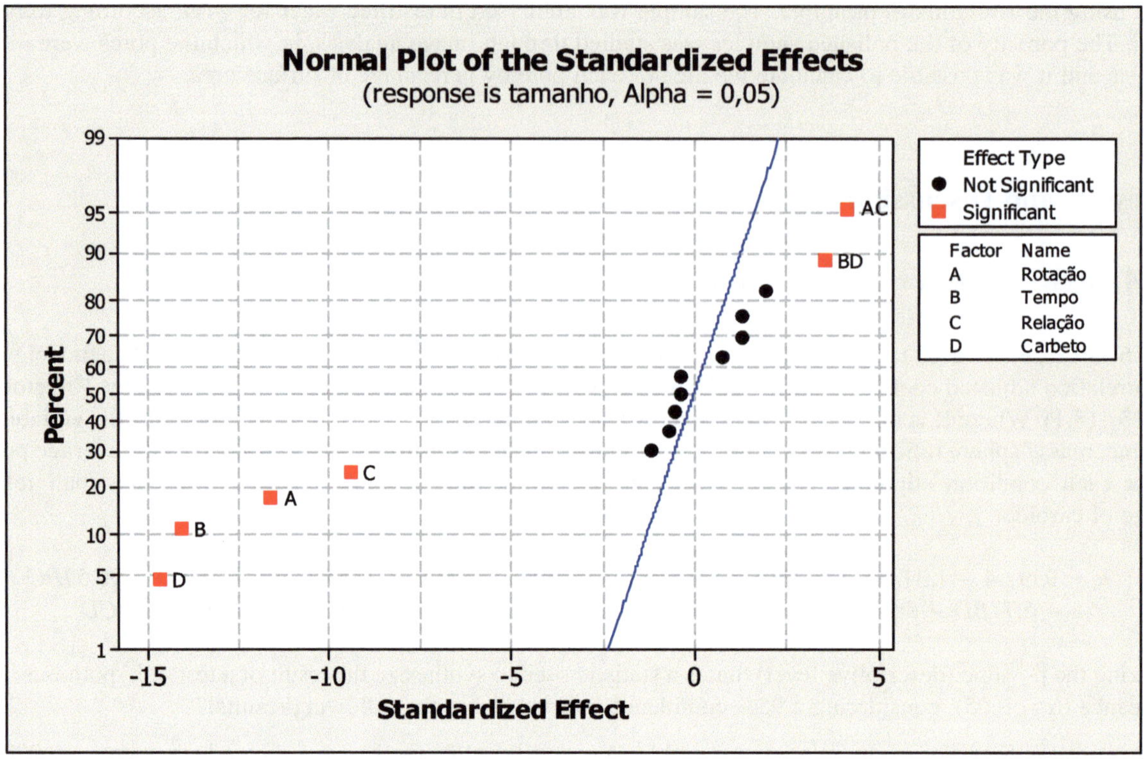

Fig. 8.1 Normal distribution for the factorial experiment with 4 factors

time and carbide interactions, and finally the interactions between mass / ball ratio and rotation. The percentage of carbide is more influential in the milling process; this may be due to the vanadium carbide having a high hardness, high resistance that in contact with the stainless steel chips, at the time of milling together with the balls break the chip, Increasing process efficiency.

The plot of normal probability of residues for particle size (Fig. 8.1) shows that set of observed values are close to the values of the theoretical distribution. Generally, the experimental points are reasonably aligned, suggesting a normal distribution. The normal probability plot is just a graph of the cumulative distribution of the residuals in the normal probability paper that is graphically with ordinate scaling so that the cumulative normal distribution is plotted to a straight line [15]. On the other hand, the observed residues are plotted against the adjusted response, which indicates a random dispersion of residues (Fig. 8.2).

8.3.2 Microstructural Characterization of the Powder Particles

The initial characterization of the UNS S31803 steel chips processed by machining, Fig. 8.3, have an average size of 2000 μm with plastic deformations on the surface originated by the machining tool.

For the analyzes of the stainless steel microstructures of the experimental design of the conditions with rotation of 350 rpm, 50 h and mass / ball ratio of 1/20, of the samples with and without carbide. Scanning electron microscopy analyzes of the samples subjected to milling conditions of 350 rpm, mass/ball relation of 1/20 for the time of 50 h (Fig. 8.4), it was verified that the material acquired an irregular morphology with average size of 100–250 μm.

Figure 8.5 shows the morphology of the stainless steel particles subjected to milling for 50 h with the addition of 3% vanadium carbide, the irregular shape of the duplex stainless steel particles having a size ranging from 10 to 80 μm is observed. The value obtained for the mean particle size distribution is 48.91 μm. The characteristic of the milling process is ductile-brittle, where the UNS S31803 steel has ductile behavior and the VC being considered a ceramic material, have characteristics of fragile materials.

The cumulative particle size analysis curve seen in Fig. 8.6 shows the limits of D10, D50 and D90. The milling with the parameters of rotation of 250 rpm, mass / ball ratio of 1/10 and time of 50 h, were that they presented bigger particle size,

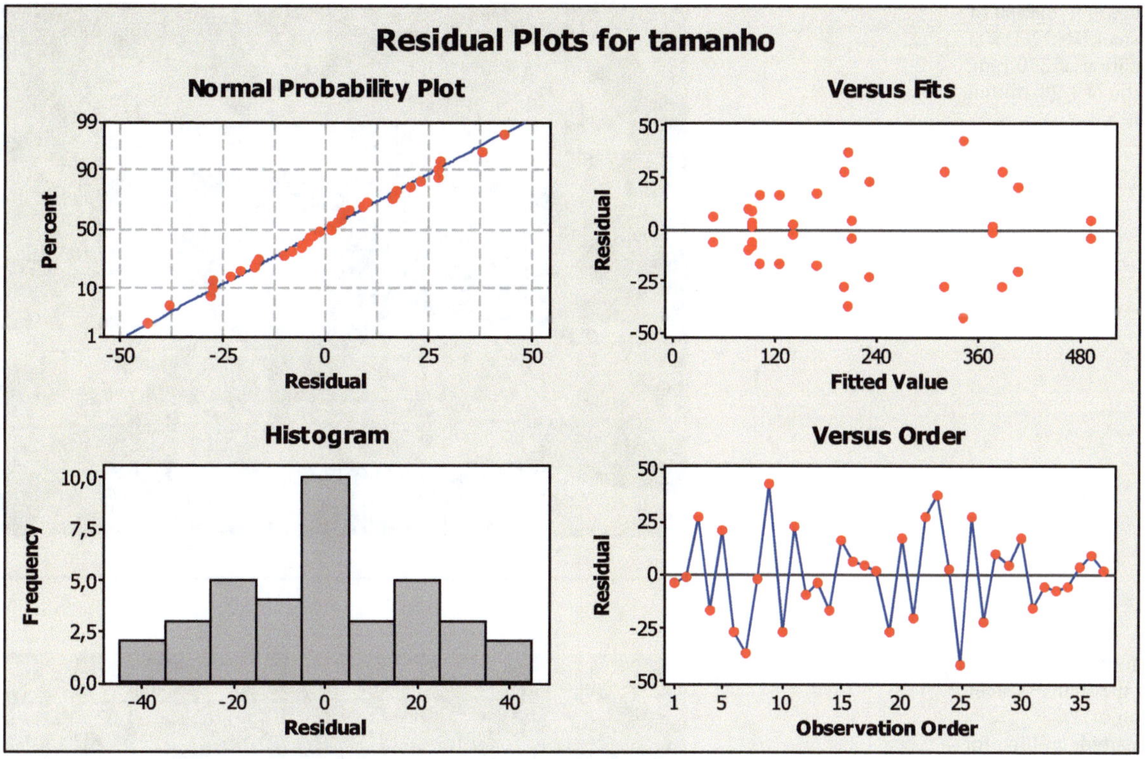

Fig. 8.2 Analysis of the residuals of the variables of the milling process: Rotation, milling time, mass/ball relation and percentage of carbide

Fig. 8.3 Duplex stainless steel chips

with the value of D90 of 559, 5 μm, representing 90% of the sizes of Particles are below this value. Increasing the variable rotation from 250 rpm to 350 rpm and mass / ball ratio from 1/10 to 1/20, it was observed that the D90 values are below 300 μm. The lower values for milling performed with the rotational parameters of 350 rpm, mass/ball ratio of 1/20 and time of 50 h and addition of 3% of carbide, showed D90 less than 100 μm.

Table 8.3 shows the particle size distribution of samples submitted to high energy milling with and without the addition of 3% vanadium carbide. Parameters D10, D50 and D90 represent the particle sizes which are respectively 10%, 50% and 90% of the total mass of the material.

Fig. 8.4 Photomicrograph of the duplex steel UNS S31803, milling condition of 350 rpm, mass/ball ratio of 1/20, milling time of 50 h

Fig. 8.5 Morphology of the particles of the stainless steel with addition of 3% Vanadium carbide milling for 50 h

Figure 8.7 shows the magnetic behavior of the duplex stainless steel under the following conditions: (a) as received; Milling stainless steel in the time of 50 h under the conditions: (b) rotation 350 rpm-ratio 1/20; (C) rotation 350 rpm-1/20 ratio with addition of 3% VC.

Analyzing the results of Fig. 8.7, it can be noticed that samples of the stainless steel after milling present an increase of the saturation magnetization, this means that there was the formation of martensite, and this increase is a function of the milling time. The values obtained for the four milling conditions showed the same values of saturation magnetization, around 90 emu / g, indicating that varying the milling parameters mass/ball relation, rotation, and the percentage of carbide from 0 to 3% in time of 50 h milling, there was no change in the magnetization values and the formation of the martensite induced by deformation.

The saturation magnetization depends only on the volume fraction of the magnetic phase that is present in the steels, therefore, studies have related the increase in the value of saturation magnetization with the transformation of austenite into deformation-induced martensitic [16, 17].

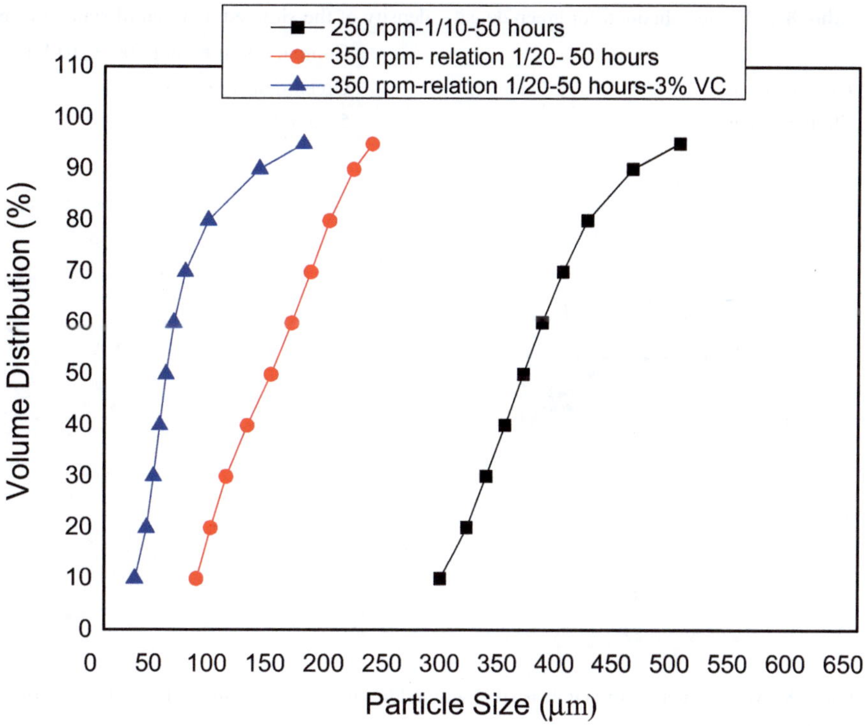

Fig. 8.6 Analysis of the volumetric distribution as a function of particle size for different milling parameters

Table 8.3 Parameters D10, D50 and D90 and a medium particle sizes of stainless steel with the addition of 3% Vanadium Carbide

Milling parameters	D10 (μm)	D50 (μm)	D90 (μm)	Medium particle size (μm)
Milling of a stainless steel- rotation 250 rpm- relation 1/10–50 h	273,6	415,9	559,5	415,8
Milling of a stainless steel- rotation 350 rpm-relation 1/20–50 h	91,27	168,3	282,5	179,5
Milling of a stainless steel- rotation 350 rpm-relation 1/20–50 h- 3% de VC	12,54	42,37	90,65	48,91

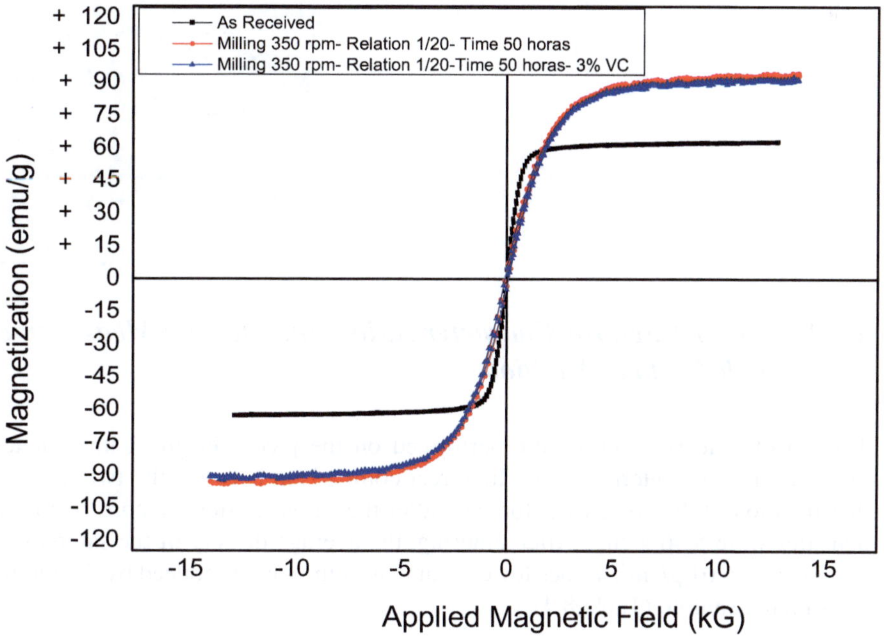

Fig. 8.7 Magnetization as a function of the magnetic field applied to different milling parameters

Table 8.4 Values obtained for green density, density of the sintered material of stainless steel

	Compaction pressing 700 ± 10 Mpa	
Composition	Green density (g/cm^3)	Apparent density (g/cm^3)
Stainless steel	5.46 ± 0,11	6.01 ± 0,08

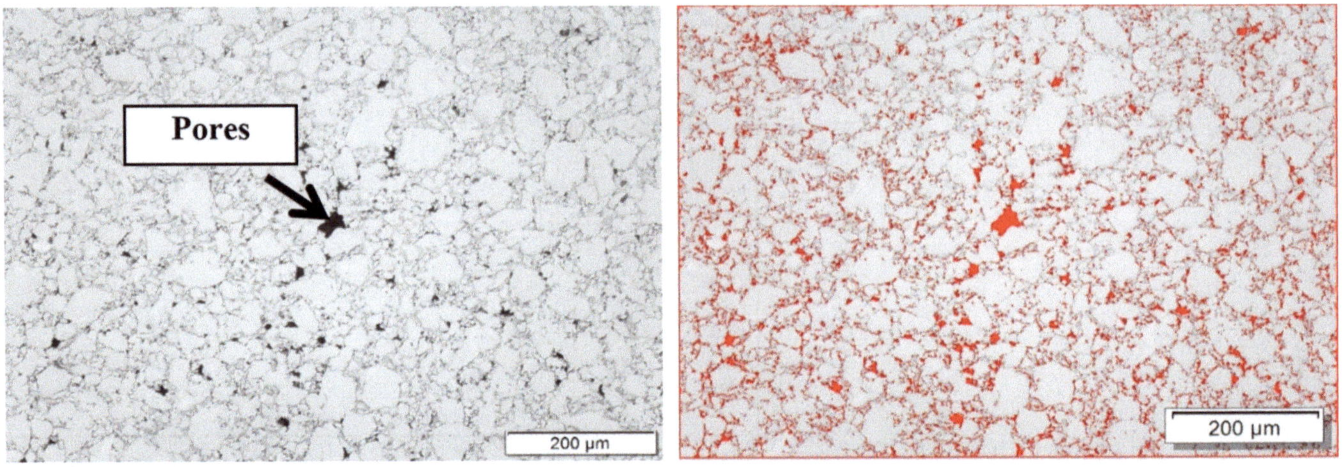

Fig. 8.8 Microstructure of stainless steel with addition of 3% VC sintered at 1200 ° C for 1 h. Regions in *red* indicate porosity

Fig. 8.9 Micrograph of a stainless steel sintered at 1200 ° C for 1 h. Images obtained via scanning electron microscopy

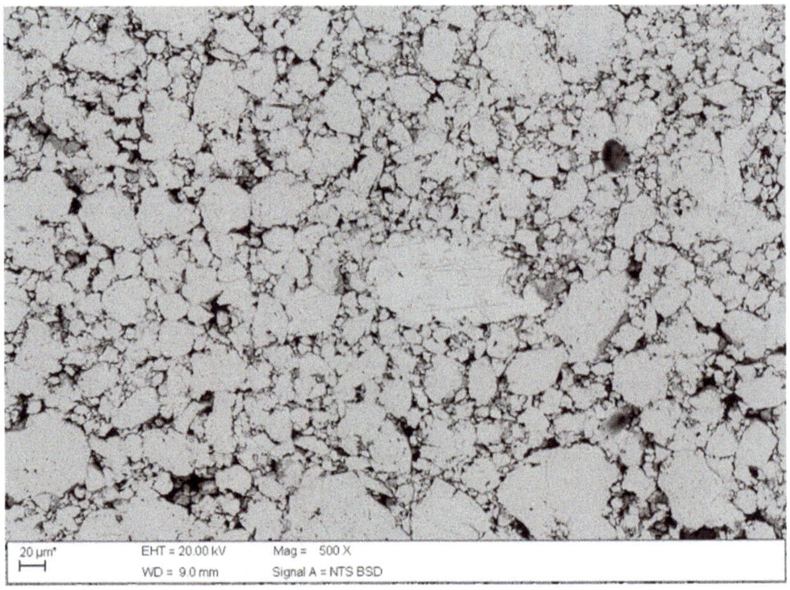

8.3.3 Microstructural Characterization and Density Measurements of Sintered Stainless Steel with Carbide Addition

The density measurements were performed on the pieces in green immediately after pressing, and by the Archimedes method after the sintering step. The direct comparison between the two measurements had the purpose of evaluating the densification of the compact before and after the sintering step. According to Table 8.4, it can be seen that the mean green densities were 5.46 g/cm^3. After sintering, the average density of the samples produced with addition of 3% of vanadium carbide was 6.01 g/cm^3, values lower than that of the steel produced by the melting process, due to the existence of pores in the sintered sample (Table 8.4).

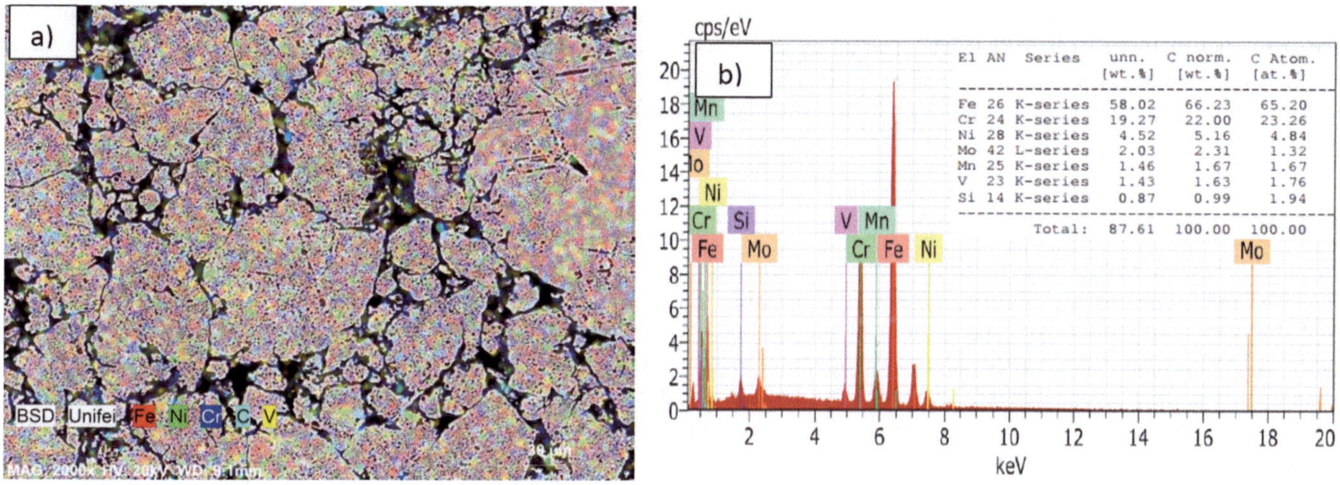

Fig. 8.10 Distribution analysis of the main elements and the vanadium carbide in the microstructure steel sintered at 1,200 °C. (**A**) Mapping analysis. (**B**) EDS analysis

Fig. 8.11 Diffraction of x rays. Where: (●) Austenite; (*) Ferrite; (◊) Martensite

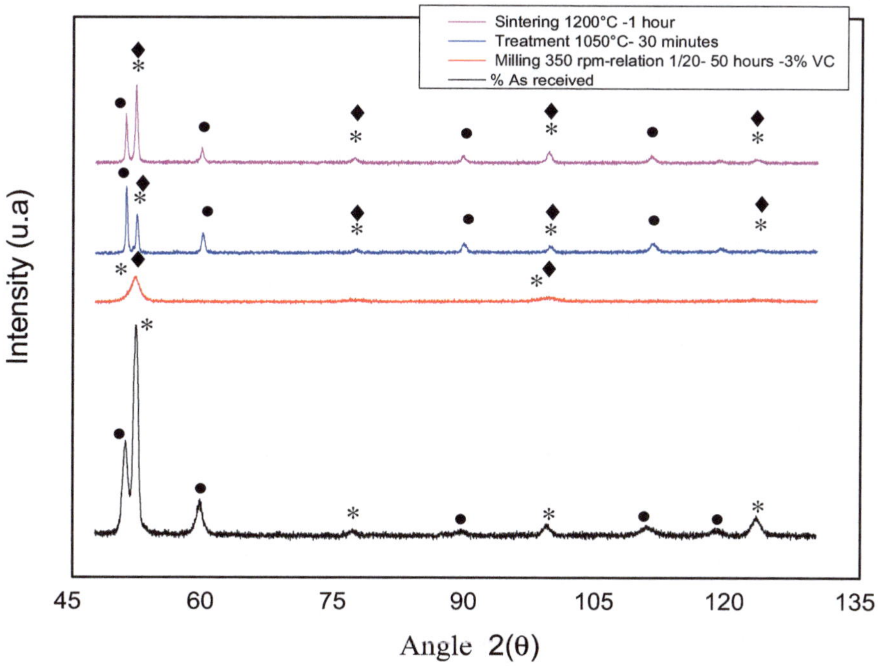

The results of the porosity of the samples with the addition of vanadium carbide are shown in Fig. 8.8. It can be seen from optical microscopy that the samples have a low residual porosity and are concentrated between the contours of the particles and that in some regions If very intense sintering occurs, material becomes homogeneous. The porosity obtained through the image analysis is 10.5%.

The micrograph of the sample of a stainless steel sintered at 1200 ° C for 1 h is shown in Fig. 8.9, it is observed that the samples have a low residual porosity and concentrated between the contours of the particles, and that in some regions a very intense sintering occurred, material becomes homogeneous.

Carbide distribution was performed through the BSD and EDS modes (Fig. 8.10a). It has been observed that the VC particles are located homogeneously on the surface of the steel particles. The VC particles were identified in (Fig. 8.10b) by their chemical elements (carbon and vanadium), as well as the presence of iron, chromium and nickel elements, constituent elements of duplex stainless steel.

The results of the X-ray diffraction analysis are shown in Fig. 8.11, the presence of the austenite and ferrite phases in the material as received is observed. The chipping was submitted to the milling process, it was observed that the peaks decreased in intensity, occurring a widening of the peak, and even disappeared, with a tendency to amorphized and the appearance of the martensitic phase, a phase that was formed by the process of plastic deformation.

The stress relief process performed at 1050 ° C observed the resurgence of the ferritic and austenitic phase and the deformation induced martensitic phase, which remained even after the sintering step at 1200 ° C.

8.4 Conclusion

Through the high energy milling process, it was possible to recycle the machining chips. With the experimental procedure and the presented parameters, the milling of the stainless steel duplex UNS S31803 is more effective when there is the addition of carbides. After thermal treatment, the ferritic and austenitic phases were observed, characterizing again a structure of duplex stainless steel. There was the presence of deformation induced martensite, even after the sintering step. For the characterization of a homogeneous final structure, it is also possible to present an order of importance in the influence of the milling process, starting with the percentage followed by the milling time, rotation and finally mass / ball ratio, in this way a low residual porosity and the density of about 77% of the melting process were also obtained.

8.5 Acknowledges

The authors are grateful to the Brazilian agencies, CAPES and FAPEMIG for support and incentives in the development of this research.

References

1. Nomani, J., Pramanik, A., Hilditch, T., Littlefair, G.: Machinability study of firsy generation duplex (2205), second generation duplex (2507) and austenite stainless steel during drilling process. Wear. **304**(1–2), 20–28 (2013)
2. Chen, T.H., Weng, K.L., Yang, J.R.: The effect of high-temperature exposure on the microstructural stability and toughness property in a 2205 duplex stainless steel. Mater. Sci. Eng. A. **338**(1–2), 259–270 (2002)
3. Breda, M., Brunelli, K., Grazzi, F., Scherillo, A., Calliari, I.: Effects of cold rolling and strain-induced martensite formation in a SAF 2205 duplex stainless steel. Metall. Mater. Trans. A. **46A**, 577–586 (2015)
4. García, C., Martin, F., Blanco, Y.: Effect of sintering cooling rate on corrosion resistance of powder metallurgy austenitic, ferritic and duplex stainless steels sintered in nitrogen. Corros. Sci. **61**, 45–52 (2012)
5. Dobrzański, L.A., Brytan, Z., Actis Grande, M., Rosso, M.: Properties of duplex stainless steels made by powder metallurgy. Arch. Mater. Sci. Eng. **28**(4), 217–223 (2007)
6. Shashanka, R., Chaira, D.: Development of nano-structured duplex and ferritic stainless steels by pulverisette planetary milling followed by pressureless sintering. Mater. Charact. **89**, 220–229 (2015)
7. Padilha, A.F.: Materiais de engenharia microestrutura e propriedades, 1st edn. Hemus S.A., Curitiba (2000)
8. Delforge, D.Y.M., Ferreira, I., da Silva, C.G.R., Paris, E.A.G., Marcelo, A.L., Novares, R.H.: Sinterização de uma mistura de cavaco de aço inoxidável com pó do mesmo material. Uma nova tecnologia para a reciclagem de metáis. Rem: Rev. Esc. Minas, Ouro Preto. **60**(1), 95–100 (2007)
9. Klar, E., Samal, E.: Powder Metallurgy Stainless Steels. Processing, Microstructures, and Properties, 243p. ASM International, Materials Park, Ohio (2007)
10. Brytan, Z., Dobrzánski, L.A., Actis Grande, M., Rosso, M.: The influence of sintering time on the properties of PM duplex stainless steel. J. Achiev. Mater. Manuf. Eng. **37**(2), 387–396 (2009)
11. Yonekubo, A.E.: Caracterização Microestrutural do aço inoxidável Superduplex UNS S32520 (UR 52 N*) Processado por moagem de alta energia. Universidade Estadual de Ponta Grossa- Ponta Grossa, Dissertação de Mestrado (2010)
12. Suryanarayana, C.: Mechanical alloying and milling. Prog. Mater. Sci. **46**(1–2), 1–184 (2001)
13. Montgomery, D.C.: Design and Analysis of Experiments, 6th edn. John Wiley & Sons, New York (2005)
14. ABNT NBR 6220:2011, Materiais refratários densos conformados – Determinação do volume aparente, densidade de massa aparente porosidade aparente, absorção e densidade aparente da parte sólida. São Paulo: Norma Brasileira (2011)
15. Palanikumar, K., Dawim, J.P.: Assessment of some factors influencing tool wear on the machining of glass fibre-reinforced plastics by coated cemented carbide tools. J.Mater. Process. Technol. **209**, 511–519 (2009)
16. Baldo, S., Mészáros, I.: Effect of cold rolling on microstructure and magnetic properties in a metastable lean duplex stainless steel. J. Mater. Sci. **45**, 5339–5346 (2010)
17. Tavares, S.S.M., Silva, M.R., Pardal, J.M., Abreu, H.F.H., Gomes, A.M.: Microstructural changes produced by plastic deformation in the UNS S31803 duplex stainless steel. J. Mater. Technol. **180**, 318–322 (2006)

Chapter 9
Design of Magnetic Aluminium (A356) Based Composites through Combined 2 Method of Sinter + Forging 3

D. Katundi, L.P. Ferreira, E. Bayraktar, I. Miskioglu, and M.H. Robert

Abstract In this work, aluminium matrix composites (AMCs with scrap A356 powder given by French Aeronautical Society) were designed through combined method of powder metallurgy and thixoforming (sinter + Forging). Three different reinforcements (Magnetic iron oxide, Fe_3O_4 – Hybrid graphene nano-platelets, GNPs, Nickel, Ni) were used and preceded under the constant process parameters such as hot compaction, sinter-forging, sintering time, etc. Mechanical and physical properties of the composites were improved with the combined processing method of powder compacted specimens and reinforcement volume fractions. Static compression tests, Microhardness tests, surface scratch tests, measurement of magnetic permeability showed that the mechanical and physical properties of these composites can be improved with the optimization of process parameters. In the present work, an alternative and a low cost manufacturing process were proposed for these composites. The microstructure and damage analyses have been carried out by Scanning Electron Microscope (SEM).

Keywords AMMCs • Magnetic permeability • Wear • Mechanical properties SEM structure

9.1 Introduction

Particulate reinforced aluminum matrix composites (AMCs) have attracted considerable attention in the field of structural and functional materials. In the present work, two types of Aluminium Matrix Composites (AMCs with scrap A356 powder given by French Aeronautical Society) were designed by using different reinforcements such as magnetic iron oxide (Fe_3O_4), hybrid graphene nano-platelets, GNPs and also Nickel. These composites were produced by powder metallurgy through combined method of Sinter + Forging with low cost manufacturing of light and efficient multifunctional materials or much more interesting one is multiferroïques composite for aeronautical applications that is one of the aims of the present work [1–3].

As well known, Multiferroïques are an attractive class of materials that are instantaneously ferromagnetic and ferroelectric, and they have potential applications in multifunctional devices, transducers, actuators and sensors [6–9]. There are two types of multiferroic used in engineering applications: intrinsic magneto electrics that exist as a single-phase state, a few ones such as MnO_3, $BiMnO_3$ and $BiFeO_3$, and extrinsic multiferroic that are composites or solid solutions of ferroelectric and ferromagnetic crystals. Since the number of intrinsic multiferroïques is not so high [4], there have been many researches to manufacture multiferroïques composites [5–9]. For composites of ferromagnetic and ferroelectric crystals, the connection between the ferromagnetic and ferroelectric order parameters is through the strain order parameter. In other words, the magnetostrictive and electrostrictive (or piezoelectric) effects of the ferromagnetic and ferroelectric crystals lead to a combination between magnetization and polarization through the elastic interactions.

D. Katundi • E. Bayraktar (✉)
Supmeca/Paris, School of Mechanical and Manufacturing Engineering, Paris, France
e-mail: bayraktar@supmeca.fr

L.P. Ferreira
Supmeca/Paris, School of Mechanical and Manufacturing Engineering, Paris, France

University of Campinas, UNICAMP-FEM, SPCampinas, Brazil

I. Miskioglu (✉)
Michigan Technology University, Engineering Mechanics Department, Houghton, MI, USA
e-mail: imiski@mtu.edu

M.H. Robert
University of Campinas, UNICAMP-FEM, SPCampinas, Brazil

P.R. Thakre et al. (eds.), *Mechanics of Composite and Multi-functional Materials, Volume 6*,
Conference Proceedings of the Society for Experimental Mechanics Series, DOI 10.1007/978-3-319-63408-1_9

In the last few decades, high performance AMCs have been widely developed with high strength, high stiffness, low density, and good wear resistance capacity [1–5]. Among them, A356 aluminium is widely used as matrix. A356 is an interesting option for aeronautical, military and automobile applications as the alloy having important properties of high strength, light weight and good capacity for foundry (fluidity). This alloy can also be successfully used as matrix from fresh scrap powder to produce high quality metal matrix composites in an economic way [6–11]. In the present work, Fe_3O_4, GNPs and Ni were chosen as the basic reinforcements to the AMCs structures to improve mechanical behaviour and wear resistance of these composites [7–13].

Generally, AMCs reinforced with magnetic iron (Fe_3O_4) exhibit good physical (electrical and magnetic) properties, which make them excellent multifunctional lightweight materials. But, they have not high wear resistance. For this reason, GNPs (<10 nm, powder was added to the AMCs. The main idea for this work is to show the influence of nickel and nano graphene on the AMCs reinforced with 10% and 20% Fe_3O_4 to improve the mechanical quality of these novel composites.

This work is concentrated on the manufacturing of AMCs using scrap A356 powder as raw material. Two groups of composites were prepared. For processing, a novel combined method of sintering + forging was performed. Morphology of the composites was evaluated by Scanning Electron Microscopy (SEM). Scratch tests were conducted to evaluate wear properties.

9.2 Experimental Conditions

In the present work, two types of Aluminium Matrix Composites (AMCs with scrap A356 powder given by French Aeronautical Society) were designed by using different reinforcements such as magnetic iron oxide (Fe_3O_4), hybrid graphene nano-platelets, GNPs and also Nickel, Cu, Zn and Zn stearate. For easy wettability of cupper, certain amount of pure nano aluminium was added in the mixture. These composites were produced by powder metallurgy through a novel combined method of "Sinter + Forging" as certain advantages regarding to conventional manufacturing processes such as low cost, capability of producing products with complex shapes, and processing simplicity, etc.

Effectively, this process aims to improve the quality and mechanical and wear behaviour of the composites created in the present work. First of all, initial size of scrap A356 powder was variable up to 200 μm. After planetary ball milling with nano aluminium (d < 23 nm) during 1 h, these powders was successfully incorporated with particulate reinforcements. After that, this main structure was mixed with ball milling during 3 h to homogenize the mixture Thanks to combined process of sinter forging, porosity and other inconvenient defects were eliminated. The general composition of scrap A356 powder was given in the Table 9.1 and also, the compositions of two types of compositions were presented in two groups (Table 9.2).

Microstructural characterization was done by means of scanning electron microscope (SEM). The dispersion of reinforcement particles in the matrix and interface at matrix/reinforcements was evaluated. Microhardness tests ($HV_{0.3}$) have been carried out on the polished and etched specimens.

All the density measurements of the specimens were carried out by using Archimedes method. These values change between 2, 38 and 2, 55 for the A356-I and A356-I respectively. And then, these results were compared. 4–5 cylindrical specimens ($H/D \geq 1.5$) for each composition then tested under quasi-static compression conditions, carried out in a servo-hydraulic MTS Universal test system (model: 5500R) at an initial rate of 10 mm/min and second rate of 5 mm/min. Maximum load endpoint was 5000 N.

Certain specimens prepared from the composites manufactured in this work were impact compression tested with Split Hopkinson Pressure Bar (SHPB) in the University of Michigan Tech, Mechanic-Materials laboratory. For impact compression testing in a Split Hopkinson Pressure Bar (SHPB), two symmetrical bars are situated in series, with the sample in between. The incident bar is hit by a striker bar during testing. The striker bar is excited from a gas gun. The transmitted bar

Table 9.1 Chemical composition of scrap A356 powder (wt %)

Element	Al	Cu	Fe	Mg	Mn	Si	Ti	Zn
wt%	Balance	≤0.20	≤0.20	≤0.25	≤0.10	6.5–7.5	≤0.20	≤0.10

Table 9.2 Compositions of the composites prepared in two groups (wt %)

Composition name	Fe_3O_4	Pure Al	Ni	Cu	GNPs	Pure Zn	Zn-Stearate
A356-I	10	5	2	2	1	<1	1
A356-II	20	5	2	2	1	<1	1

collides with a momentum trap. Strain gauges are mounted on both the incident and transmitted bars. In this work, "diameter/height" ratio of the specimens was kept around 0.75. All the tests were run at 15 psi air pressure.

Wear resistance was measured by scratch wear tests at a frequency of 15 Hz. All of the compositions were tested in two different numbers of cycles, $50*10^3$, $100*10^3$ cycles. After scratch test, damaged zone was investigated by 3D optical roughness meter. Volume loss/time and maximum depth were evaluated. Electrical and magnetic measurements have been carried out by the laboratory of the French aeronautical society on two specimens for each composite and evaluated in this work.

9.3 Results and Discussion

9.3.1 Microstructural Evaluation of the Composites A356-I and A356-II

As indicated in former section, Basic idea in this work is to create a novel composite with an optimized magnetic permeability and good electrical conductivity. For this reason, a magnetic iron oxide (Fe_3O_4) was used as primary reinforcement in these composites and GNPs, Ni, Cu etc. were added as secondary reinforcements for obtaining an improved mechanical behaviour and wear resistance. As indicated in former papers published the same authors of this work, magnetic iron oxide can be easily synthesized in the laboratory in nanoscale [5, 9–12, 14].

Figure 9.1 shows the structure of magnetite Fe_3O_4 taken by SEM Microstructure after ball milling with matrix (at the left side) [11, 12, 14] and also, general crystal structure and crystallographic data of the magnetite iron oxide (Fe_3O_4 cubic, Fd3m) was shown in the same figure (at the right side); the black balls present Fe^{2+}, the green balls present Fe^{3+} and the red balls presents O^{2-} ions respectively in this picture [6–8].

As shown in Fig. 9.1, magnetite Fe_3O_4 has the face centered cubic spinel structure, based on 32 O^{2-} ions and close-packed along the [111] direction. Fe_3O_4 differs from most other iron oxides in that it contains both divalent and trivalent iron. Fe_3O_4 has a cubic inverse spinel structure that consists of a cubic close packed array of oxide ions, where all of the Fe^{2+} ions occupy half of the octahedral sites and the Fe^{3+} are split evenly across the remaining octahedral sites and the tetrahedral sites. In stoichiometric magnetite $Fe^{II}/Fe^{III} = 1/2$, and the divalent irons may be partly or fully replaced by other divalent ions (Co, Mn, Zn, etc.). Thus, Fe_3O_4 can be both an n- and p-type semiconductor. However, Fe_3O_4 has the lowest resistivity among iron oxides due to its small band gap (0.1 eV) [6–8, 12].

Again, general structure of aluminium powder used in this work was given in the Fig. 9.2 taken by SEM microstructure after ball milling (left side) and XRD analyses complete of the structure after all of the operation of sinter + forging (at right side). It is interesting to see the last phases from this diagram. It means that a typical phase, Fe_2Cu occurred during the process sinter forging of the composite designed in the present work. This phase (Fe_2Cu) exhibited single chain magnet behaviour because very strong ferromagnetic interactions exist between the Fe^3 and Cu^2 ions.

Again, general microstructures of the two types of the composites were presented in the Fig. 9.3. As an example, only the microstructure of the first composition of this series (A356–1) was given in this Fig. 9.3a. The distribution of particle

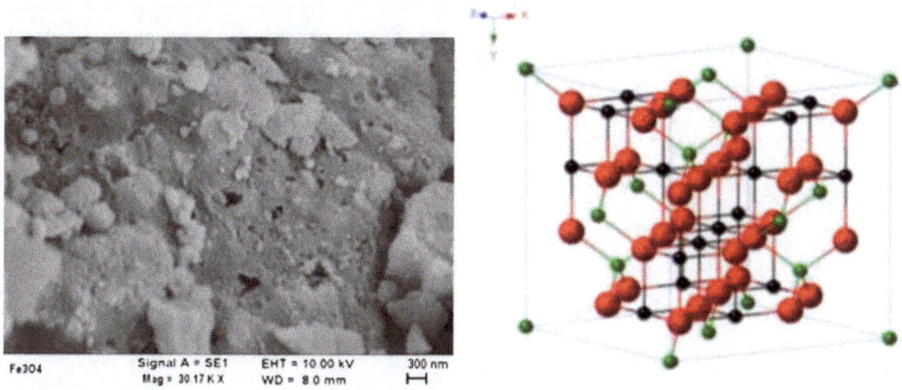

Fig. 9.1 SEM Microstructure of magnetite Fe_3O_4 after ball milling with matrix at the *left* picture [5, 10–12, 14] and general crystal structure and crystallographic data of the magnetite iron oxide (Fe_3O_4 cubic, Fd3m); the *black* ball present Fe^{2+}, the *green* ball present Fe^{3+} and the *red* ball present O^{2-} ions, at the *right* picture [6, 7]

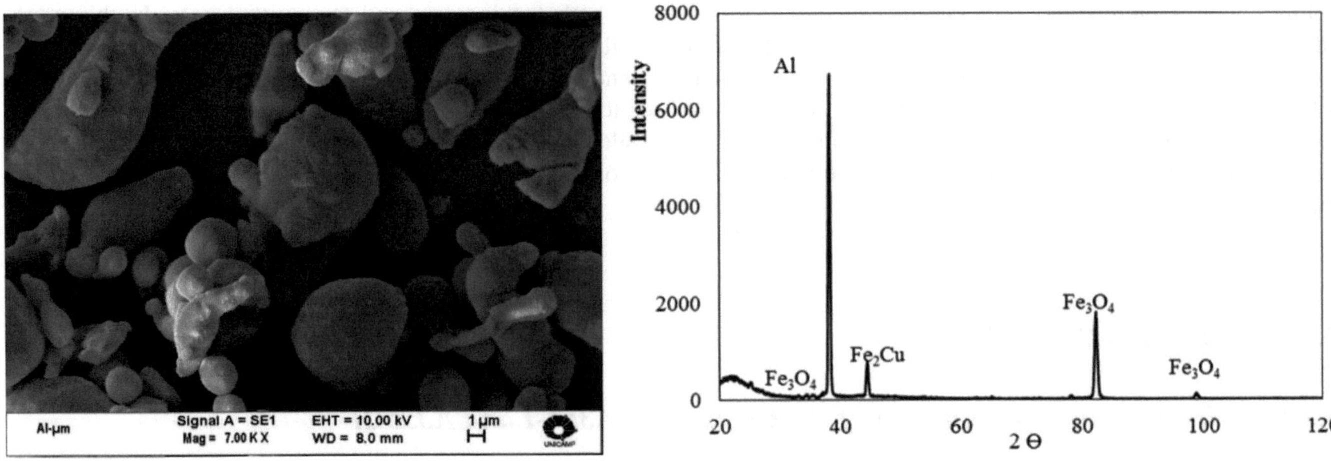

Fig. 9.2 SEM Microstructure of aluminium powder after ball milling with matrix (at the *left* picture) and XRD analyses for the final structure of the composites after sinter + forging process (at the *right* picture)

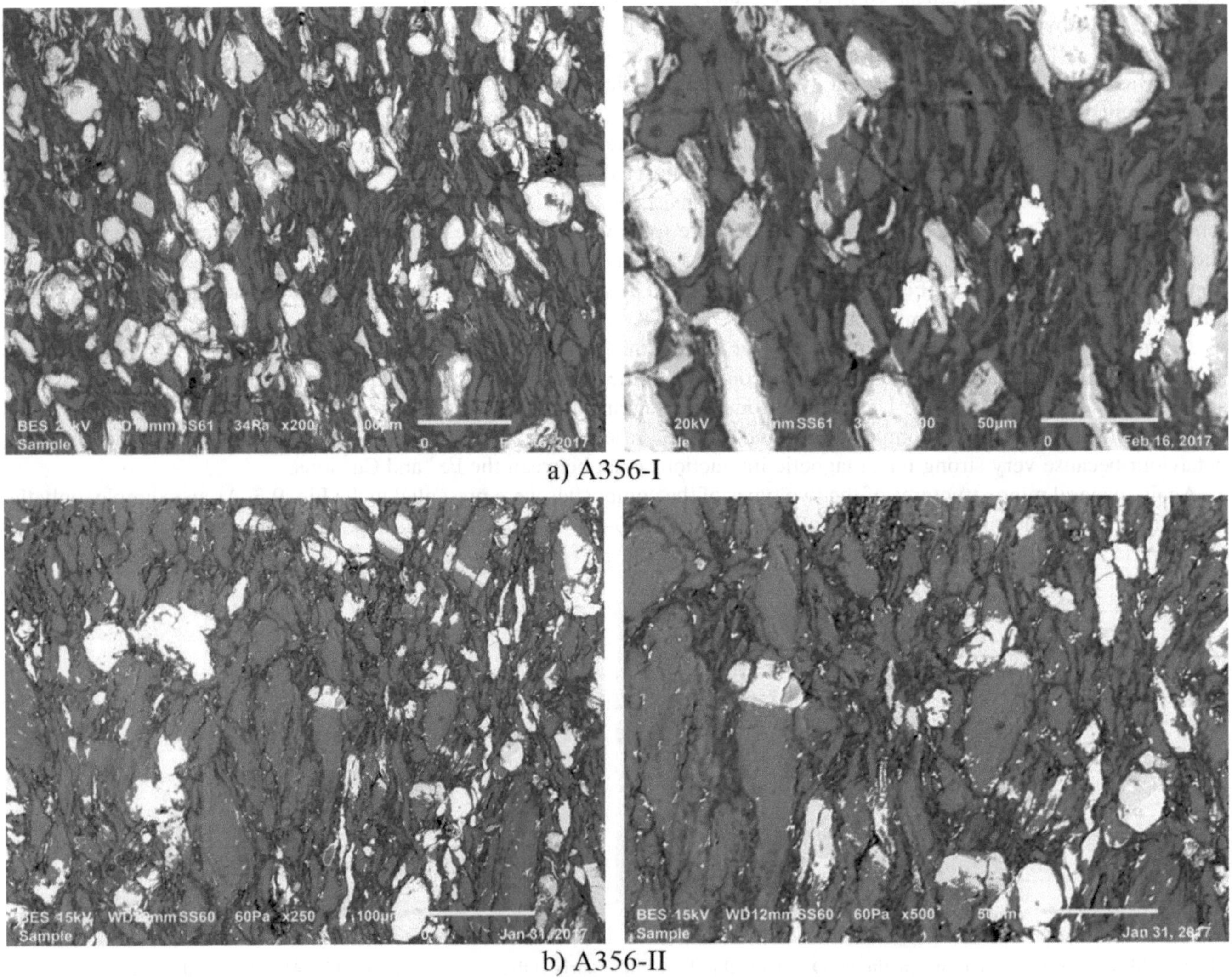

Fig. 9.3 (**a**) General microstructure of the composite A356-I and (**b**) A356-II with detail of matrix/reinforcement interface

reinforcements in the matrix were perfectly arranged thanks to the combined method of sintered + forging. In the Fig. 9.3b, microstructure of the second composition (A356-II) was presented. From these microstructures, one may observe certain details on the globular structure and perfect interface at the matrix/reinforcement and essentially distribution of fine magnetic iron oxide in the structure; eutectic structure is formed always around the globular grains and many others are observed inside of the grains. As for the big size particulate reinforcements, they are at the border of interface and/or to be struggling in the aluminium particles.

All of these evaluations of the microstructure give a clear idea about the combined process (sintered forging). This process applied for these composites is caused by bonding diffusion at the interface between matrix and reinforcement and some of the particles were forced into the grains during the forging (second) stage of this process. For this reason, very tough, solid and homogeneous structure could be obtained. Porosity and other structural defects were quasi eliminated.

9.3.2 Static Compression Test Results

The idea for novel composite design can be applied very well on the industrial parts in an economic way. Even if scrap aluminium powders are used as matrix, the microstructure can be improved by this process very well. Actually, sintered forging process is a novel process mainly called as near-net shape process for the manufacturing of the pieces [13–15]. Mainly, this process is used for bulk materials in industrial applications. In fact, low-cost sinter-forging approach to processing of particle-reinforced metal matrix composites gives always high performance applications of the industrial pieces (fatigue-creep, static and impact compression, etc.).

In the frame of this present work, only a simple static compression test was given in the laboratory scales for mechanical behaviour of these novel composites. This idea should be developed very well on the many other composites in the industrial scale. It means that very tough and strong pieces can be obtained with a microstructure by this manufacturing process similar to the thixoforming but cheaper than the other manufacturing processes.

Figure 9.4 gives static compression test results made for two compositions, A356-I and A356-II processed by simple sinter forging. These values are mean values obtained from four tests for each composition. Evidently standard deviation is variable around ±10 MPa for both of two composites. These results are around three and four times higher regarding to the specimens processed only simple sintering for the same compositions with higher toughness levels [11, 16–23]. Ultimate tensile strength values in static compression tests for the specimens processed with only sintering process have shown lower values variable between 45 70 MPa with a standard deviation around ±15 MPa due to certain micro defects – porosity in the microstructure [17, 19, 23]. All of the details about the compression test values was summarised in the Table 9.3. The same confirmation was given by micro hardness measurements as indicated in Table 9.3.

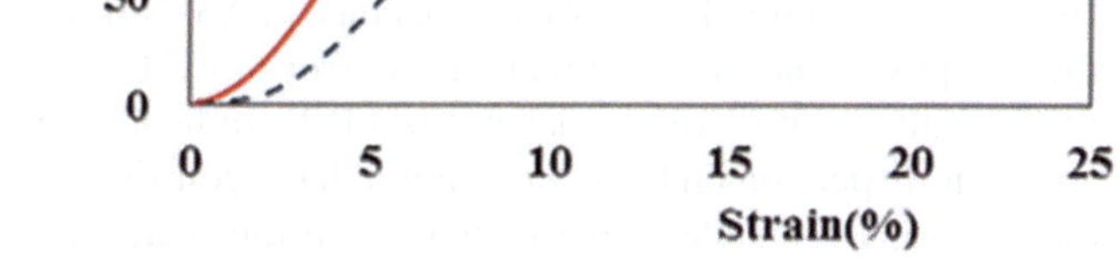

Fig. 9.4 Static compression test results for two compositions A356-I and A356-II, processed with sinter + Forging

Table 9.3 Comparison of the compression test results for two composites in case of sintering only and sinter + forging processes

Static Compression Test	Sintering	only	Sinter	+ Forging
	A356–1	A356–2	A356–1	A356–2
Yield stress (MPa)	30	35	120	130
UTS (MPa)	50 ± 15	65 ± 15	200 ± 10	225 ± 10
Microhardness ($HV_{0.3}$)	40 ± 20	46 ± 12	138 ± 15	147 ± 15

9.3.3 Impact: Compression Test Results with Split Hopkinson Pressure Bar (SHPB)

As indicated in the former section, certain values in case of impact compression tests were considered as the indicative values that have been carried out by means of Split Hopkinson Bar (SHPB) for two compositions studied here. Certain specimens prepared from the composites manufactured in this work were impact compression tested with Split Hopkinson Pressure Bar (SHPB). For the impact compression testing in a Split Hopkinson Pressure Bar (SHPB), two symmetrical bars are situated in series, with the sample in between. The incident bar is hit by a striker bar during testing. The striker bar is excited from a gas gun. The transmitted bar collides with a momentum trap. Strain gauges are mounted on both the incident and transmitted bars.

In this work, "diameter/height" ratio of the specimens was kept around 0.75. All the tests were run at 15 psi air pressure.

Figure 9.5 gives more detail information on the impact test results. Evolution of engineering strain rate is varied depending on the time. And also engineering stress strain values were given for the two types of composites tested under impact compression test by using SHPB. All of these tests seem high toughness behaviour. A356-I gives relatively higher toughness regarding to that of A356-II. A little bit variable in the striker bar was detected. Here, it seems that the striker bar velocity shows slightly a variation during the impact but it does not influence the final values.

In reality, the values are very indicatives for these composites to understand the influence of graphene platelets (GNPs) reinforcements in aces of impact conditions. It should be necessary to make much more tests with different reinforcements percentages. Many experimental results in the literature reveal significant increase of GNPs in hardness values as well as tensile strength levels of composites as compared to unreinforced aluminium [13, 24]. The improvements in properties are generally attributed to uniformly dispersed graphene platelets, an excellent interfacial bonding between graphene and aluminium matrix and grain refinement caused by the addition of graphene. In the present work, There is not so much brittleness due to graphene platelets (GNPs); may be much more percentage than 1 wt% graphene platelets can increase brittleness, it does not seems brittleness effect in the present work.

9.3.4 Wear Resistance by Scratch Test

Reduction of friction rests the main challenge for wear-resistant composites. Evaluation of wear resistance of the composites A356-I and A356-II have been carried out in two different numbers of cycles, $50*10^3$, $100*10^3$ cycles. Influence of reinforcement elements mainly GNPs and magnetic iron oxide as the major reinforcements tested and the results are presented for two composites in Figs. 9.6, 9.7, 9.8 and 9.9 respectively.

The surface, volume and the depth values are presented in the same figures for each test condition. Effects of reinforcements mainly Ni, GNP and Fe3O4 have obvious advantages over others.

As shown by the results of scratch tests, the composite reinforced with different ratios (particularly magnetic iron oxide) has presented lightly higher wear resistance in A356-II than the A356-I. Considering the studied conditions (friction between the composite and the zirconia ball, number of cycles, normal force, etc.), it seems that not so much differences in overall wear results even if wear resistance of GNPs indicated in literature [24]. In this particular case, the small size of the GNPs and their dispersion on the matrix should have contributed to improve the wear resistance. But it does not seem real effect of these nano GNPs under experimental conditions carried out in the present work. Influences of magnetic iron oxide powders have slightly improved wear resistance of these composites contrary to former results obtained by the same authors of this work [17–19, 21, 23].

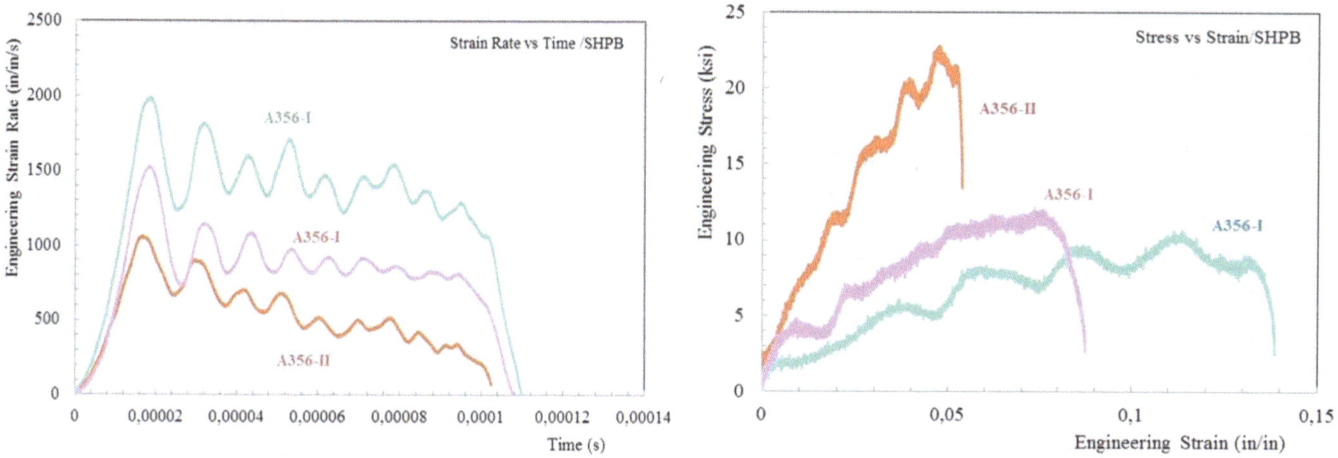

Fig. 9.5 Evolution of engineering strain rate depending on the time (at the *left* side) and engineering stress strain values (at the *right* side) for the two types of composites tested under impact compression test by using SHPB

Parameters (A356-1)	N° of Cycle 50.10^3	
	At the bottom	At the top
Surface (mm^2)	0.465	0.0166
Volume (μm^3)	1879933	4476
Depth - max. (μm)	11.04	3.90
Depth - mean (μm)	4.04	0.270

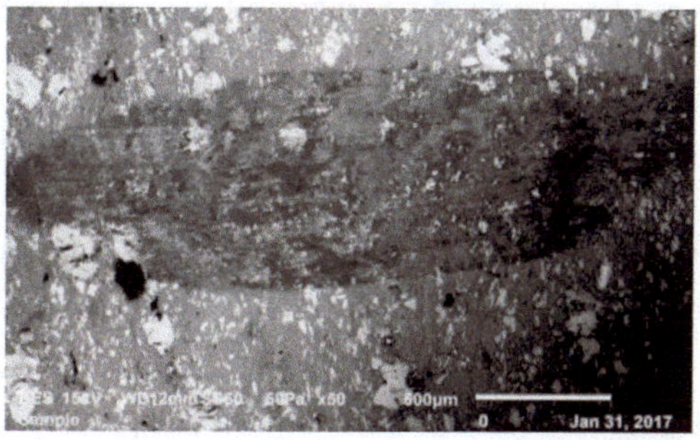

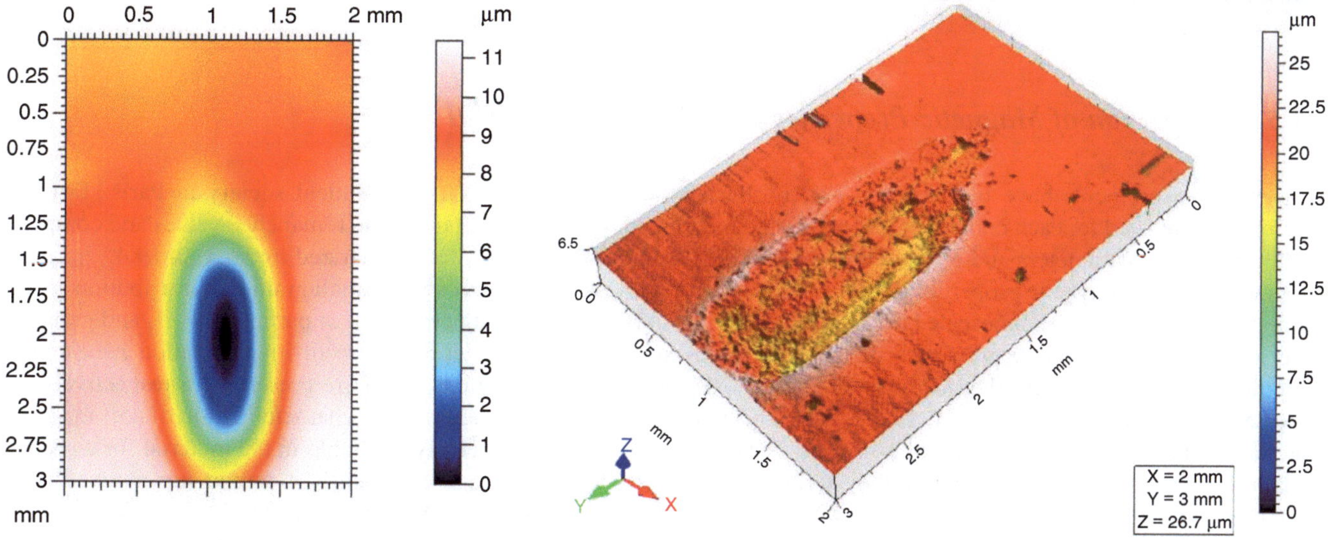

Fig. 9.6 Images of surface damage zones produced by scratch wear test for A356–1 specimen in the cycle of 50.10^3 and comparison of damage zone with SEM pictures defined with surface and volume lost during the test, etc

Parameter (A356-2)s	N° of Cycle 50.10^3	
	At the bottom	At the top
Surface (mm^2)	1.35	0.0382
Volume (μm^3)	4635266	9551
Depth - max. (μm)	8.74	2.02
Depth - mean (μm)	3.44	0.250

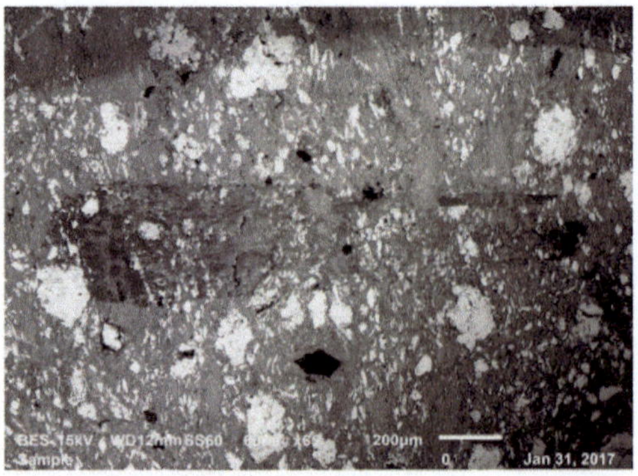

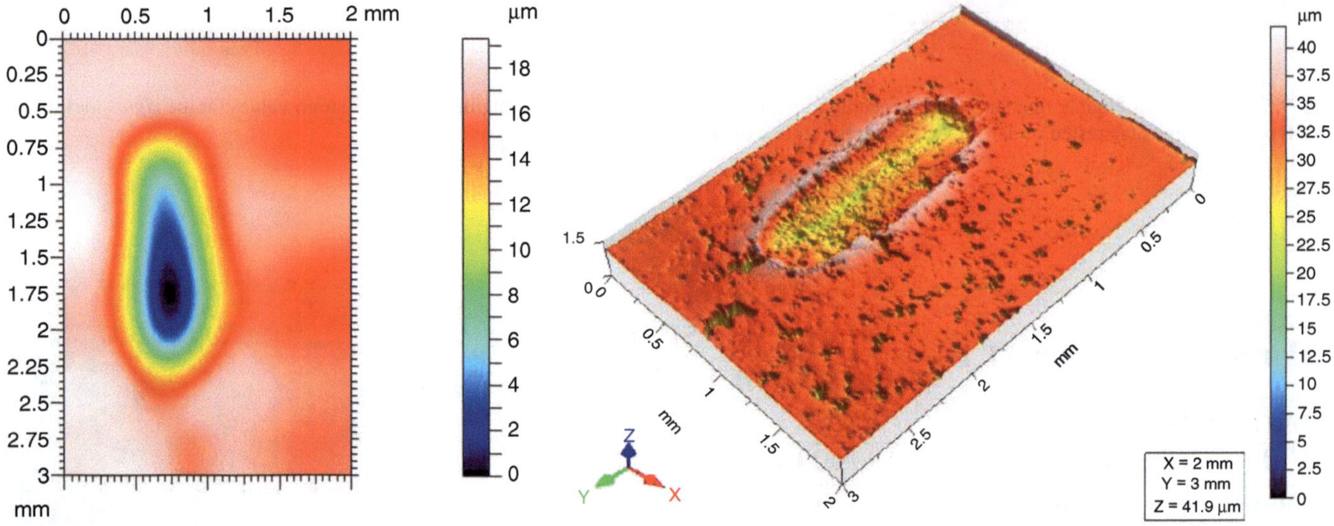

Fig. 9.7 Images of surface damage zones produced by scratch wear test for A356–2 specimen in the cycle of 50.10^3 and comparison of damage zone with SEM pictures defined with surface and volume lost during the test, etc

9.3.5 Evaluation of Magnetic Properties for A356-I and A356-II

Magnetic measurements have been carried out by the laboratory of the French aeronautical society in Paris. Two test specimens were used for each composite and evaluated by drawing the magnetic saturation, remanence values and compared certain hysteresis parameters and presented in the Fig. 9.10 and critical values are summarized in the Table 9.4.

As well known, magnetic properties of the composite structures can be improved with small grain size distribution well in the matrix especially in the nanoscale [17, 21, 23]. As shown in the Fig. 9.9, the percentage of particles Fe_3O_4 and graphene have improved the magnetic properties of the composites studied. The field dependence of magnetization measured is presented in the same figure; the magnetic saturation (Ms) and remanence values, (Mr) are presented in respectively. As indicated in the former papers [5, 11, 17, 21, 23], a basic and important parameter in the characterization of soft magnetic materials is the power loss; this kind of power gives a measure of the energy density available in the material for a specific application. Data from the analysis of the hysteresis curve as the saturation magnetization (M_s), remanence magnetization (M_r), coercive field (H_c), the relationship between $M_{r.}$ and $M_{s.}$ and susceptibility (X_m) are presented in Table 9.4.

It seems that composites A356-II exhibit slightly better magnetic properties. As well known, the value of $M_{s.}$ must be maximized as much as possible to confirm a better response under the application of the pieces. These results are the first time published and original for these two composites and other measurements for these composites are going on in the frame of the common research project.

Parameters (A356-1)	N° of Cycle 100.10^3	
	At the bottom	At the top
Surface (mm^2)	0.895	0.0299
Volume (μm^3)	6669921	23584
Depth - max. (μm)	20.9	4.73
Depth - mean (μm)	7.46	0.789

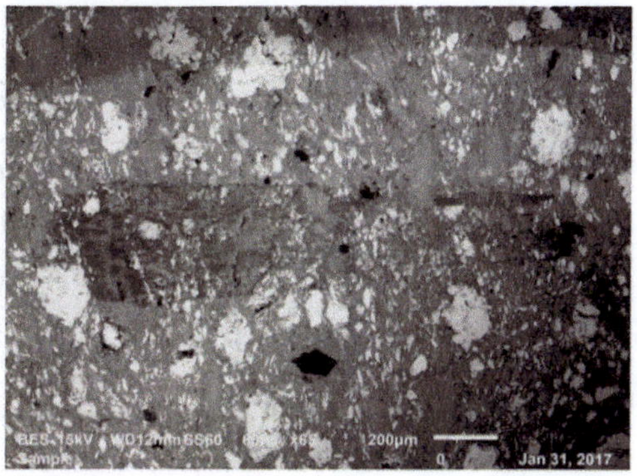

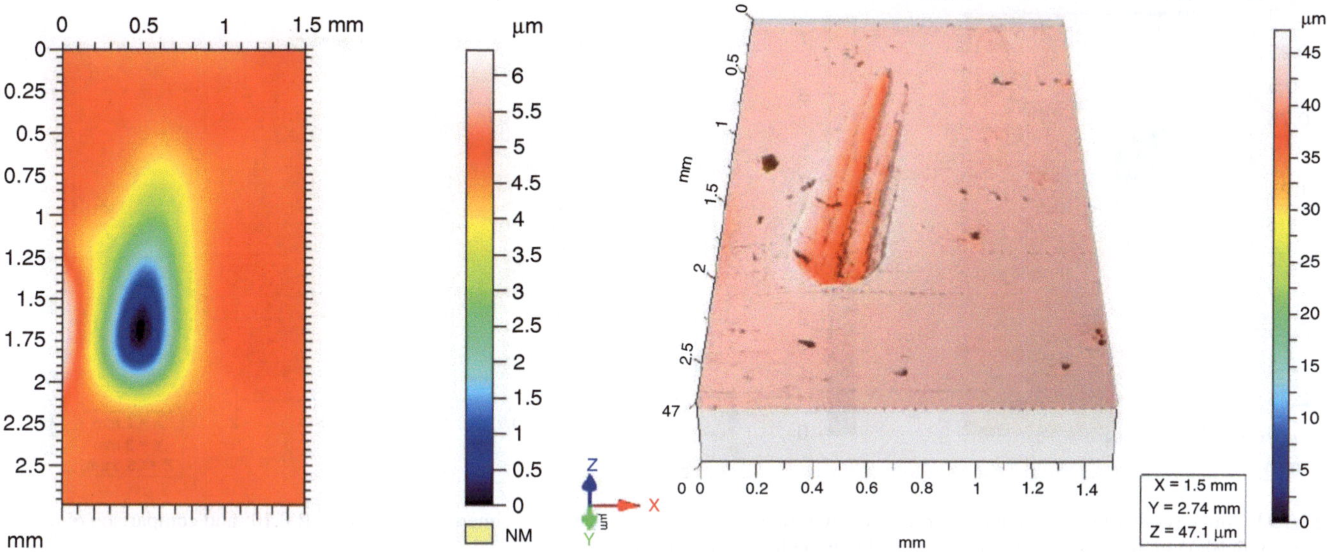

Fig. 9.8 Images of surface damage zones produced by scratch wear test for A356–1 specimen in the cycle of 100.10^3 and comparison of damage zone with SEM pictures defined with surface and volume lost during the test, etc

9.3.6 Measurements of Electrical Properties

Electrical properties (resistivity and conductivity) were measured with an Agilent 4338B milliohm meter. Five specimens were used for each composition and the mean values of conductivity and resistivity are presented in Table 9.5. For measurements, DC regulated power supply voltage and current were set as 20 V and 20A respectively. Data acquisition Card "NI9234" was connected in parallel with the output of the power to acquire the voltage data (voltage input accuracy was 24 bits). A high precision multi-meter "Agilent U1253 N" was connected in series to measuring the current intensity (A).

All the data from the measurements above were displayed and stored with Lab View program. It is worth to mention that in the present work these results were obtained under laboratory conditions. Naturally, the samples for the two composites studied were prepared carefully according to the standards specified by the manufacturers of the measurement devices, but the form and other effects such as surface conditions of each sample can easily influence the measurements. That is the reason that all the samples were produced under the same conditions and polished carefully on both of their surfaces to keep a parallelism. These composites are listed in the same table with pure aluminum in order to make a simple comparison under laboratory conditions. These results evidently should be considered as preliminary results and give an indicative data. It seems that form and surface conditions should be improved again.

Parameters (A356-2)	N° of Cycle 100.10^3	
	At the bottom	At the top
Surface (mm^2)	0.927	0.0972
Volume (μm^3)	3707875	52448
Depth - max. (µm)	14.9	5.49
Depth - mean (µm)	4.00	0.540

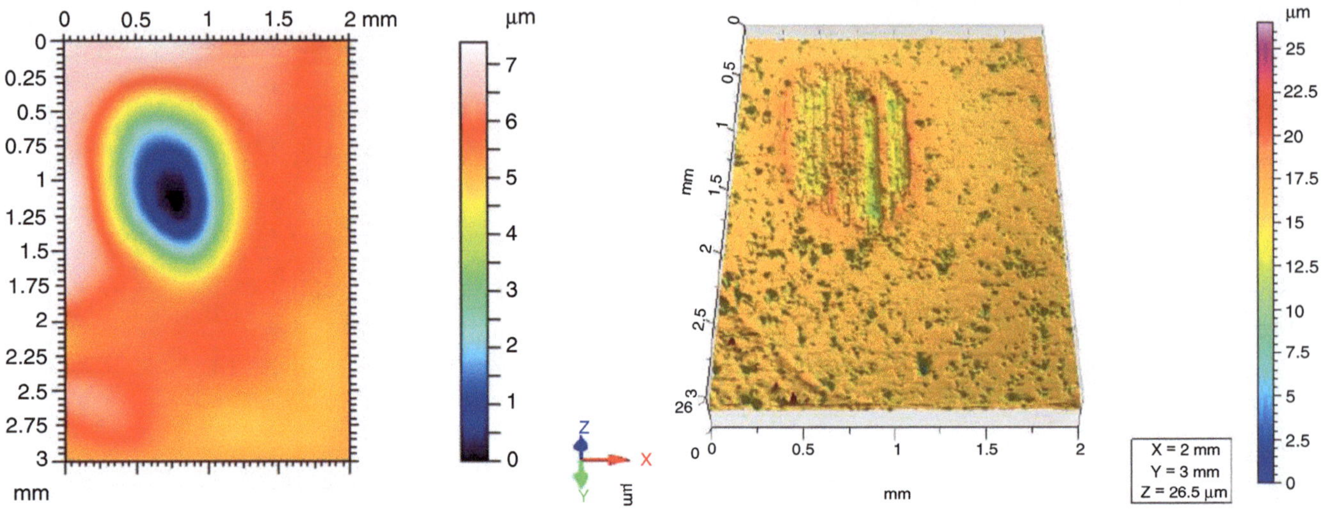

Fig. 9.9 Images of surface damage zones produced by scratch wear test for A356–2 specimen in the cycle of 100.10^3 and comparison of damage zone with SEM pictures defined with surface and volume lost during the test, etc

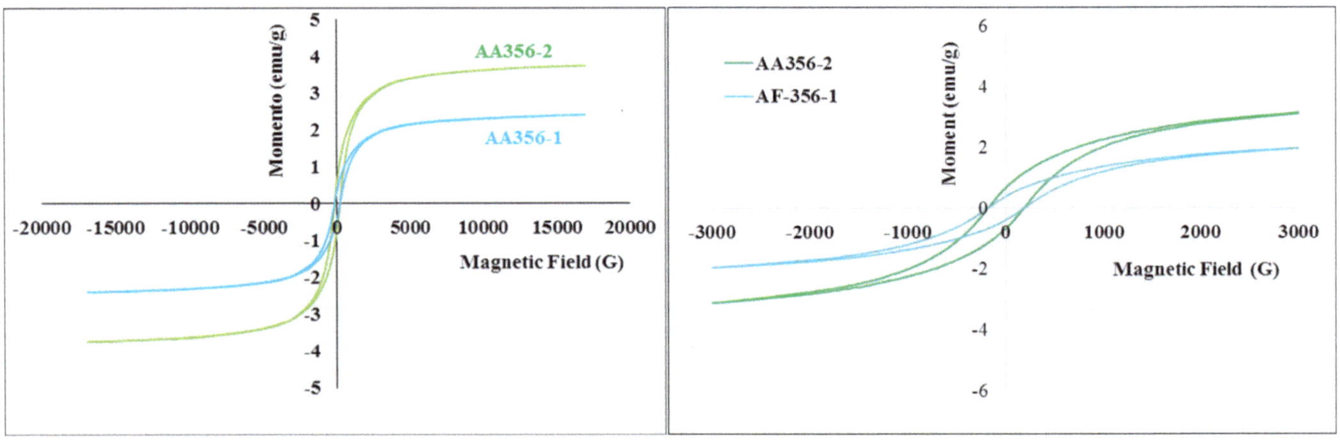

Fig. 9.10 Magnetic properties; magnetic saturation *curves* at *left* side and the hysteresis *curves* at *right* side, for two composites, A356-I and A356-II respectively

Table 9.4 Hysteresis parameters of two composites, A356-I and II after Sinter + Forging process

Sample	M_s (emu/g)	M_r (emu/g)	H_c (G)
A356–1	3,86 ± 0,03	0,0650 ± 0,0025	189,446 ± 50,03
A356-II	2,49 ± 0,05	0,0643 ± 0,002	189,422 ± 50,09

Table 9.5 Electrical properties (resistivity & conductivity) measured for three compositions

Composition	Resistivity (ohm.m)	Conductivity (S/m)
Pure Aluminium	0, 000192	11,997
A356-I	0,000581	5974
A356-II	0,000435	6804

9.4 Conclusion

New designed Magnetic Aluminum Based metal matrix composite reinforced with nano graphene platelets (GNPs) and nano Fe_30_4 magnetic particles can potentially be a useful material for aeronautic and automotive applications. In the frame of this common project, novel composites have been developed from scrap aluminium A356 powder reinforced with magnetic iron oxide (Fe_30_4) and GNPs as main reinforcements in an economic way. Low cost manufacturing of these composites have been successfully managed through the combined method of sinter forging.

Microstructural analysis has shown that a good bonding at interface of matrix-reinforcement thanks to combined process sinter-forging, a tough and sound microstructure was obtained without porosity. Wear resistance and ductility of these structures should be improved with doping process and good preparation conditions; ball milling in longer time is needed for helping the fine and homogeneous distribution of the particles in the matrix.

Mechanical behaviours of these composites are better than the former magnetic composites developed with only sintering processes. Electrical properties (resistivity and conductivity) and magnetic properties measured for two compositions seems very optimistic values for future work Optimizations of the certain parameters (processing parameters, reinforcement content, etc.) require much more experimental work to create magnetic parts in an industrial scale. Here, only limited measurements at room temperature were presented as they are indicative parameters for better understanding the effect of the reinforcements on the optimization of the mechanical, magnetic and electrical properties of the aluminum based composites that indicated in the present work.

Acknowledgments This work has been carried out on the frame of collaboration between SUPMECA/PARIS, UNICAMP/SP/CAMPINAS and MICHIGAN TECHNICAL UNIVERSITY/ HOUGHTON-MI-USA.

Some of the magnetic measurements in the laboratory of French aeronautical society in PARIS, Authors acknowledge and appreciate so much for valuable technical helps and discussions.

References

1. Chawla, N.: Industrial paper, "MMCs in automotive applications". Adv. Mat. Process. **1**, 29–31 (2006)
2. Din, T., Campbell, J.: High strength aerospace aluminum casting alloys a comparative study. Mater. Sci. Technol. **12**, 644–650 (1996)
3. Chainarong, S., Muangjunburee, P., Suthummanon, S.: Friction stir processing of SSM356 Aluminium alloy, 12th global congress on manufacturing and management, GCMM 2014. Procedia Eng. **97**, 732–740 (2014)
4. Choi, D.-H., et al.: Microstructure and mechanical property of A356 based composite by friction stir processing. Trans. Nonferrous Metals Soc. China. **23**, 335–340 (2013)
5. Bayraktar, E., Katundi, D.: New aluminum matrix composites reinforced with iron oxide journal of achievements in materials and manufacturing engineering. JAMME. **38**(1), 7–14 (2010)
6. Wu, W., Wu, Z., Yu, T., Jiang, C., Kim, W.S.: Review: Recent progress on magnetic iron oxide nanoparticles: synthesis, surface functional strategies and biomedical applications. Sci. Technol.Adv. Mater. **16**, 43p (2015)
7. Zhuang, P.-F., Luo, L., Liu, T., Luo, Y., Xie, X.-H., Zheng, H., Liang, Z., Jiao, C.-Q., Wang, J.-L., Hu, J.-X., Cheng, H., Duan, C.-Y.: Cyano-bridged Fe_2Cu clusters: Control of magnetic properties through cis–*trans* arrangement. Inorg. Chem. Commun. **48**, 8–11 (2014)
8. Dong, D.-P., Liu, T., Zheng, H., Liang, Z., Zhuang, P.-F., Cheng, H., Duan, C.-Y.: Synthesis, structures and single chain magnet behavior of a cyano-bridged $\{Fe_2Cu\}$ chain. Inorg. Chem. Commun. **24**, 153–156 (2012)
9. Srinivasu, R., Sambasiva, A., Madhusudhan, G., Srinivasa, K.: Friction stir surfacing of cast A356 aluminum end silicon alloy with boron carbide and molybdenum disulphide powders. Defense Technol. **11**, 140–146 (2015)
10. Ayari, F., Katundi, D., Bayraktar, E.: Damage of aluminium matrix composite reinforced with iron oxide (Fe3O4): experimental and numerical study AIP Conference Proceedings, pp. 1315–1328, 27 (2011)

11. Ferreira, L.-M.-P., Bayraktar, E., Robert, M.-H.: Magnetic and electrical properties of aluminium matrix composite reinforced with magnetic nano iron oxide. J. Adv Mater. Process. Technol. **2**(1), 165–173., Taylor & Francis-USA (2016)
12. Bayraktar, E., Ayari, F., Tan, M.-J., Tosun Bayraktar, A., Katundi, D.: Manufacturing of aluminum matrix composites reinforced with iron-oxide nanoparticles: microstructural and mechanical properties. Metall. Mater. Trans. B. **45B**(26), 352–362. Laughlin, David E. (ed.), ASM-TMS/USA (2014)
13. Ferreira, L.M.P.: Maria Helena Robert, Emin Bayraktar, Diana Zaimova new design of aluminium based composites through combined method of powder metallurgy and thixoforming advanced materials research. AMR. **939**(1), 68–75 (2014)
14. Ferreira, L.-M.-P., Bayraktar, E., Robert, M.-H., Miskioglu, I.: Optimization of magnetic and electrical properties of new Aluminium matrix composite reinforced with magnetic Nano iron oxide (Fe_3O_4) mechanics of composite and multifunctional materials. SEM-Springer-USA. **7** (1), 11–18 (2015)
15. Santella, M.-L., Engstrom, T., Storjohann, D., Pan, T.-Y.: Effects of friction stir processing on mechanical properties of the cast aluminum alloys A319 and A356. Scr. Mater. **53**, 201–206 (2005)
16. Kleemann, W., Borisov, P., Bedanta, S., Shvartsman, V.V.: Multiferroic and magneto electric materials: Multiferroic and magneto electric materials: novel developments and perspectives IEEE transactions on ultrasonic. Ferroelect. Freq. Control. **57**(10), 2228–2232 (2010)
17. Ferreira, L.F.P., Bayraktar, E., Miskioglu, I., Robert, M.H.: Recycle of aluminium (A356) for processing of new composites reinforced with magnetic nano iron oxide and molybdenum, mechanics of composites and multifunctional materials, chapter 18, vol. 7, pp. 153–161, 2016, SEM-Springer-USA, ISBN 978-3-319-41766-0 (2016)
18. Ferreira, L.F.P., Bayraktar, E., Miskioglu, I., Katundi, D.: Design of hybrid composites from scrap aluminum bronze chips, pp. 131–138, chapter 15, SEM-Springer-USA, ISBN 978-3-319-41766-0 (2016)
19. Ferreira, L.F.P., Bayraktar, E., Robert, M.H., Miskioglu, I.: Particles reinforced scrap aluminum based composites by combined processing sintering + Thixoforming, pp. 145–152, chapter 17, SEM-Springer-USA, ISBN 978-3-319-41766-0 (2016)
20. Kursun, A., Ferreira, L.F.P., Bayraktar, E., Miskioglu, I.: Design of hybrid composites from scrap aluminum reinforced with (SiC+TiO2+Gr +Ti+B), pp. 225–232, chapter 27, SEM-Springer-USA, ISBN 978-3-319-41766-0 (2016)
21. Ferreira, L.F.P., Gatamorta, F., Bayraktar, E., Robert, M.H.: Manufacturing of low cost composites with porous structures from scrap aluminium (AA2014) chips, pp. 233–240, chapter 28, SEM-Springer-USA, ISBN 978-3-319-41766-0 (2016)
22. Leibholz, R., Robert, M.H., Leibholz, H., Bayraktar, E.; Development of functionally graded nodular cast iron reinforced with recycled WC particles, pp. 241–250, chapter 29, SEM-Springer-USA, ISBN 978-3-319-41766-0 (2016)
23. Ferreira, L.F.P., Miskioglu, I., Bayraktar, E., Robert, M.H.: Aluminium matrix composites reinforced by Nano Fe3O4 doped with TiO2 by thermomechanical process, pp. 251–259, chapter 30, SEM-Springer-USA, ISBN 978-3-319-41766-0 (2016)
24. Garg, P., Gupta, P., Kumar, D., Parkash, O.: Structural and mechanical properties of graphene reinforced aluminum matrix composites. J. Mater. Environ. Sci. **7**(5), 1461–1473 (2016.) ISSN: 2028-2508

Chapter 10
Design of Low Composites from Recycled Copper + Aluminium Chips for Tribological Applications

F. Gatamorta, E. Bayraktar, I. Miskioglu, D. Katundi, and M.H. Robert

Abstract In this work, a special copper aluminium matrix composite (ACMMCs) obtained from the fresh scrap – chips of AA7075 and- pure electrolytic copper were designed through combined method of powder metallurgy and sinter + Forging. First of all, Al-Cu matrix was doped with ZnO after the ball milling with two basic reinforcements (Nb_2Al– SiC, etc.) was carried out during the 4 h. A basic composition was prepared depending on the doping percentage of ZnO as 30 wt%. Mechanical and physical properties of this composite designed here can be improved with the doping process followed by combined method of powder compacted specimens and doping volume fractions. The surface scratch tests and micro-hardness results were compared according to the optimization conditions of the doping and the reinforcement. Static compression and impact-drop weight tests were carried out. The microstructure and damage analyses have been carried out by Scanning Electron Microscope (SEM).

Keywords ACMMCs • Sintered-forging • Wear • Impact • Compression • SEM analyze

10.1 Introduction

Recent industrial applications find more attractive particulate reinforced metal matrix composites that have made considerable progress in the field of structural and functional materials. In the present work, special copper aluminium Matrix Composite (ACMMCs with scrap AA7075 powders atomized for using as fine powder) was designed by using different reinforcements such resistance against tribological damages such as friction of two pieces and/or new types electrical brushes, etc.). This composite was produced by powder metallurgy through combined method of Sinter + Forging with low cost manufacturing of light and efficient multifunctional materials or much more interesting one is multiferroïques composite for industrial applications that is one of the aims of the present work [1–3].

As well known, for tribological applications, certain types of electric motors or generators to function, the coils of the rotor must be connected to complete an electrical circuit. To accomplish this, for example a copper or brass commutator or 'slip rings' are affixed to the shaft, and springs press braided copper wire 'brushes' onto the rings which conduct the current. Such types of brushes provided poor commutation as they moved from one commutator segment to the next. For this reason, as an alternative composite can be produced a high resistance brushes' made from copper aluminium reinforced with high resistant reinforcements such as Nb_2Al and SiC, etc.). Since the traditional brushes such as graphite – copper based structures wear out very easily during their functions coupled with other pieces; they can be replaced in products with other non-traditional structures that one of these structures is proposed in the present work.

In the last few decades, high performance MMCs have been widely developed with high strength, high stiffness, low density, and good wear resistance capacity [1–15]. Among them, AA7075 aluminium is widely used as matrix. AA7075 is an interesting option for aeronautical, military and automobile applications as the alloy having important properties of high

F. Gatamorta • D. Katundi
University of Campinas, UNICAMP-FEM, SP, Campinas, Brazil

E. Bayraktar (✉)
University of Campinas, UNICAMP-FEM, SP, Campinas, Brazil

Supmeca/Paris, School of Mechanical and Manufacturing Engineering, Paris, France
e-mail: bayraktar@supmeca.fr

I. Miskioglu (✉)
Michigan Technology University, Engineering Mechanics Department, Houghton, MI, USA
e-mail: imiski@mtu.edu

M.H. Robert
Supmeca/Paris, School of Mechanical and Manufacturing Engineering, Paris, France

P.R. Thakre et al. (eds.), *Mechanics of Composite and Multi-functional Materials, Volume 6*,
Conference Proceedings of the Society for Experimental Mechanics Series, DOI 10.1007/978-3-319-63408-1_10

strength, light weight and good capacity for foundry (fluidity). This alloy can also be successfully used as matrix from fresh scrap powder to produce high quality metal matrix composites in an economic way [6–14].

In the present work, Nb_2Al and SiC were chosen as the basic reinforcements to the ACMMCs structure to improve mechanical behaviour and wear resistance of this composite [7–12]. This work is concentrated on the manufacturing of ACMC using scrap AA7075 powder as raw material. For processing, a novel combined method of sintering + forging was performed. Morphology of the composites was evaluated by Scanning Electron Microscopy (SEM). Compression and impact-drop weight tests, scratch tests were conducted to evaluate wear properties.

10.2 Experimental Conditions

In the present work, special Copper-Aluminium Matrix Composites ((ACMMCs with scrap AA7075 powders atomized for using as fine powder) was designed by using different reinforcements such as Nb_2Al and SiC, Zn and Zn stearate. For easy wettability of electrolytic copper, certain amount of pure nano aluminium was added in the mixture. These composites were produced by powder metallurgy through a novel combined method of "Sinter + Forging" as certain advantages regarding to conventional manufacturing processes such as low cost, capability of producing products with complex shapes, and processing simplicity, etc. Effectively, this process aims to improve the quality and mechanical and wear behaviour of the composites created in the present work.

First of all, initial size of scrap AA7075 atomized powder was milled for using as fine powder and mixed with certain amount of pure nano aluminium (5%). After addition of electrolytic copper powder in this mixture, planetary ball milling was carried out during 1 h. Finally, this mixture was doped with ZnO and incorporated with particulate reinforcements. After that, this main structure was mixed with ball milling during 3 h to homogenize the mixture Thanks to combined process of sinter forging, porosity and other inconvenient defects were eliminated. The general composition of this composite was given in the Table 10.1.

Microstructural characterization was done by means of scanning electron microscope (SEM). The dispersion of reinforcement particles in the matrix and interface at matrix/reinforcements was evaluated. Microhardness tests ($HV_{0.3}$) have been carried out on the polished and etched specimens.

All the density measurements of the specimens were carried out by using *Archimedes* method. These values change between 2.85 and 3.5. For each condition; sintering and sinter + forging (called after here F16-S and F16-S + F respectively), 4 cylindrical specimens ($H/D \geq 1.5$) were then tested under quasi-static compression conditions in order to see difference between only sintered and sintered forging specimens.

These tests were carried out in a servo-hydraulic MTS Universal test system (model: 5500R) at an initial rate of 10 mm/min and second rate of 5 mm/min. Maximum load endpoint was 5000 N. Certain specimens prepared from the composite manufactured in this work were impact –drop weight test compression tested on a universal drop weight test device (Dynatup Model 8200 machine) with a total weight of 1.9 kg, punch height of 1000 mm and with an impact velocity of 3 m/s.

Wear resistance was measured by scratch wear tests at a frequency of 15 Hz. All of the specimens were tested in two different numbers of cycles, $50*10^3$, $100*10^3$ cycles. After scratch test, damaged zone was investigated by 3D optical roughness meter. Volume loss/time and maximum depth were evaluated. Electrical conductivity and resistivity measurements have been carried out on the laboratory scale at the French aeronautical society for each condition (sintering and also sinter + forging) and evaluated in this work.

Table 10.1 Compositions of the composite prepared in this work (wt %)

Composition Name	Matrix: Mixture and doping of Cu + Al + ZnO (2:2:1)	Nb_2Al	SiC	Pure Zn	Zn-Stearate
F-16	Balance	10	10	2	3

10.3 Results and Discussion

10.3.1 Microstructural Evaluation of the Composite under Different Manufacturing Conditions

As indicated in former section, Basic idea in this work is to create a novel composite with an optimized wear resistance and good electrical conductivity. For this reason, electrolytic copper and aluminium was processed together as a matrix doped with ZnO and then two primary reinforcements such as Nb_2Al and SiC, Zn and Zn stearate etc., were added for obtaining an improved mechanical behaviour and wear resistance.

Figure 10.1 shows the SEM microstructure of this composite produced under two different processes; only sintering (F16-S) and sinter + forging (F16-S + F) respectively.

Again, all of the distributions of particle reinforcements in the matrix were perfectly arranged for two manufacturing conditions. However, the better microstructural evolution without micro porosity was obtained thanks to the combined method of sintered + forging. From these microstructures, one may observe certain details on the globular structure and perfect interface at the matrix/reinforcement and essentially distribution of hard reinforcement particles in the structure; eutectic structure is formed very often and many particles are observed inside of the grains. As for the big size particulate reinforcements, they are at the border of interface and/or to be struggling in the matrix.

As made microstructural analyses for this composite, Fig. 10.2 gives a general EDS results carried out on these structures.

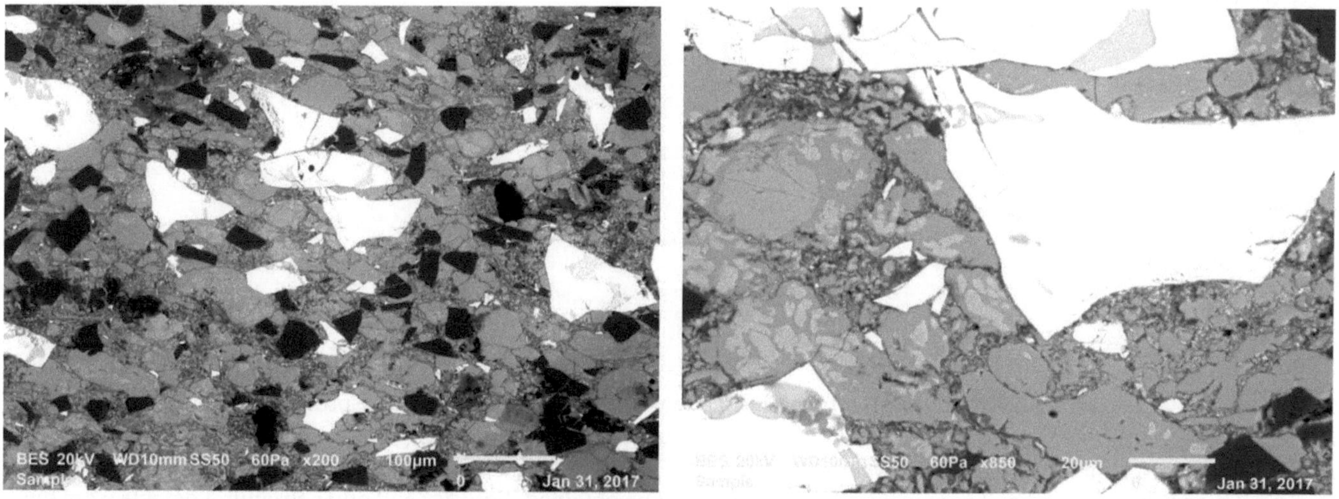

a) F16-S (microstructure)

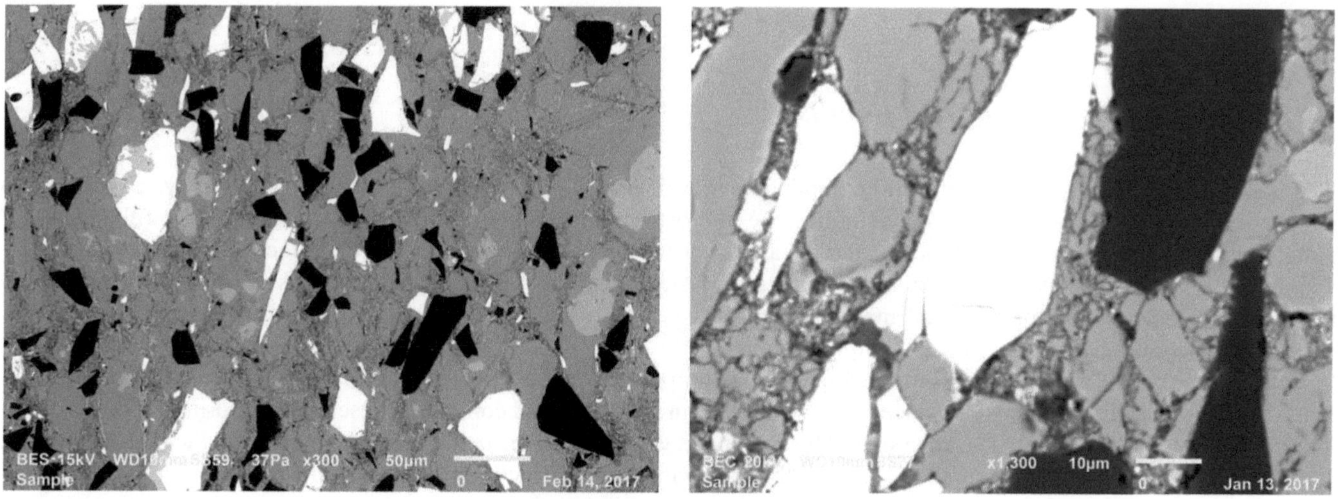

b) F16 (S+F microstructure)

Fig. 10.1 (**a**) General microstructure of the sintering composite and (**b**) sinter + forging with detail of matrix/reinforcement interface

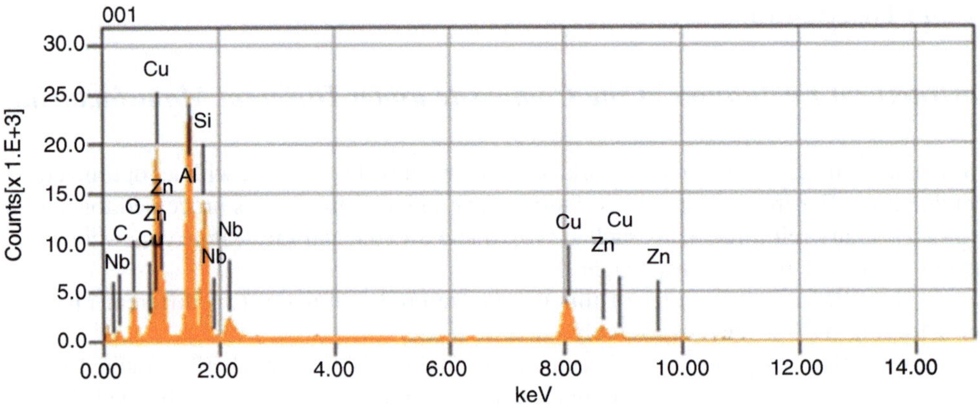

Fig. 10.2 General microstructural analyses for the composite F16

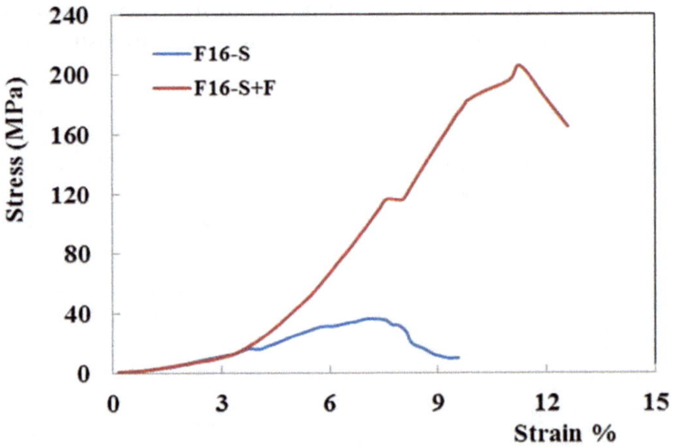

Fig. 10.3 Static compression test results for two manufacturing conditions, processed with sintering and sinter + Forging

All of these evaluations of the microstructure give a clear idea about the combined process (sintered forging). This process applied for these types of composites is caused by bonding diffusion at the interface between matrix and reinforcement and some of the particles were forced into the grains during the forging (second) stage of this process. For this reason, very tough, solid and homogeneous structure could be obtained. Porosity and other structural defects were quasi eliminated.

10.3.2 Static Compression Test Results

The idea for novel composite design can be applied very well on the industrial parts in an economic way. Even if scrap copper and aluminium powders are used as matrix, the microstructure can be improved by this process very well. Actually, sintered forging process is a novel process mainly called as near-net shape process for the manufacturing of the pieces [12–15]. Mainly, this process is used for bulk materials in industrial applications. In fact, low-cost sinter-forging approach to processing of particle-reinforced metal matrix composites gives always high performance applications of the industrial pieces (fatigue-creep, static and impact compression, etc.).

In the frame of this present work, only a simple static compression test was given in the laboratory scales for mechanical behaviour of this novel composite produced with two different processes and compared these manufacturing processes (only sintering and sinter + forging). This idea should be developed very well on the many other composites in the industrial scale. It means that very tough and strong pieces can be obtained with a microstructure by this manufacturing process but cheaper than the other manufacturing processes.

Figure 10.3 gives static compression test results obtained on the specimens made for two manufacturing processes; only sintering and sinter + forging. These values are mean values obtained from four tests for each manufacturing condition.

Table 10.2 Comparison of the compression test results in case of sintering only and sinter + forging processes

Static compression Test results (mean values)	F16-S Sintering only (MPa)	F16-S + F Sinter + Forging (MPa)
Yield stress (MPa)	17	113
UTS (MPa)	38	210
Microhardness ($HV_{0.3}$)	257	398

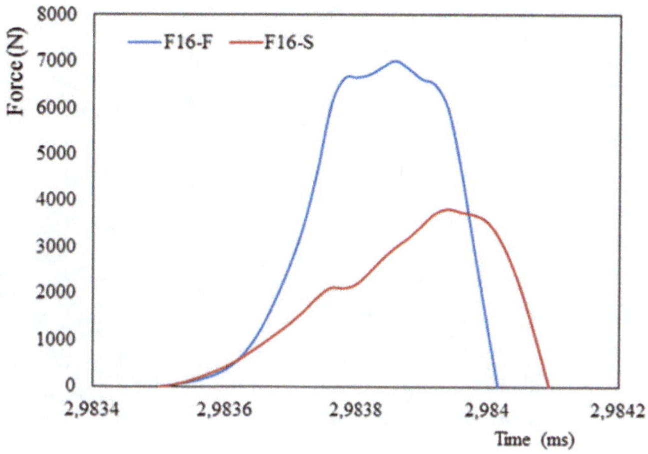

Fig. 10.4 Impact behaviour for two different processes; sintering and sinter + forging obtained by dynamic compression test (F16-S and F16-S + F) respectively

Evidently standard deviation is variable around ±15-20 MPa for both of two processes. One may observe that sinter + forging process gives always very high resistance regarding to the simple sintering process. Ultimate tensile strength values in static compression tests for the specimens processed with only sintering process have shown lower values variable between 35–40 MPa with a standard deviation around ±20 MPa due to certain micro defects – porosity in the microstructure. For the sinter + forging process these values have found at least ten times higher than simple sintering process (around 200–220 MPa). All of the details about the compression test values was summarised in the Table 10.2. The same confirmation was given by micro hardness measurements as indicated in Table 10.2.

10.3.3 Dynamic Compression (Impact-Drop-Weight) Testing

Low velocity or dynamic compression tests results were given in the Fig. 10.4. Maximum force was evaluated there by the values from both support data points. Here, a series of impact tests were performed at room temperature at the centre of cylindrical specimens using the instrumented drop weight test device as explained in detail in 2nd the section (experimental conditions). Three specimens were tested for each composition (Fig. 10.4).

First of all, the effect of simple sintering and also combined effect of sinter + forging processing on the impact resistance of the specimens can be detected very clearly from these graphs.

It seems from these graphs, impact resistance is related to absorbed energy. In fact both of the two different types of specimens have shown that the most part of the impact force is used to maintain the balance with the inertia force, and only a small portion of the impact force is actually used to deform and fracture the specimen. Absorbed energy should be related with the process used here that this energy increases considerably in the structure obtained with sinter + forging. These results are only obtained in laboratory scales and should be improved with detail analyses.

10.3.4 Wear Resistance by Scratch Test

Reduction of friction rests the main challenge for wear-resistant composites. Evaluation of the wear resistance of this composite developed in this work have been carried out in two different numbers of cycles, $50*10^3$, $100*10^3$ cycles.

a

Parameters (F16S)	No of Cycle 50.10^3	
	At the bottom	At the top
Surface (mm^2)	0.201	0.0793
Volume (μm^3)	386110	49820
Depth - max. (μm)	28.8	7.16
Depth - mean (μm)	1.92	0.628

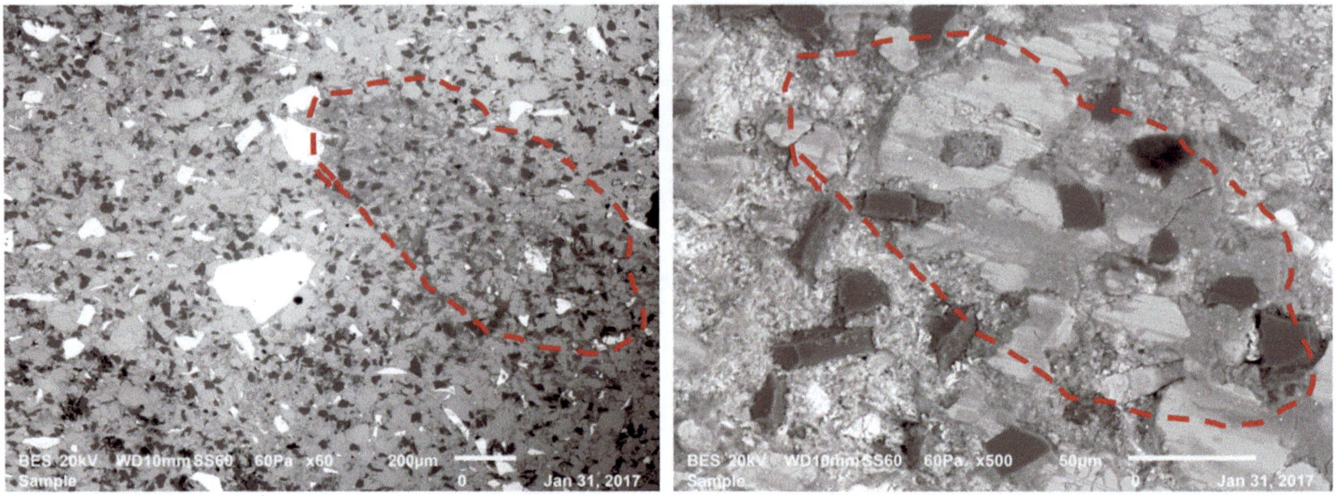

b

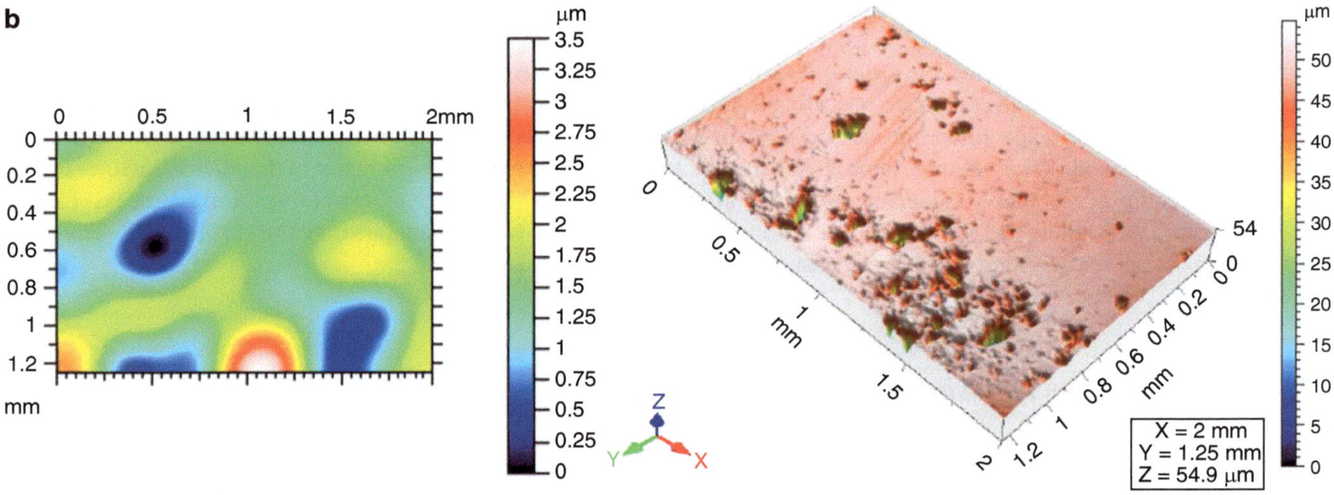

Fig. 10.5 (**a**) Comparison of damage zone with SEM pictures after the scratch test F16S ($N = 50.10^3$ cycles) defined with surface and volume lost during the test shown schematically on the pictures, (**b**) Images of surface damage zones produced by scratch wear test for F16S specimen in the cycle of 50.10^3 and comparison of damage zone with SEM pictures defined with surface and volume lost during the test, etc

Influence of reinforcement elements and essentially influence of two manufacturing processes are presented for two different specimens only sintering and sinter + forging in Figs. 10.5, 10.6, 10.7 and 10.8 respectively.

The surface, volume and the depth values are presented in the same figures for each test condition. Effects of reinforcements mainly Nb_2Al and SiC, and evidently effects of process have obvious advantages over others.

As shown by the results of scratch tests, the composite reinforced and processed with sinter + forging has presented higher wear resistance regarding to the simple sintering. In this particular case, the size of the intermetallics (Nb_2Al) and their dispersion on the matrix should have contributed to improve the wear resistance under experimental conditions carried out in the present work. However, it was very difficult to detect the damaged traces on the specimens produced with sinter + forging. For this reason, we could not supply these pictures for F16-S + F specimens.

a

Parameter (F16S)s	No of Cycle 100.10^3	
	At the bottom	At the top
Surface (mm^2)	0.226	0.0456
Volume (μm^3)	185764	13427
Depth - max. (μm)	8.09	2.37
Depth.- mean (μm)	0.827	0.295

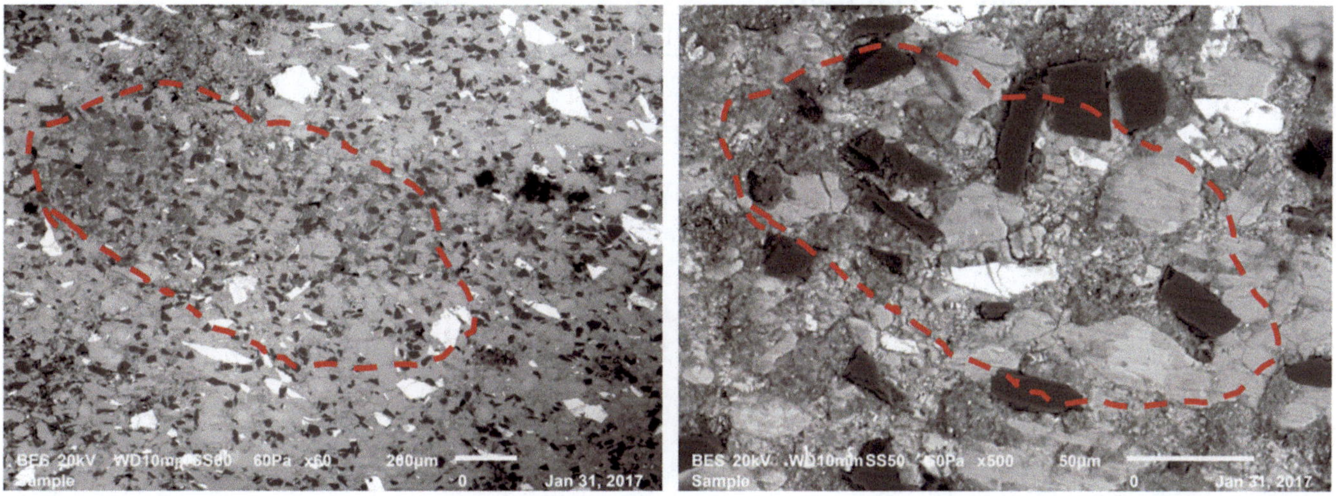

b

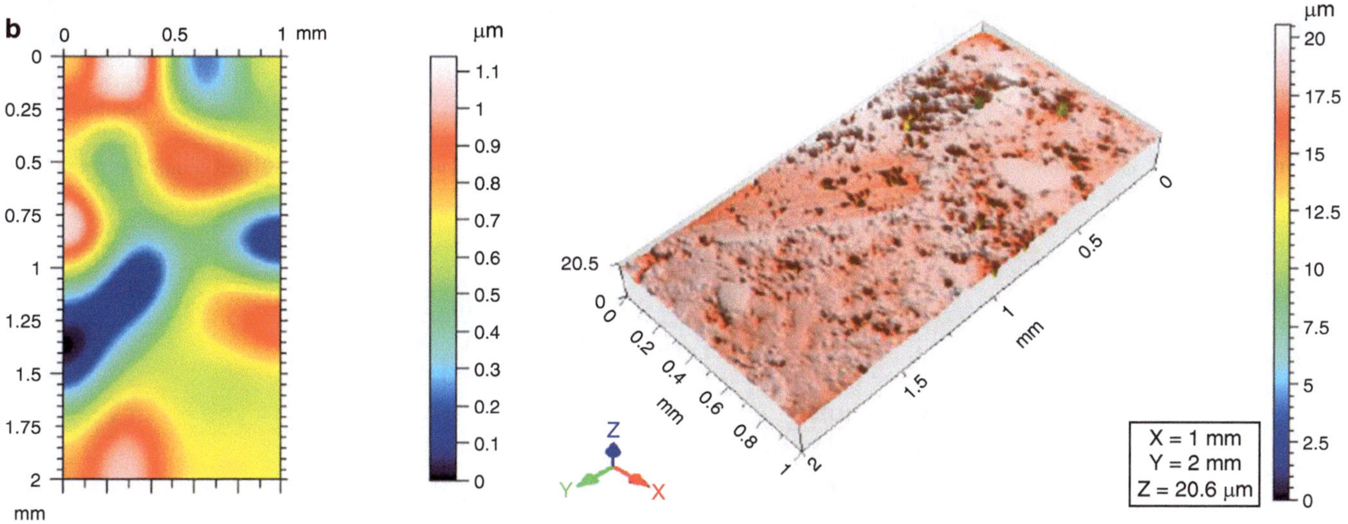

Fig. 10.6 (**a**) Comparison of damage zone with SEM pictures after the scratch test F16S (N = 100.10^3 cycles) defined with surface and volume lost during the test shown schematically on the pictures. (**b**) Images of surface damage zones produced by scratch wear test for F16S specimen in the cycle of 100.10^3 and comparison of damage zone with SEM pictures defined with surface and volume lost during the test, etc

10.3.5 Measurements of Electrical Properties

Electrical properties (resistivity and conductivity) were measured with an Agilent 4338B milliohm meter. Five specimens were used for each manufacturing condition (simple sintering and sinter + forging) and then, the mean values of conductivity and resistivity are presented in Table 10.3.

For measurements, DC regulated power supply voltage and current were set as 20 V and 20A respectively. Data acquisition Card "NI9234" was connected in parallel with the output of the power to acquire the voltage data (voltage input accuracy was 24 bits). A high precision multi-meter "Agilent U1253 N" was connected in series to measuring the current intensity (A).

Parameters (F16F)	N° of Cycle 50.10³	
	At the bottom	At the top
Surface (mm²)	0.360	0.0252
Volume (μm³)	563850	18497
Depth - max. (μm)	10.7	5.92
Depth - mean (μm)	1.57	0.734

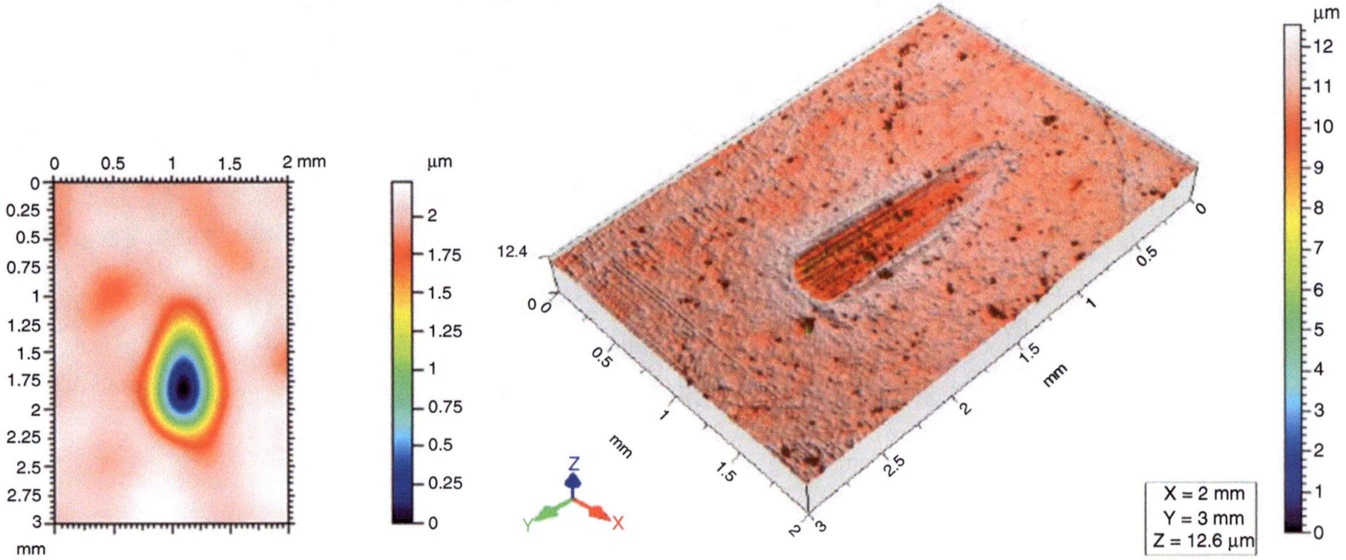

Fig. 10.7 Images of surface damage zones produced by scratch wear test for F16-S + F specimen in the cycle of 50.10^3 and comparison of damage zone defined with surface and volume lost during the test, etc

Parameter (F16F)s	N° of Cycle 100.10³	
	At the bottom	At the top
Surface (mm²)	0.449	0.0321
Volume (μm³)	545835	7848
Depth - max. (μm)	7.18	1.45
Depth. - mean (μm)	1.22	0.244

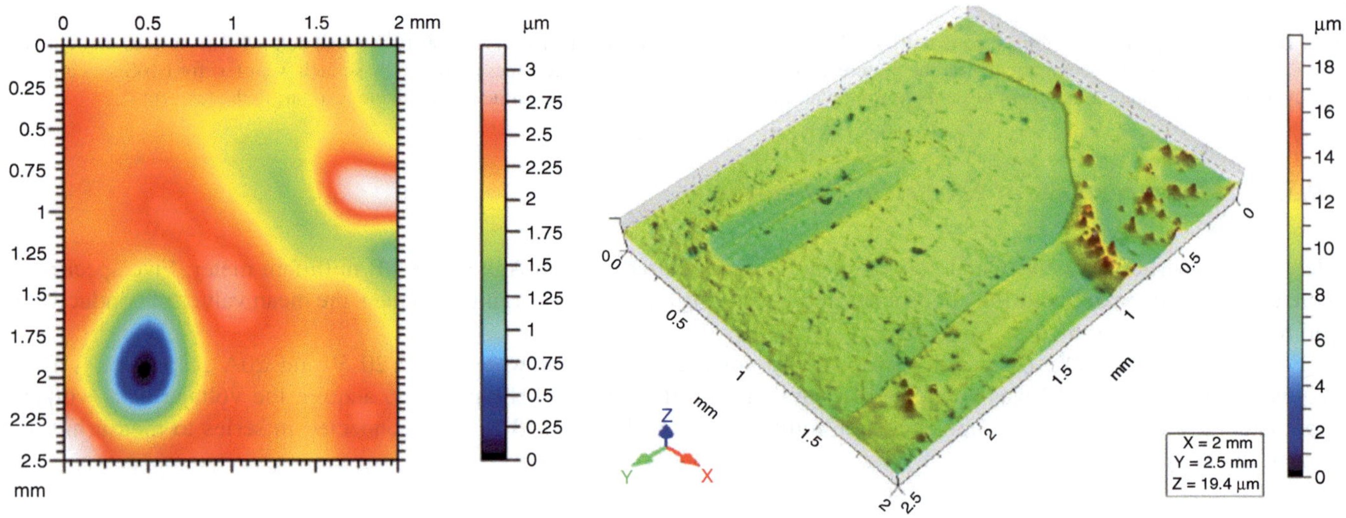

Fig. 10.8 Images of surface damage zones produced by scratch wear test for F16-S + F specimen in the cycle of 100.10^3 and comparison of damage zone defined with surface and volume lost during the test, etc

Table 10.3 Electrical properties (resistivity & conductivity) measured for the composition of F16

Composition	Resistivity (ohm.m)	Conductivity (S/m)
Pure Aluminium	0, 000192	11,997
F16-S	0,000181	10,792
F16-S + F	0,000175	10,984

All the data from the measurements above were displayed and stored with Lab View program. It is worth to mention that in the present work these results were obtained under laboratory conditions. Naturally, the samples for each manufacturing condition (simple sintering and sinter + forging) were prepared carefully according to the standards specified by the manufacturers of the measurement devices, but the form and other effects such as surface conditions of each sample can easily influence the measurements. That is the reason that all the samples were produced under the same conditions and polished carefully on both of their surfaces to keep a parallelism. These specimens are listed in the same table with pure aluminum in order to make a simple comparison under laboratory conditions. These results evidently should be considered as preliminary results and give an indicative data. It seems that form and surface conditions should be improved again.

10.4 Conclusion

New designed copper aluminum based metal matrix composite reinforced mainly with Nb_2Al and SiC particles can potentially be a useful material for aeronautic and automotive applications where tribological problems are important such as in electrical motor where friction problem is known on the electrical brushes etc. This novel composite was designed as alternative composite against the conventional pieces used in these areas. In the frame of this common project, novel composites have been developed from scrap copper + aluminium AA7075 powder reinforced with Nb_2Al and SiC particles as main reinforcements in an economic way. Low cost manufacturing of these composites have been successfully managed through the combined method of sinter + forging.

Microstructural analysis has shown that a good bonding at interface of matrix-reinforcement essentially in the specimens manufactured with combined process sinter + forging; a tough and sound microstructure was obtained without porosity. Wear resistance and ductility of these structures should be improved with doping process and good preparation conditions; ball milling in longer time is needed for helping the fine and homogeneous distribution of the particles in the matrix.

Mechanical behaviours of these specimens produced with sinter + forging process are better than the specimens produced with only sintering processes. Electrical properties (resistivity and conductivity) measured for two different processes seems very optimistic values for future work of the production of alternative pieces used in electrical motor and other tribological applications. Optimizations of the certain parameters such as processing parameters, reinforcement content, etc. require much more experimental work to create real parts in the industrial scales. Here, only limited measurements at room temperature were presented as they are indicative parameters for better understanding the effect of the reinforcements on the optimization of the mechanical, wear and electrical properties of this type of composite produced in the present work.

Acknowledgments This work has been carried out on the frame of collaboration between SUPMECA/PARIS, UNICAMP/SP/CAMPINAS and MICHIGAN TECHNICAL UNIVERSITY/ HOUGHTON-MI-USA. Authors acknowledge and appreciate so much for valuable technical helps and discussions.

References

1. Chawla, N.: Industrial paper, "MMCs in automotive applications". Adv. Mat. Process. **1**, 29–31 (2006)
2. Din, T., Campbell, J.: High strength aerospace aluminum casting alloys a comparative study. Mater. Sci. Technol. **12**, 644–650 (1996)
3. Lo, S.H.J., Dionne, S., Sahoo, M., Hawthorne, H.M.: Mechanical and tribological properties of zinc-aluminium metal-matrix composites. J. Mater. Sci. **27**(21), 5681–5691 (1992)
4. Choi, D.-H., et al.: Microstructure and mechanical property of AA7075 based composite by friction stir processing. Trans. Nonferrous Met. Soc. China. **23**, 335–340 (2013)
5. Kaczmar, J.W., Granat, K., Grodzka, E., Kurzawa, A.: Tribological properties of Cu based composite materials strengthened with Al_2O_3 particles. Archive of Foundry Engineering. **12**(2), 33–36 (2012)
6. Kumar, P., Srivastava, V.K.: Wear characterization of mica-loaded Al-Cu dual matrix particulate composites. Tribol. Trans. **59**(6), 1134–1114 (2015)

7. Selvakumar, N., Gangatharan, K.: Research article, electrical resistivity, tribological behaviour of multiwall carbon nanotubes and nano boron carbide particles reinforced copper hybrid composites for pantograph application. Adv. Mater. Sci. Eng. **2016**, 1–18 (2016)
8. Casstevens, J.M.: Friction and wear characteristics of power metallurgy copper–graphite brushes at high sliding speeds. Wear. **49**, 169–178 (1978)
9. Zhao, H., Barber, G.C., Liu, J.: Friction and wear in high speed sliding with and without electrical current. Wear. **249**, 409–914 (2001)
10. Lhmn, C.: Tribological properties of unidirectional poly – Phenylene sulfide-carbon fiber laminated composites. Wear. **107**, 147–159 (1987)
11. Groth, K., Heidenfelder, F., Holinski, R.: Advancements of tribological performance of carbon brushes in electrical motors. Industrial Lubrication Tribology. **53**(1), 5–9 (2001)
12. Esezobor, D., Oladoye, A.: Improvementon the tribological characteristics of particulate copper silicon carbide composites, TMS – EPD, Congress pp. 827–834, 2011
13. Vettivel, S.C., Selvakumar, N., Leema, N., Lenin, A.H.: Electrical resistivity, wear map and modeling of extruded tungsten reinforced copper composite. Mater. Des. **56**, 791–806 (2014)
14. Kaya, H.: Dependence of electrical resistivity on temperature and composition of Al–Cu alloys. Mater. Res. Innov. **16**(3), 224–229 (2012)
15. Lim, S.C., Gupta, M., Ren, L., Kok, J.K.M.: Tribological properties of Al-Cu/SiCp metal matrix composites fabricated using the rheo-casting technique. J. Mater. Process. Technol. **89–90**, 591–596 (1999)

Chapter 11
Liquid Metal Dispersions for Stretchable Electronics

A.S. Koh, G.A. Slipher, and R.A. Mrozek

Abstract The mechanical effects of dispersing solid particles within a continuous polymer matrix have been well studied for a variety of polymeric systems. These dispersions are of particular interest for electronically conducting polymer applications where the combined mechanical and electrical performance of the material is important, especially in the presence of large deformations. A key weakness of these systems, however, is apparent in soft polymer systems where the polymer is chosen for its elastic qualities, which are often negated by the presence of solid particulate fillers. To overcome this challenge liquid metal, in this work the eutectic Gallium-Indium-Tin (GalInStan), was dispersed in polydimethyl-siloxane (PDMS), and the resulting composited material electromechanical properties were evaluated with respect to particle size and loading. Mechanical properties were compared to rigid Ni and soda-lime silica particles of similar sizes. Data presented will also illustrate the "fluid" dispersion viscosity behavior associated with GalInStan loading, and the effect of viscosity on dispersed liquid metal particle size, which is a key parameter in controlling material electronic performance. This work widens the potential scope of electronically conductive filled elastomers beyond the low-strain regime and promises to greatly increase the fatigue life of fabricated stretchable electronic devices based on conductive elastomers.

Keywords Stretchable robotics • Dispersion • Galinstan • Liquid metal

11.1 Introduction

Developing stretchable electronics is a significant area of interest for personal, medical, and military applications [1]. By creating materials able to achieve high strains while maintaining desired electrical properties, it becomes possible to create soft robotic systems better able to interface directly with biology (e.g. brain-computer interfaces) [2]. Strain-tolerant electronically conductive materials have been widely studied in the areas of artificial muscles [3], sensors [4], and woven batteries [5]. In order to create stretchable electronic material a variety of strategies can be used including the use of conjugated conductive polymers, the embedding of conductive materials into an elastic polymer in defined shapes or channels, and the blending of non-conductive polymers with conductive materials [6].

This work will focus on the blending of a non-conductive polymer polydimethyl siloxane (PDMS) with the liquid metal Galinstan (GaInSn). Little work has been done to-date to understand the processing, material, and electronic benefits and challenges of directly mixing liquid metal with polymers. Data presented here will begin to focus on the effects of polymer viscosity, Galinstan loading, and particle size on the final properties of a Galinstan/PDMS blend. While significant literature exists studying the effects of solid particle loading into polymers, little more than theoretical work has been published directly understanding the material effects of liquid metal inclusions in a polymer matrix [7]. This work will also begin to look at the differences between solid and liquid metal loading.

11.2 Experimental

Blends of metal and PDMS were created by using a high shear overhead mixer at 2000 rpm for 2 h. Metal loading was varied from 10% to 70 wt%. The polymer matrix was a 1:1 vol:vol blend of vinyl and tri-methyl terminated silicones with a viscosity of ~1000 cP or ~200 cP. Blend rheology was measured to determine strain and frequency dependence at 1 Hz and 1% strain respectively. Particle size and morphology were determined through confocal microscopy.

A.S. Koh (✉) • G.A. Slipher • R.A. Mrozek
U.S. Army Research Laboratory, Aberdeen Proving Ground, Aberdeen, MD 21005, USA
e-mail: amanda.s.koh.ctr@mail.mil

P.R. Thakre et al. (eds.), *Mechanics of Composite and Multi-functional Materials, Volume 6*,
Conference Proceedings of the Society for Experimental Mechanics Series, DOI 10.1007/978-3-319-63408-1_11

11.3 Results and Discussion

Figure 11.1 shows the confocal images of 10% and 50% Galinstan loading in high and low viscosity PDMS blended for 2 h. From dispersion literature it would be expected that low viscosity systems would blend to create smaller particles and lower loadings would blend to form smaller particles as there is more energy per unit volume of dispersed material. Within the lower PDMS viscosity dispersion, lower loadings do produce significantly smaller particles, but the trend is reversed for the higher viscosity PDMS dispersion. Particles formed at 50% loading are at least ten times as large when blended in low viscosity PDMS. This suggests that how the energy of mixing is dissipated through the sample is not the only processing parameter determining the morphology of the final blend. It is important to note that flocculation of metal particles is present at low loading samples, particularly in the high viscosity PDMS, which may impact electrical percolation. Strong interactions between metal particle interfaces will also contribute to the processing, stability, and final material properties of the dispersion.

Figure 11.2 shows the dispersion viscosity of the low and high matrix viscosity blends across strain and frequency ramps as well as the corresponding data from the base polymer matrix. The 10% loading samples show no strain or frequency dependence while this dependence develops at the higher loading. From Fig. 11.2 it is clear that the 10% loading does not drastically alter the bulk viscosity for either matrix, while the 50% loading samples jump 60x for the high viscosity matrix and 30x for the low viscosity matrix. Similarly to the particle size images, this suggests the importance of the matrix viscosity for the final material properties.

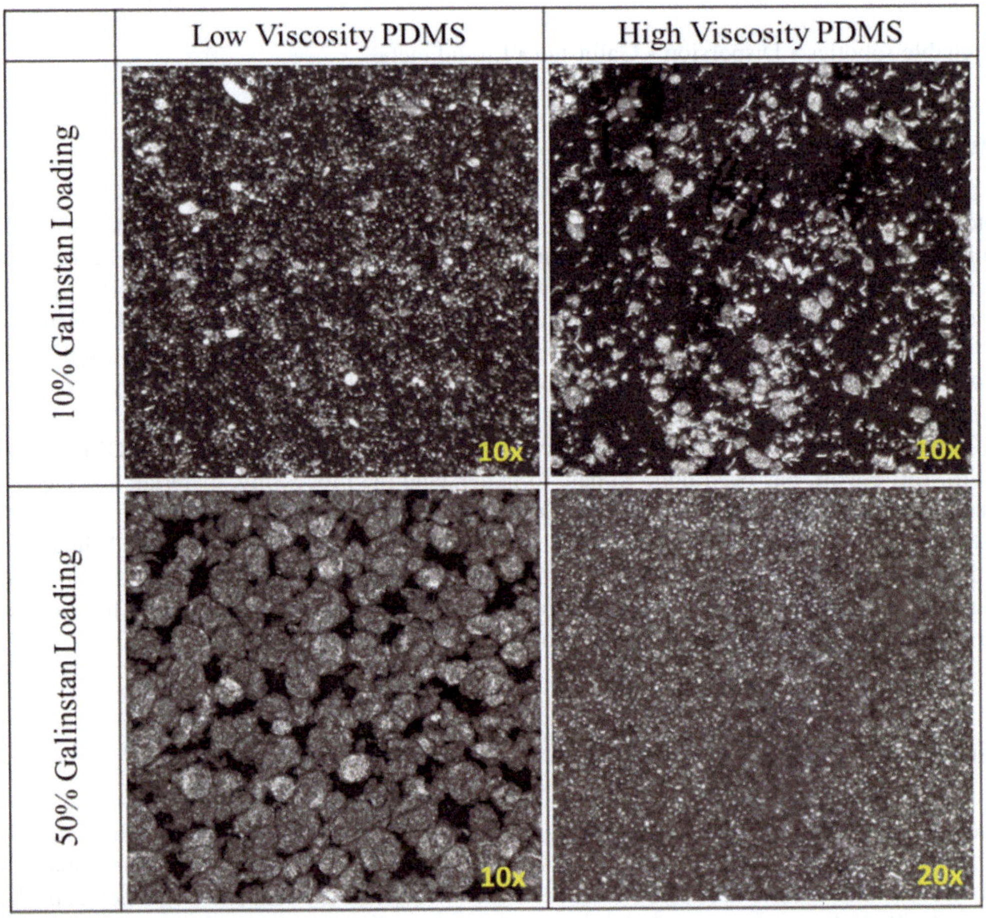

Fig. 11.1 Confocal images of PDMS/Galinstan dispersions

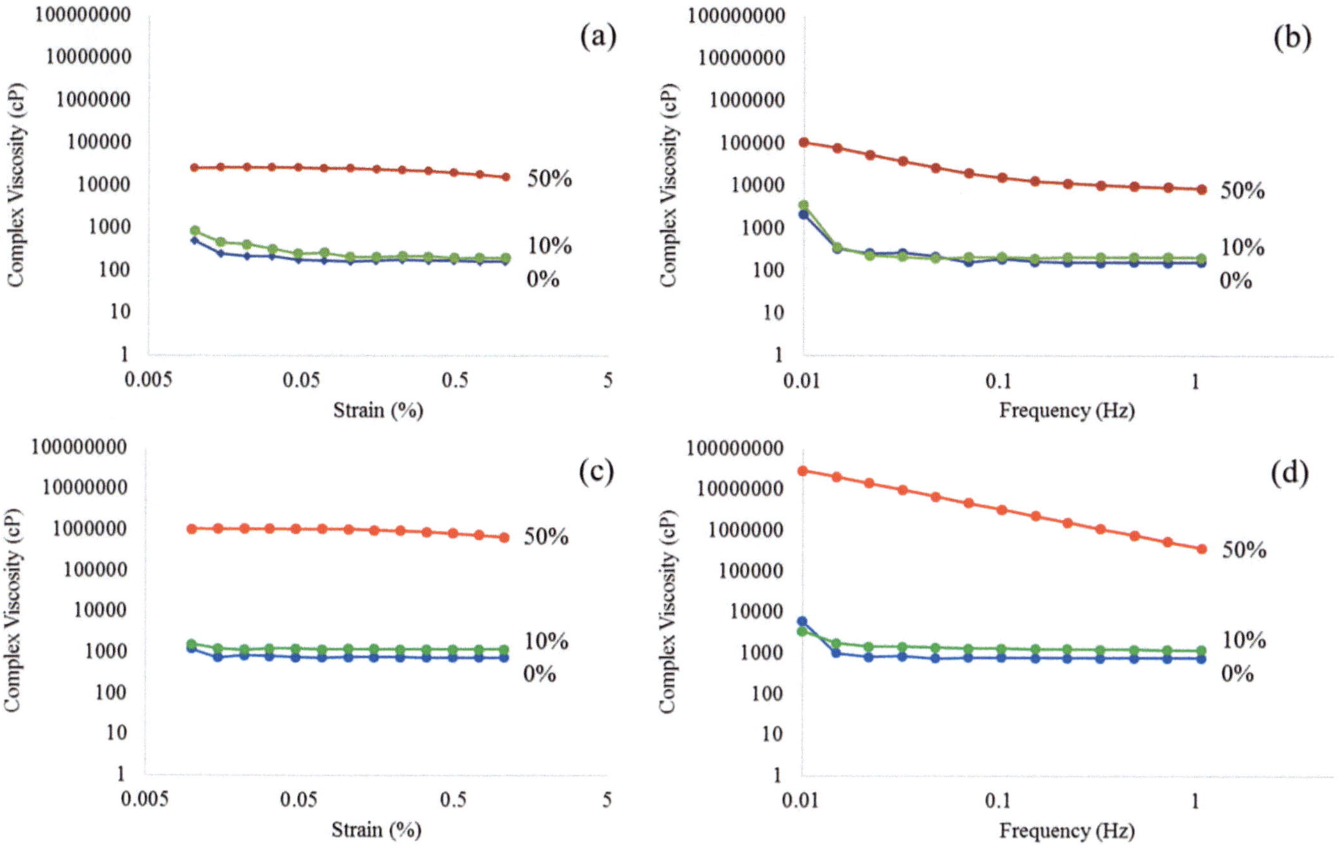

Fig. 11.2 Strain and frequency sweeps of (**a, b**) low viscosity PDMS and (**c, d**) high viscosity PDMS Galinstan dispersions. Galinstan loadings are marked for each curve

11.4 Conclusions

Data presented in this work describes the successful blending of PDMS and Galinstan in the liquid (uncured) state. Liquid metal loading and PDMS matrix viscosity were critical to the particle size of the blended liquid metal as well as the dispersion viscosity. To understand the benefits of this Galinstan/PDMS blend it is necessary to do electrical measurements of the different loading blends in addition to a comparison with solid particle systems. The blend, however, does show promise as an easily prepared and processable system for designing stretchable electronics for future soft robotic applications without the need for hard metals, lithography, or high modulus polymers.

References

1. Song, J.: Mechanics of stretchab le electronics. Curr. Opinion Solid State Mater. Sci. **19**, 160–170 (2015)
2. Majidi, C.: Soft robotics: a perspective—current trends and prospects for the future. Soft Rob. **1**, 5–11 (2014)
3. Madden, J., Vandesteeg, N., Madden, P., Takshi, A., Zimet, R., Anquetil, P., Lafontaine, S., Wieringa, P., Hunter, I.: IEEE J. Ocean. Eng. **29**, 706–728 (2004)
4. Majidi, C., Kramer, R., Wood, R.: A non-differential elastomer curvature sensor for softer-than-skin electronics. Smart Mater. Struct. **20**, 1–7 (2011)
5. Gaikwad, A., Zamarayeva, A., Rousseau, J., Chu, H., Derin, I., Steingart, D.: Highly stretchable alkaline batteries based on embedded conductive fabric. Adv. Mater. **24**, 5071–5076 (2012)
6. Park, M., Park, J., Jeon, U.: Design of conductive composite elastomers for stretchable electronics. Nano Today. **9**, 244–260 (2014)
7. Style, R., Wettlaufer, J., Dufresne, E.: Surface tension and the mechanics of liquid inclusions in compliant solids. Soft Matter. **11**, 672–679 (2014)

Chapter 12
Laser Cutting of the TiN +Al_2O_3 Reinforced Aluminium Matrix Composites Through Semisolid Sintering

Sonia Ezeddini, D. Katundi, Emin Bayraktar, and I. Miskioglu

Abstract Laser cutting is commonly used for manufacturing of strong composites as an advance machining process. It is suitable for geometrically complex profile cutting in the metals used in manufacturing engineering. In the present work, laser cutting process of aluminium matrix composites obtained from scrap chips reinforced with hard ceramics (TiN + Al_2O_3) were carried out. They are being widely used in the aerospace and automobile industries such as aircraft structure, internal combustion engines, plain bearings, wheels, pistons, brake rotors for high speed trains, etc. In fact, TiN does not react with aluminium. It keeps its thermal stability up to 3000 °C and chemically inert to most of the common acids. It is industrially important due to its high hardness, good electrical and optical properties, etc. TiN is presently being used in cutting tools, solar-control films and other micro-electronic applications. For this reason, it was interesting to develop a new composite from scrap aluminium chips reinforced with TiN +Al_2O_3 ceramic powders and to examine its tailored beahviour of these composites. The thermal effects of laser cutting and effects of main operating parameters such as laser power, and cutting speed on the cutting edge and on the cutting surface were examined. The evolution of the microhardness underneath the cutting surface due to laser power is also examined. The composite used in this study was produced through combined method of powder metallurgical (P/M) and thixoforming (Semi solid). Microstructure of cutting edge and cutting surfaces are investigated in detail by scanning electron microscopy (SEM). Cutting surfaces have been analyzed with 3D optical surface roughness-meter (3D–SurfaScan).

Roughness evaluations were taken as optimization criteria as a function of the cutting surface and cutting parameters (power, speed, gas pressure etc.) that have been carried out by Taguchi method. A simple and useful tool was proposed for using in real manufacturing environment.

Keywords Recycled composite • Laser cutting • Taguchi method • SEM, Tailored behaviour

12.1 Introduction

Composite materials are very suitable for their tailored properties that are accomplished by combining two or more materials to obtain the crucial scale of properties. The most characteristic feature of composites is that, while the individual constituents retain their properties, they synergize to give properties which may not be found in any one of them alone [1–6]. Recently, extensive research study has been carried out, and it was shown that there is tremendous promise of ceramic reinforced metal-matrix composites (MMCs) [6–9].

Various processing techniques have been developed to engineering composites for various fields of applications [4, 5, 11] such as aerospace, defense, automobile, and sports sectors. In MMCs, the addition of a small amount of second-phase materials with high shear strength imparts unique properties to the base materials [4–7]. Metal matrix composites (MMCs) have been developed to respond the demand for lighter materials with high specific strength, stiffness and wear resistance [1, 2, 7, 8, 10, 11]. Aluminium is preferred as matrix material in MMCs because of its low density, and manufacturing facilities.

It means that high performance indices due to their light weight, high strength-to-weight ratio, ease in melting and casting, and good engineering properties [3]. However, particulate reinforced aluminium matrix composites (AMCs) are much more attractive MMC materials due to their strength, ductility and toughness as well as their ability to be processed by

S. Ezeddini (✉) • D. Katundi • E. Bayraktar (✉)
Supmeca/Paris, School of Mechanical and Manufacturing Engineering, St-Ouen, France
e-mail: bayraktar@supmeca.fr

I. Miskioglu (✉)
Michigan Technological University, ME-EM Department, Houghton, MI, USA
e-mail: imiski@mtu.edu

P.R. Thakre et al. (eds.), *Mechanics of Composite and Multi-functional Materials, Volume 6*,
Conference Proceedings of the Society for Experimental Mechanics Series, DOI 10.1007/978-3-319-63408-1_12

non-conventional methods; they are relatively low costs and characteristic isotropic properties. Therefore, AMCs have been applied successfully to structural components, especially in the automotive and aviation industries, and the number of applications is expected to increase with the development of low-cost processing methods [10, 11]. To improve the structure and properties of particulate reinforced MMCs various processing techniques have evolved over the last 20 years. The potential for forming processes based upon semi-solid metal alloys was first recognized in the early 1970s. The microstructure of a suitable alloy comprises spheroidal particles of solid surrounded by liquid phase of a lower melting point, rather than the interlocking tree-like dendrites of conventionally cast alloy. It is this microstructure that gives the material its thixotropic properties, i.e. when sheared the material flows, but when allowed to stand, it thickens. Thixoforming is one member of semi-solid forming processes and it possesses characteristics of both casting and forging [4, 7, 8].

For this reason, it was interesting to develop a new composite from scrap aluminium chips reinforced with TiN +Al_2O_3 ceramic powders and to examine its tailored beahviour of these composites. In the present work, laser cutting process of aluminium matrix composites obtained from scrap chips reinforced with hard ceramics (TiN + Al_2O_3) were carried out. The thermal effects of laser cutting and effects of main operating parameters such as laser power, and cutting speed on the cutting edge and on the cutting surface were examined. The evolution of the microhardness underneath the cutting surface due to laser power is also examined.

The composite used in this study was produced through combined method of powder metallurgical (P/M) and thixoforming (Semi solid). Microstructure of cutting edge and cutting surfaces are investigated in detail by scanning electron microscopy (SEM). Cutting surfaces have been analyzed with 3D optical surface roughness-meter (3D–SurfaScan). Roughness evaluations were taken as optimization criteria as a function of the cutting surface and cutting parameters (power, speed, gas pressure etc.) that have been carried out by Taguchi method. A simple and useful tool was proposed for using in real manufacturing environment.

12.2 Experimental Conditions

12.2.1 Manufacturing of Composites and Microstructural Evaluation

Composition of the specimens designed in this work was arranged as follows (Table 12.1): basic reinforcement used here are Al_20_3 and TiN is variable between 10% and 30%. As secondary reinforcements, Ti, Cu and Ni were used. A little amount h-BN was added to the microstructure because Boron Nitride (BN) is a very suitable ceramic powder because of their high intrinsic thermal conductivity. It is not sensitive at all against moisture and it is very suitable for lubrication effect during mixing and compacting of composition. All of three types of composites will be called here after TiN-I, TiN-II and TiN-III depending on the amount of TiN respectively.

Processing route has been carried out through combined process of powder metallurgical (P/M) and thixoforming. Firstly, all of the element of this composition have been mixed and milled in a pulverize device during 4 h by using a protection and successfully ball milled (4000 rpm) during 1 h and compacted by CIP (Cold Isostatic Pressure) at a pressure of 250 MPa and after that thixo forming sintering after heating at the temperature of 700 °C for 45 min under argon inert atmosphere, then it was cooled in the oven.

After manufacturing of the composites, microstructural evaluation has been carried out for each composite. The dispersion of reinforcement particles in the matrix and interface at matrix/reinforcements was evaluated. Microhardness tests have been carried out on the polished and etched specimens. Certain basic mechanical parameters have been given here as an indicative results of these composites.

Table 12.1 Chemical compositions of three different composites designed in this work

Specimen name	Al Matrix	Al_2O_3 (wt%)	TiN (wt%)	Ti (wt%)	Cu (wt%)	Ni (wt%)	BN
TiN-I	B	10	10	5	5	5	1
TiN-II	B	10	20	5	5	5	1
TiN-III	B	10	30	5	5	5	1

Table 12.2 Basic parameters of CO_2 laser cutting process

	Levels		
Cutting parameters	Min	Medium	Max
Power (W)	3000	3500	4000
Cutting speed (m/min)	0.048	0.144	0.192
Gas pressure (nitrogen) (bar)	4	8	14

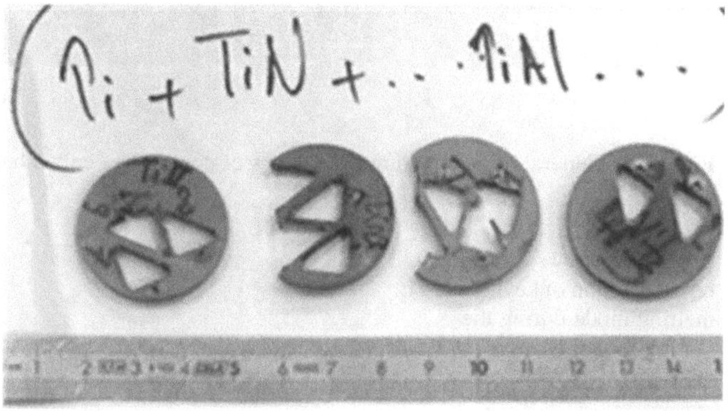

Fig. 12.1 Experimental specimens cut using CO_2 laser cutting process

12.2.2 Laser Cutting Process and Cutting Parameters

All of the laser cutting processes have been carried out on a CO_2 laser machine called "4000-TLF TURBO" Basic parameters of laser cutting process have been given in the Table 12.2. In the present work, three levels of powers (Pw) and three levels of cutting speeds (S), with three different gas pressures (Gp) have been chosen in order to examine their impact on the basic performance parameters. Nine samples were cut for each composition. Totally, 27 Laser-cut-samples were used and these samples have been prepared for damage evaluation (Fig. 12.1).

After achievement of the cutting process, the widths of resulting cuts were measured and SEM micro-photography of the cutting surfaces was carried out. To measure the striation depths and widths, the surface roughness measurement was carried out using a 3D optical roughness meter (Surf-Scan). For intensive evaluation, measurements were taken in different zones for each composition (Remelted Zone, Heat Affected Zone (HAZ) and the centre of the specimen) by using SEM.

Optimization process for the experimental matrix with variables was employed to develop for each experimental test. Taguchi method and L9 Orthogonal Array were used to reduce number of the experiments. The design of experiments (DOE) and measurements of the average roughness values "Ra" have been carried out with Minitab 17 software. All of the optimization parameters were evaluated for each composition with these diagrams obtained by software.

12.3 Results and Discussions

12.3.1 Microstructure of (TiN +Al_2O_3) Reinforced Aluminium Matrix Composites

It is well known that the structures of the ceramic particulates reinforcement aluminum matrix composites are complex hybrid characters. Distribution of TiN in the matrix is relatively difficult regarding to other ceramics (i.e. Al_2O_3, TiB_2, etc.) due to low wettability (Fig. 12.2). In this work, TiN particulates (size of particles is variable between 3 and 20 μm) were treated before mixing with matrix to activate surface of powder and a little amount of BN was added to improve its cohesion with matrix. The presence of the intermetallic phases is confirmed and estimated that solid state phase reactions start during the treatment of mechanofusion. Additionally, Al_2Cu was also observed after a heat treatment at 750 K during sintering process.

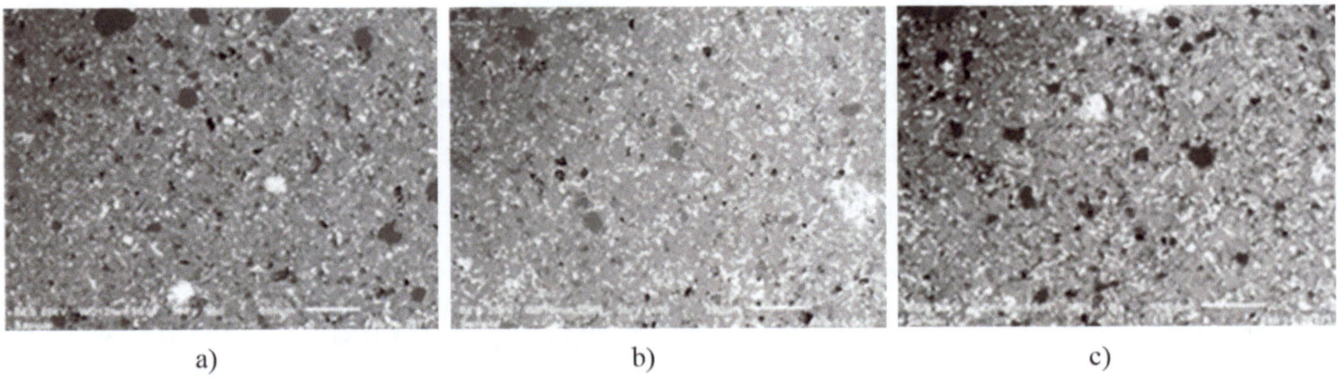

Fig. 12.2 General microstructure three types of composites called here (**a**) TiN1, (**b**) TiN2 and (**c**) TiN3

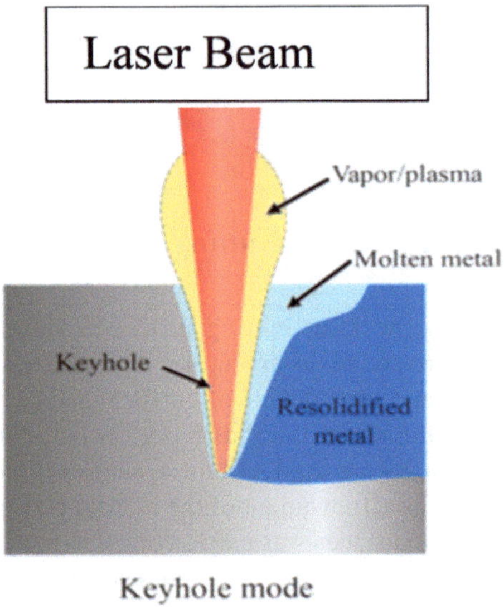

Fig. 12.3 Shows a schematic representation of keyhole laser melting mode during the cutting process

12.3.2 Evaluation of the Results for Laser Cutting Process

Laser gas assisted cutting process of (TiN +Al_2O_3) reinforced composite pieces have been carried out. The effect of laser output power, gas pressure and cutting speed on the surface striations and surface roughness evolution of the cutting edge surfaces were analyzed. Profiles of the laser cutting surfaces were also evaluated depending on the process parameters (power, gas pressure and cutting speed, etc.).

In this work, laser beam was conducted with different conditions. As well known, when higher power densities are used, some amount of material is vaporized, forming a cavity called keyhole containing vapour, plasma or both of them. Around this cavity, there is a molten metal. The pressure inside the cavity can give stability to the liquid pressure and keep the cavity opened during the process. Figure 12.3 shows a schematic representation of keyhole laser melting mode during the cutting process.

Generally, the keyhole mode is used for cutting processes of the pieces because laser absorption in keyhole mode is much higher than in the conduction mode. Additionally, the larger depth to width ratio of the melt pool is a benefit for most cutting and very often for welding processes. For the tailored pieces, mainly is very suitable to use this mode in manufacturing engineering. But, if certain parameters could not be optimized, certain instabilities of the cavity can cause in the formation of porosity due to gas bubbles that are imprisoned inside the remelted zone of the material after solidification.

Figures 12.4, 12.5 and 12.6 show surface roughness evolution of the workpieces for two laser output power settings (3 and 4 kWs) at three cutting speeds (0.048, 0.144 and 0.192 m/min). Two different gas pressures (4 and 14 bar) were used of these cutting processes. These surface roughness evolutions after cutting were presented at the left column. In the same figures, 3D surface topographies for the cutting edge surfaces were presented at the right column under the same conditions. As can be

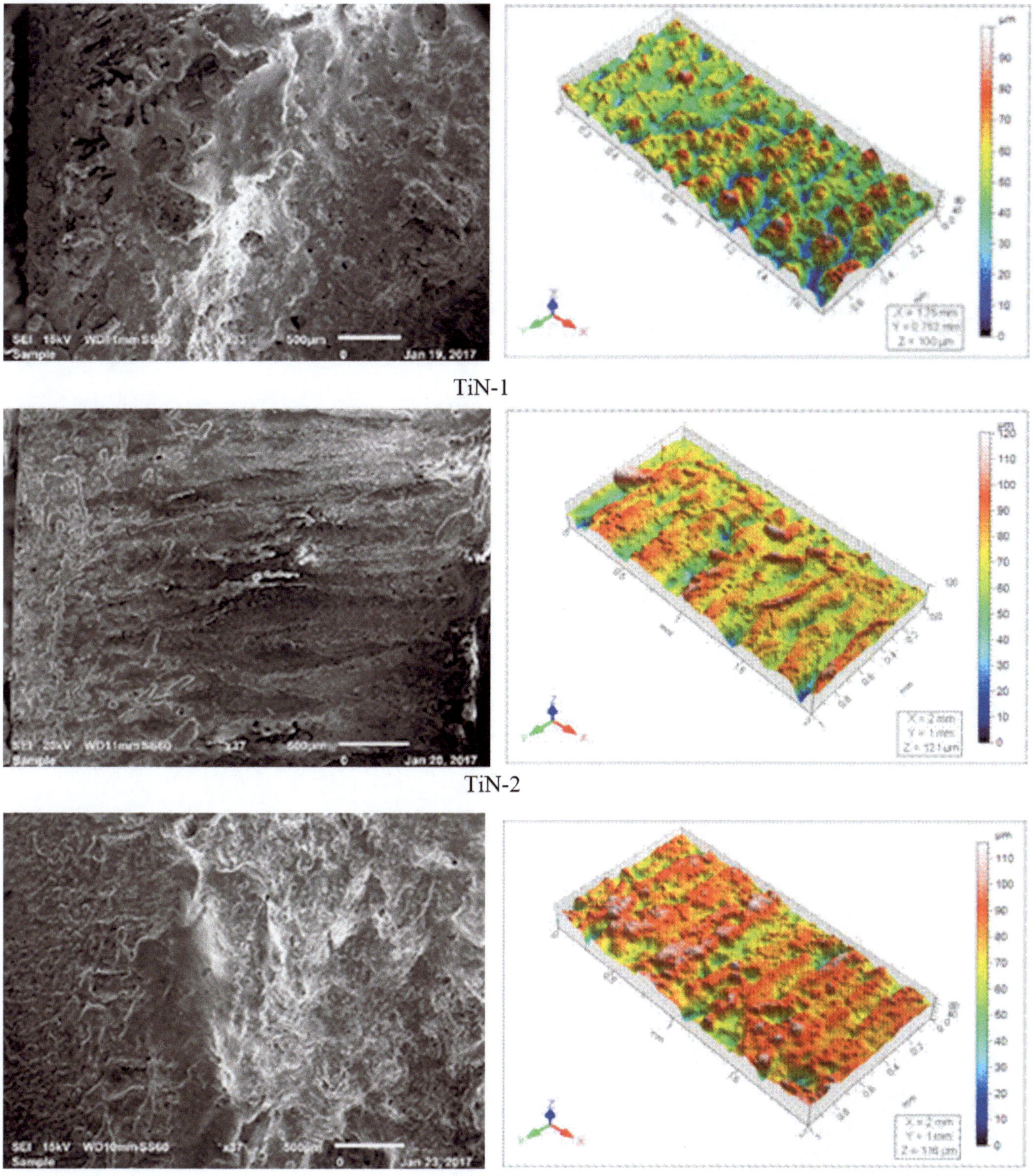

Fig. 12.4 (**a**) SEM image showing the variation of the laser cutting surface morphology of the composite TiN1 at the left column under the cutting conditions: P = 4 kW, S = 0.048 m/min and P = 14 bar. And 3D surface topography of the laser cutting surface of the composite TiN1 at the right column under the same cutting conditions. (**b**) SEM image showing the variation of the laser cutting surface morphology of the composite TiN2 at the left column under the cutting conditions: P = 4 kW, S = 0.048 m/min and P = 14 bar. And 3D surface topography of the laser cutting surface of the composite TiN2 at the right column under the same cutting conditions. (**c**) SEM image showing the variation of the laser cutting surface morphology of the composite TiN3 at the left column under the cutting conditions: P = 4 kW, S = 0.048 m/min and P = 14 bar. And 3D surface topography of the laser cutting surface of the composite TiN3 at the right column under the same cutting conditions

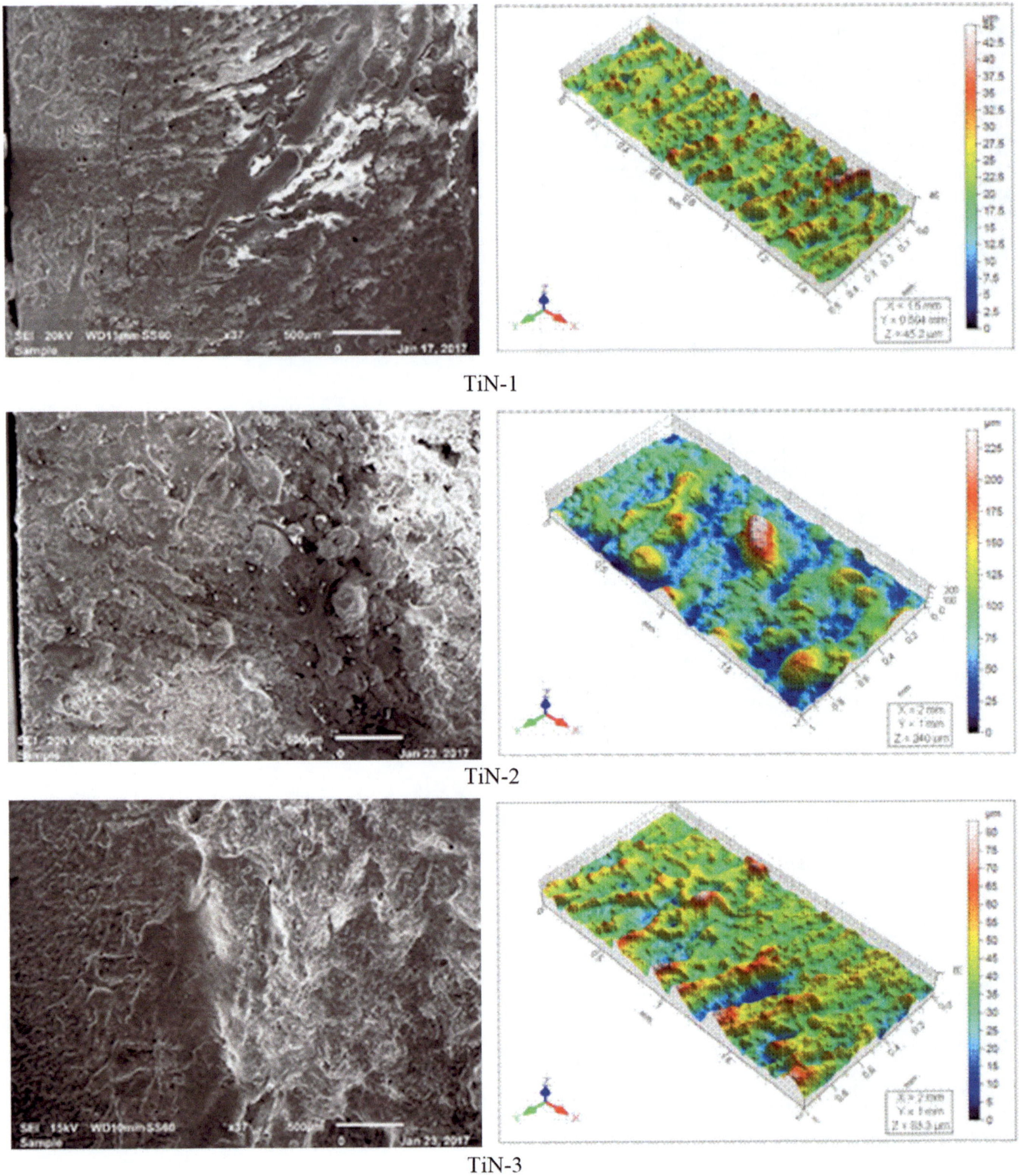

Fig. 12.5 (**a**) SEM image showing the variation of the laser cutting surface morphology of the composite TiN1 at the left column under the cutting conditions: P = 3 kW, S = 0.192 m/min and P = 14 bar. And 3D surface topography of the laser cutting surface of the composite TiN1 at the right column under the same cutting conditions. (**b**) SEM image showing the variation of the laser cutting surface morphology of the composite TiN2 at the left column under the cutting conditions: P = 3 kW, S = 0.192 m/min and P = 14 bar. And 3D surface topography of the laser cutting surface of the composite TiN2 at the right column under the same cutting conditions. (**c**) SEM image showing the variation of the laser cutting surface morphology of the composite TiN3 at the left column under the cutting conditions: P = 3 kW, S = 0.192 m/min and P = 14 bar. And 3D surface topography of the laser cutting surface of the composite TiN3 at the right column under the same cutting conditions

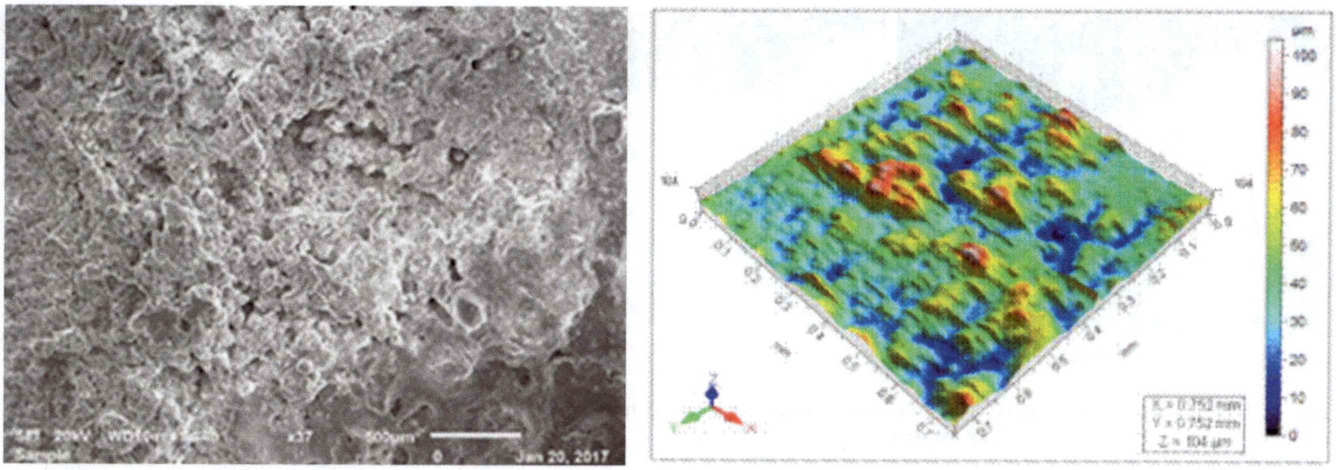

TiN-1

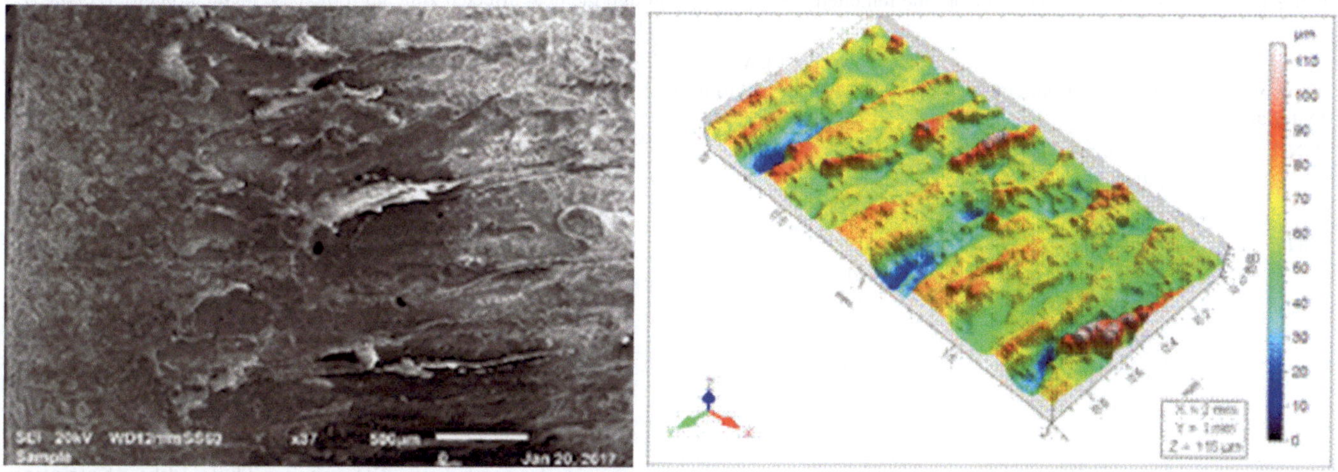

TiN-2

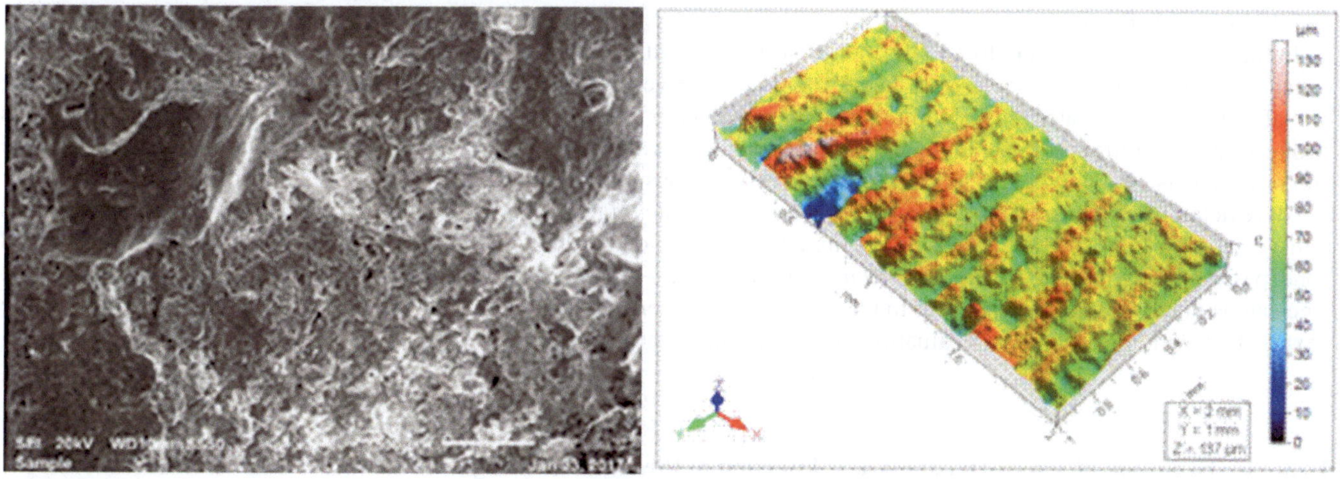

TiN-3

Fig. 12.6 (**a**) SEM image showing the variation of the laser cutting surface morphology of the composite TiN1 at the left column under the cutting conditions: P = 4 kW, S = 0.144 m/min and P = 4 bar. And 3D surface topography of the laser cutting surface of the composite TiN1 at the right column under the same cutting conditions. (**b**) SEM image showing the variation of the laser cutting surface morphology of the composite TiN2 at the left column under the cutting conditions: P = 4 kW, S = 0.144 m/min and P = 4 bar. And 3D surface topography of the laser cutting surface of the composite TiN2 at the right column under the same cutting conditions. (**c**) SEM image showing the variation of the laser cutting surface morphology of the composites TiN3 at the left column under the cutting conditions: P = 4 kW, S = 0.144 m/min and P = 4 bar. And 3D surface topography of the laser cutting surface of the composite TiN3 at the right column under the same cutting conditions

Fig. 12.7 Microstructure with transversal view to the surface under the cutting edge. Microstructure evolution from the remelted zone, passing through the heat affected zone until the substrate - base metal structure

Table 12.3 Microhardness evolution from the remelted zone, passing through the heat affected zone until the base metal structure (HV0.05)

Type of materials	Distance from cutting edge (μm)	Power = 4 kW Speed = 0.048 m/min Pressure = 14 bar	Power = 3 kW Speed = 0.192 m/min Pressure = 14 bar	Power = 4 kW Speed = 0.144 m/min Pressure = 4 bar
TiN 1	30–40 μm	111	143	138
	80-90 μm	61	64	63
	300 μm	102	109	185
TiN 2	30–40 μm	119	126	139
	80–90 μm	101	119	112
	300 μm	118	124	141
TiN 3	30–40 μm	140	156	144
	80–90 μm	90	82	85
	300 μm	123	178	182

observed from these SEM pictures, the cutting edge surfaces seem similar with little differences. The widths of striations vary with flow of liquid metal depending on the cutting parameters. And also surface topographies justify striations defined mainly by remelted zone.

In fact, heat input is bigger close to the melted surface, facilitating diffusion and very often cause in grain growth. As the distance from cutting edge increases, a darker region is observed that occurred after heat treatment with very high cooling rate after high heating rate. This is nature phenomenon of laser cutting processes. Thickness of the remelted zone depends strongly on the cutting speed that influences surface roughness value much higher than other parameters such as gas pressure and power. This case will be discussed in the next two sections see later in 12.3.2.1 and 12.3.2.2 respectively.

More detail information on this case, some pictures were shown in the Fig. 12.7 for three types of compositions taken as transversal view to the surface under the cutting edge. These pictures show the microstructure evolution from the remelted zone, passing through the heat affected zone until the substrate – base metal structure. This zone is not affected from heating. One can be noted that the region closer to the remelted zone seem as recrystallized zone. This idea should be justified in the next section with microhardness evaluation in the different zones from cutting edge to the base metal.

12.3.2.1 Evolution of Micro-Hardness as a Function of the Laser Cutting Conditions

More detail information on the microstructure evolution from the remelted zone, passing through the heat affected zone until the base metal structure was given by using microhardness (HV0.05) measurements All of the microhardness measurements for three types of compositions taken as transversal view to the surface under the cutting edge was indicated in the Table 12.3. One can be noted that the values measured in the region closer to the remelted zone seem as recrystallized zone. This zone was identified as distance of 80–90 μm from cutting edge. It is estimated as Heat Affected Zone (HAZ) and finally third zone is the base metal that was not affected with heating during cutting process. Microhardness evaluation in these three different zones from the remelted zone, passing through the heat affected zone until the base metal structure show that the microstructure influences strongly with heat treatment. Evidently, the second zone (HAZ) gives lower microhardness values for all of the three cutting conditions. Most probably, second zone (HAZ) undergone a softening due to the recrystallization.

12.3.2.2 Evolution of Surface Roughness as Function of Cutting Conditions

After achievement of experimental evaluation of the cutting process, a detail optimization process was carried out. In a simple way, an experimental matrix with variables was employed to develop for each experimental test. Essentially, design of experiments (DOE) and measurements of the average roughness values "Ra" have been carried out with Minitab 17 software. All of the optimization parameters were evaluated for each composition with these diagrams obtained by software. First results were given in the Fig. 12.8a–c. In these figures Evolution of surface roughness values (Ra) as a function of gas pressure, power and cutting speed respectively.

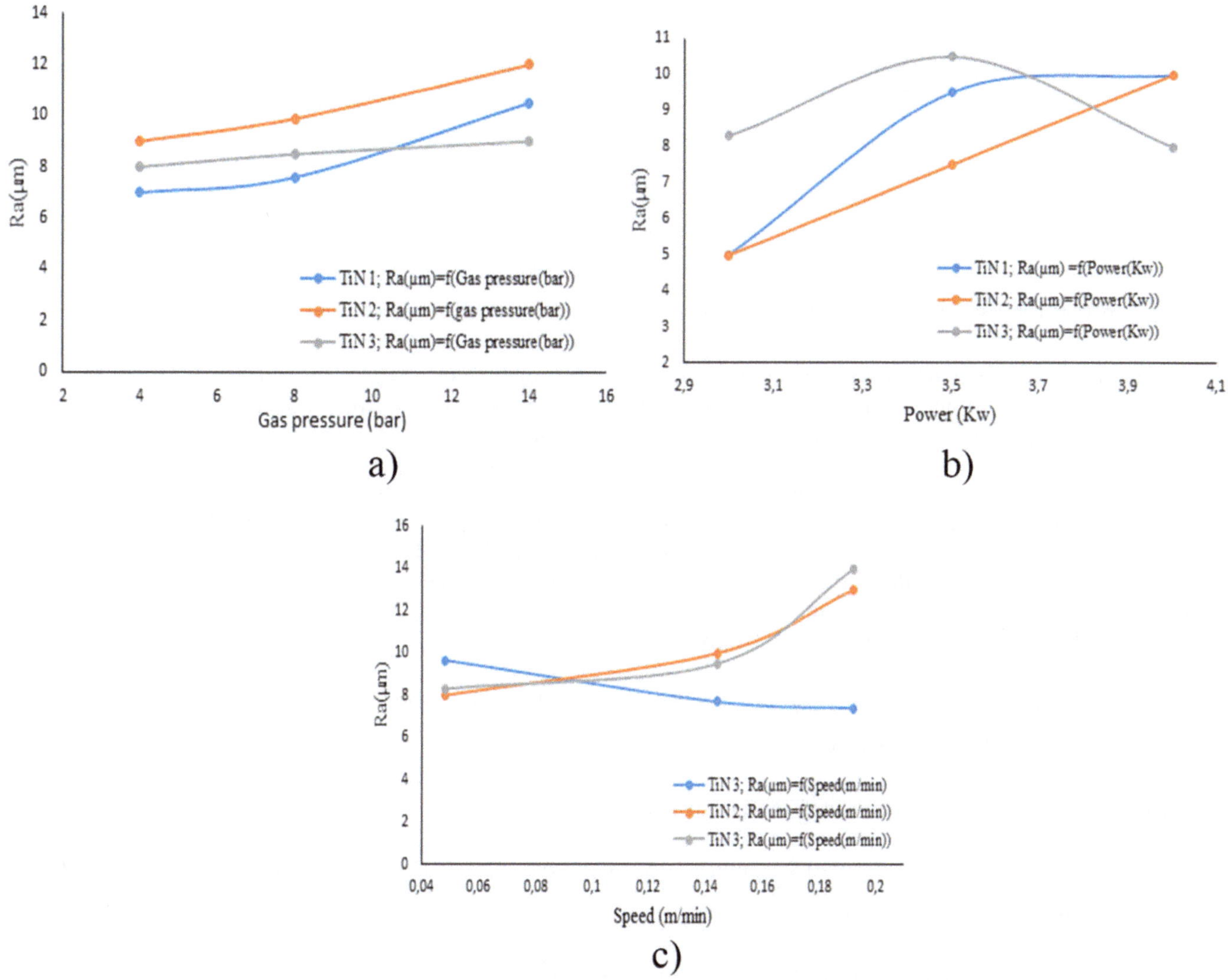

Fig. 12.8 Evolution of surface roughness values as a function of (**a**) gas pressure, (**b**) power and (**c**) cutting speed respectively for three compositions of TiN1 TiN2 and TiN3

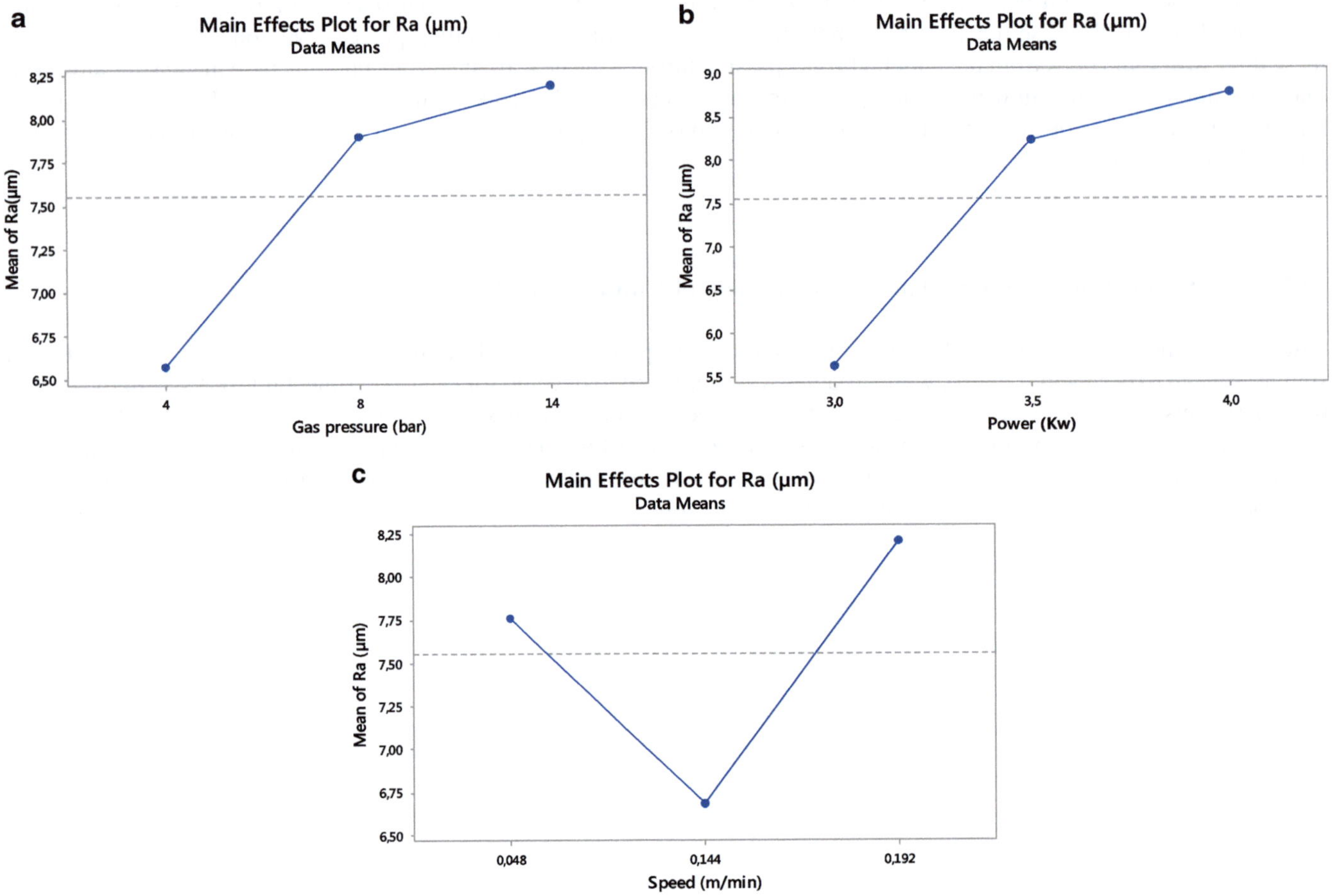

Fig. 12.9 Surface roughness values as a function of (**a**) gas pressure, (**b**) power and (**c**) cutting speed respectively for the composite TiN1

For simplifying evolution of surface roughness values (Ra), a new optimization was given as the mean values depending on the power, gas pressure and cutting speed in the Fig. 12.9a–c respectively for the composite TiN1.

In the Fig. 12.10, three cutting parameters were related for the composite TiN1, as interactions of the cutting parameters with each other. For example, (a) effect of Gas pressure on the surface roughness was explained depending on the power, (b) effect of cutting speed on the surface roughness was explained depending on the power and finally (c) effect of Gas pressure on the surface roughness was explained depending on the cutting speed. In the same way, all of the optimization results were also given for the composites TiN2 and TiN3 in the Figs. 12.11, 12.12, 12.13 and 12.14 respectively. One can follow from the evolutions of the surface roughness values depending on the cutting conditions and can be proposed "Regression Equations" as a useful tool for TiN1, TiN2 and TiN3 by considering cutting parameters.

Regression Equation for the composite TiN1is given as follows;

$$Ra(\mu m) = 6.3 + 0.4\,P(Kw) - 36\,S(m/\text{min}) + 227\,Gp(bar) + 9.7\,Power(Kw)^{*}S(m/\text{min}) - 33P(Kw)^{*}Gp(bar) - 620\,S(m/\text{min})^{*}Gp(bar)$$

Regression Equation for the composite TiN 2 is given as follows;

$$Ra(\mu m) = -137.1 + 40.0\,P(Kw) + 990\,S(m/\text{min}) - 1550\,Gp(bar) - 266\,P(Kw)^{*}S\,(m/\text{min}) + 241\,P(Kw)^{*}Gp(bar) + 4560\,S(m/\text{min})^{*}Gp(bar)$$

Regression Equation for the composite TiN 3 is given as follows;

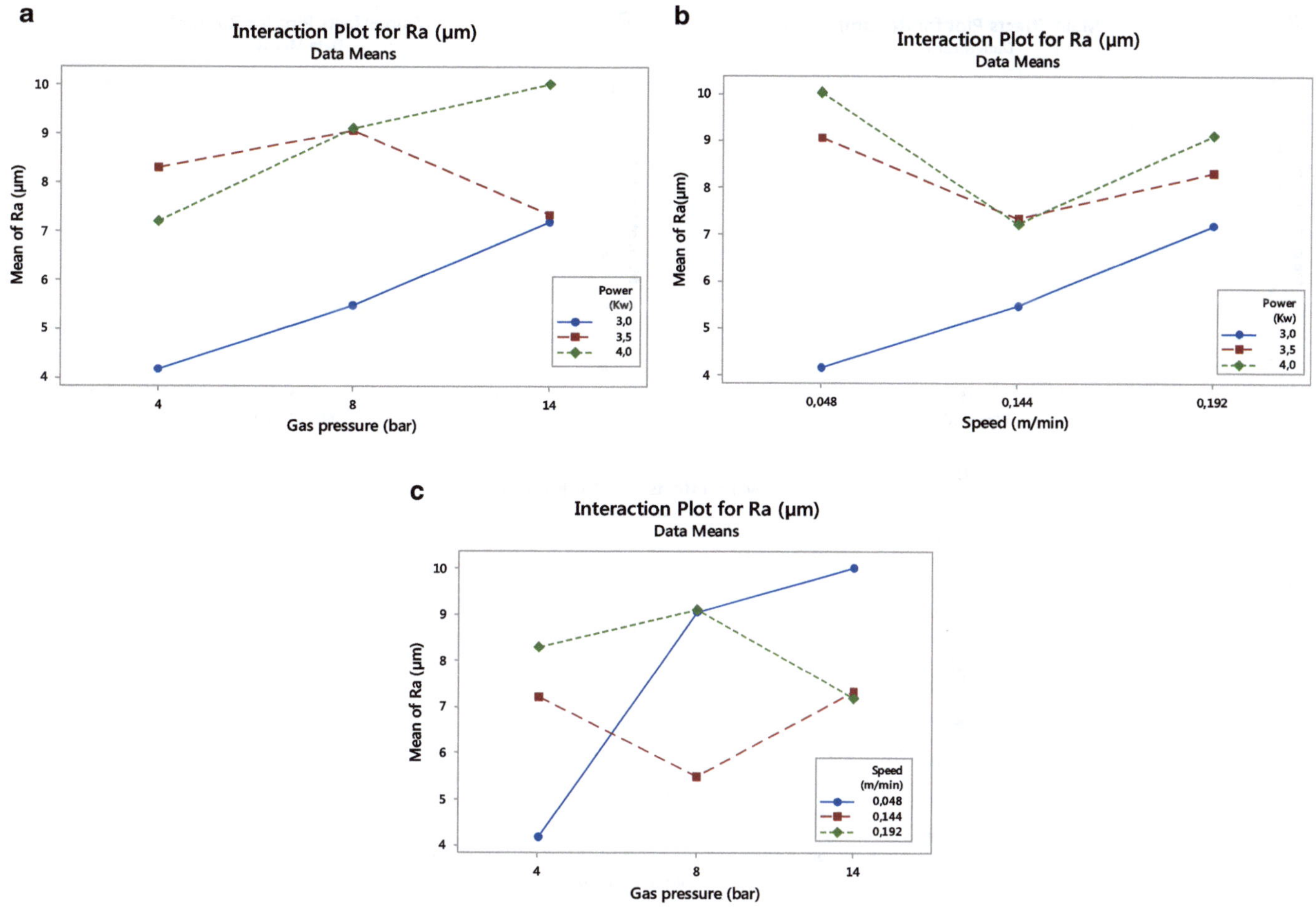

Fig. 12.10 Effects of interaction of power, speed and gas pressure on the surface roughness Ra value (TiN1)

$$Ra(\mu m) = 99.1 - 24.8\,P(Kw) - 517\,S(m/\min) + 1717Gp\,(bar) + 139.0\,P(Kw)^{*}S(m/\min) - 377\,P(Kw)^{*}Gp\,(bar) - 2275\,S(m/\min)^{*}Gp(bar)$$

12.4 Conclusions

In this work, a new composite from scrap aluminium chips reinforced with TiN +Al_2O_3 ceramic powders was developed in different compositions and was examined its tailored behaviour of these composites. The thermal effects of laser cutting process and effects of the operating parameters such as laser power, gas pressure and cutting speed on the cutting edge and on the cutting surface were examined. The evolution of the microhardness underneath the cutting surface is also examined. The composite used in this study was produced through combined method of powder metallurgical (P/M) and thixoforming (semi solid). Microstructure of cutting edge and cutting surfaces are investigated by scanning electron microscopy (SEM). Cutting surfaces have been analyzed with 3D optical surface roughness-meter (3D–SurfaScan).

Roughness evaluations were taken as optimization criteria as a function of the cutting surface and cutting parameters (power, speed, gas pressure etc.) that have been carried out by Taguchi method. A simple and useful tool was proposed for using in real manufacturing environment. These results give partial results of the common research that is going on.

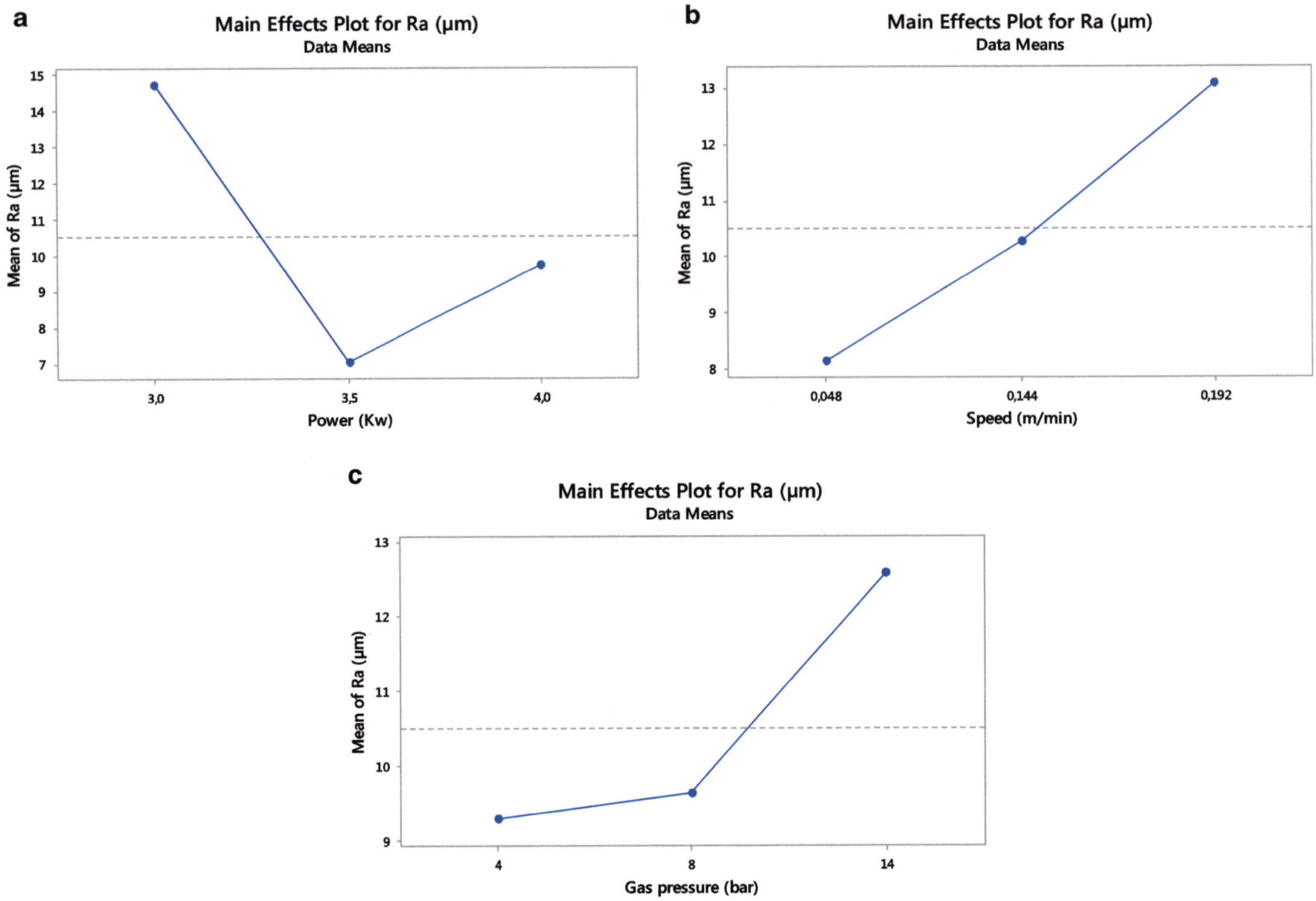

Fig. 12.11 Effects of power, speed and gas pressure on the surface roughness Ra value (TiN2)

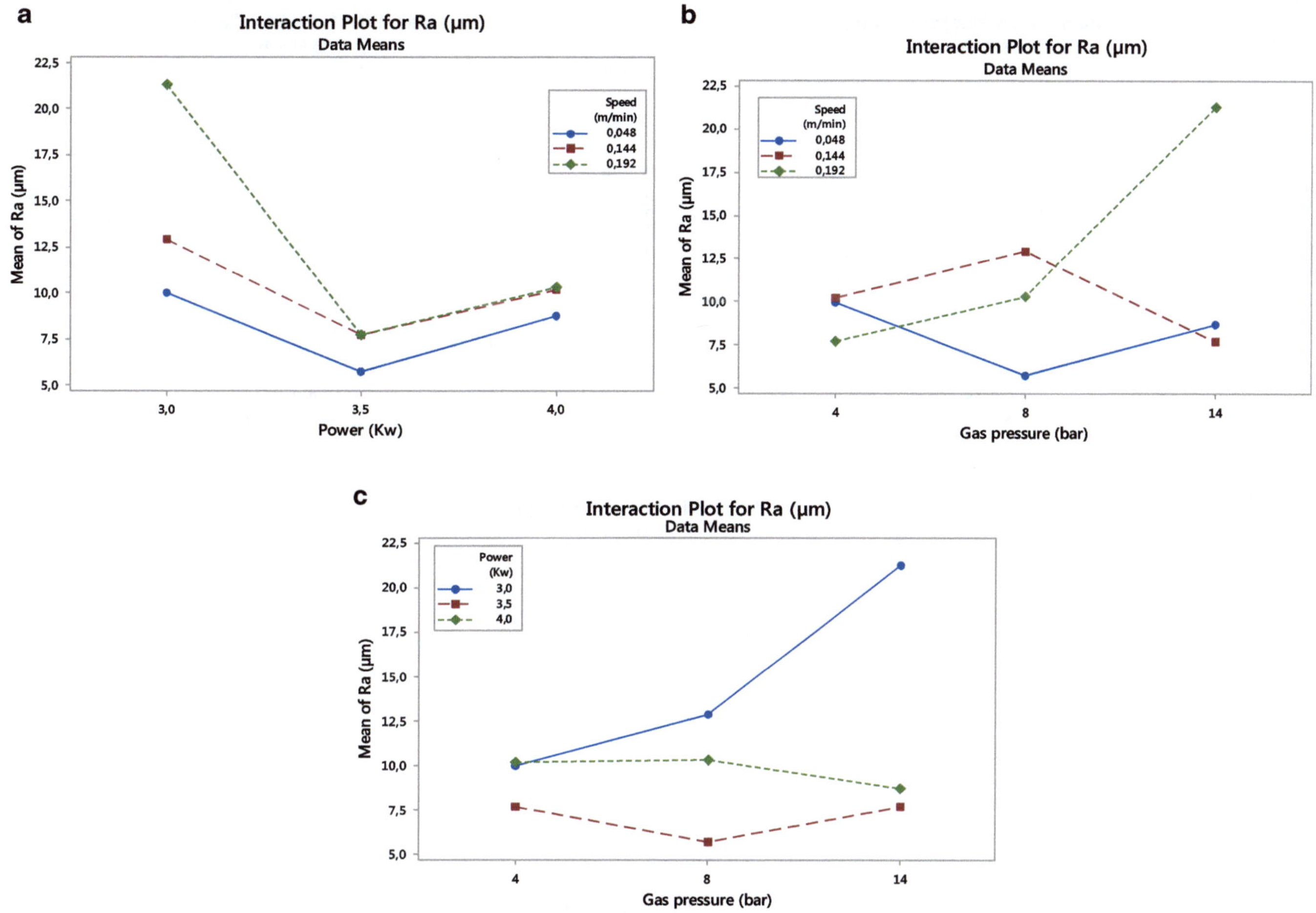

Fig. 12.12 Effects of interaction of power, speed and gas pressure on the surface roughness Ra value (TiN2)

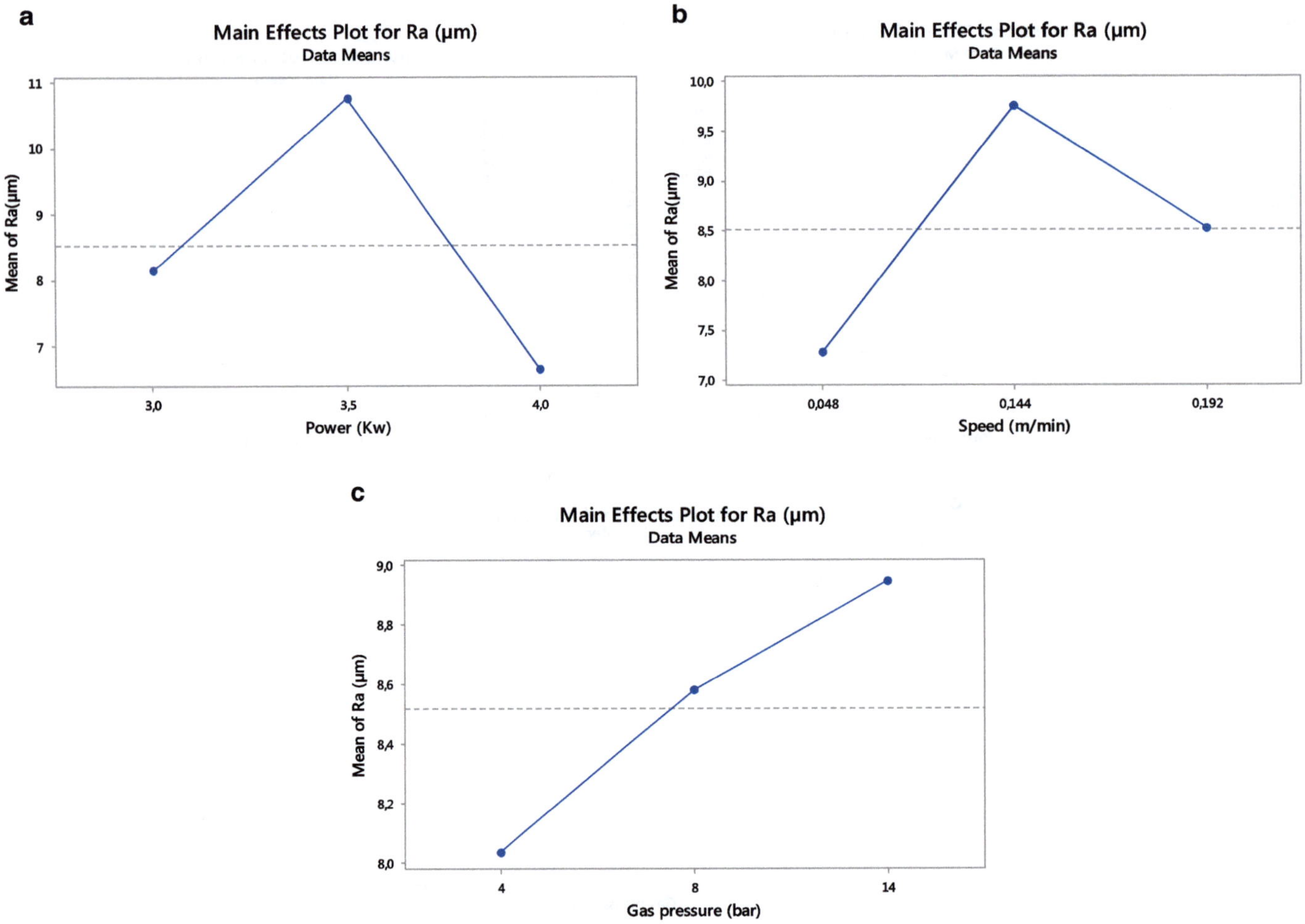

Fig. 12.13 Effects of power, pressure and speed on the surface roughness Ra value (TiN3)

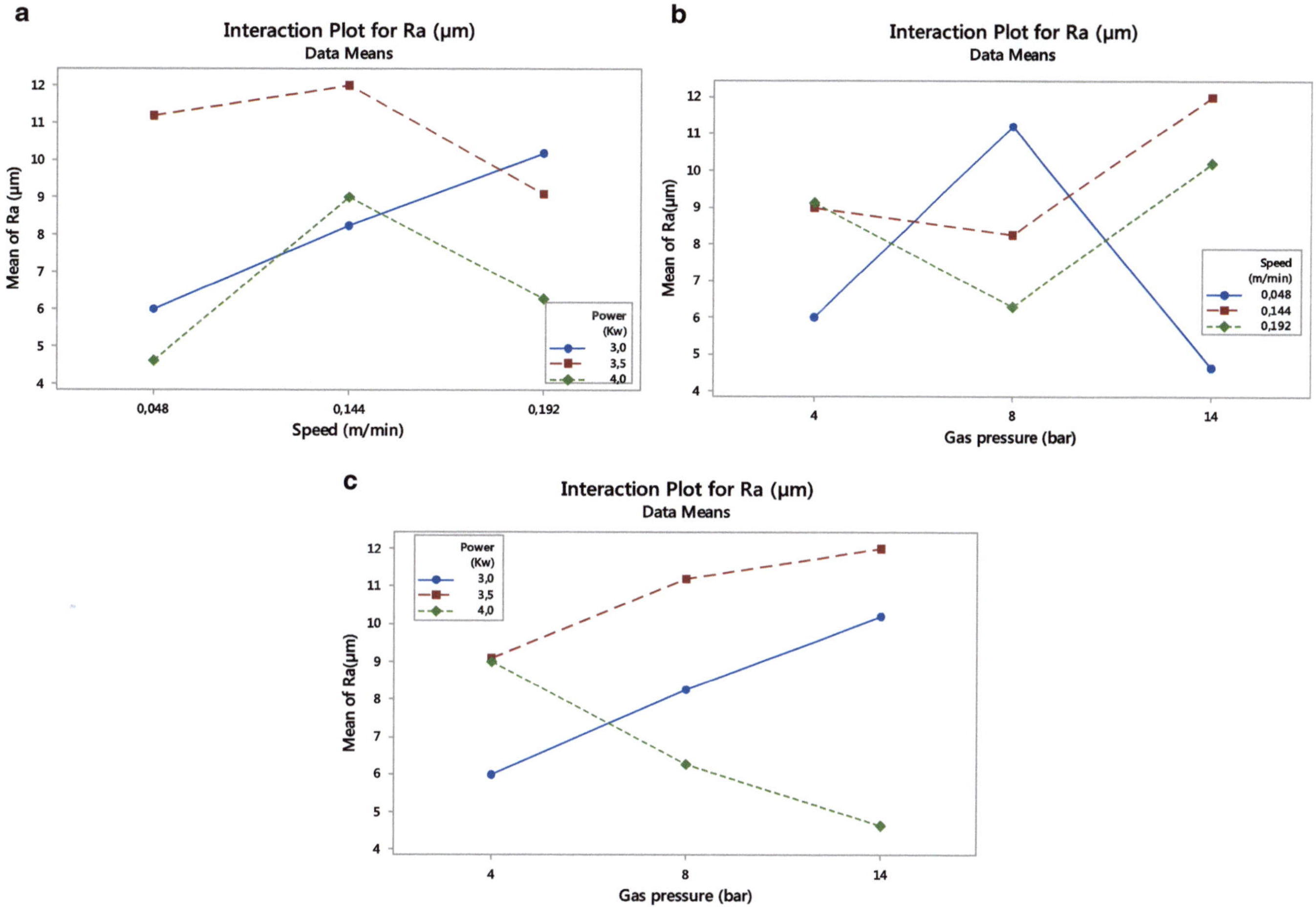

Fig. 12.14 Effects of interaction of power, speed and pressure on surface roughness Ra (TiN3)

References

1. Ray, A.K., Venkateswarlu, K., Chaudhury, S.K., Das, S.K., Kumar, B.R., Pathak, L.C.: Fabrication of TiN reinforced aluminium metal matrix composites through a powder metallurgical route. Mater. Sci. Eng. A. **338**(1–2), 160–165 (2002)
2. Shyu, R.F., Weng, F.T., Ho, C.T.: In situ reacted titanium nitride-reinforced aluminum alloy composite. J. Mater. Process. Technol. **122**(2–3), 301–304 (2002)
3. Reddy, V.K., Reddy, A.C.: Role of reinforcing particle size in the wear behavior of AA6061- Titanium Nitride composites. In: 5th International Conference on Modern Materials and Manufacturing, pp. 394–398. Bangalore (2013)
4. Katundi, D., Ezeddini, S., Bayraktar, E., Miskioglu, I.: Design of new Al – TiN composites through combined method of powder metallurgy and thixoforming. In: AMPT-2016, Kuala Lumpur, 8–11 November 2016
5. Policarpio Ferreira, L.-M., ROBERT, M.-H., Bayraktar, E.: Production of aluminum / SiC / $NiAl_2O_4$ MMCs by thixoforming of recycled chips. Int. J. Solid State Phenom. **217–218**(1), 286–293 (2015)
6. Miraoui, I., Boujelbene, M., Bayraktar, E.: Analysis of cut surface quality of sheet metals obtained by laser machining: Thermal effects. AMPT, J. Adv. Mater. Process. Technol. Taylor & Francis-USA. **1**(3–4), 633–642 (2015)
7. Kursun, A., Bayraktar, E., Enginsoy, H.M.: Experimental and numerical study of alumina reinforced aluminum matrix composites: Processing, microstructural aspects and properties. Compos. Part B. **90**(1), 302–314 (2016)
8. Bayraktar, E., Ayari, F., Tan, M.-J., Tosun Bayraktar, A., Katundi, D.: Manufacturing of aluminum matrix composites reinforced with iron-oxide nanoparticles: Microstructural and mechanical properties. Metall. Mater. Trans. B. **45B**(26), 352–362 (2014)
9. Lopez, V.H., Scoles, A., Kennedy, A.R.: The thermal stability of TiC particles in an Al7 wt%Si alloy, mater. Sci. Eng. **A356**, 316–325 (2003)
10. Kursun, A., Bayraktar, E., Helena, M.: Robert, low cost manufacturing of aluminium-alumina composites, JAMPT, journal of advances in materials and processing technologies. JAMPT, J. Adv. Mater. Process. Technol. Taylor & Francis-USA. **1**(3–4), 515–528 (2015)
11. Policarpio Ferreira, L.-M., Robert, M.-H., Bayraktar, E., Zaimova, D.: New design of aluminium based composites through combined method of powder metallurgy and thixoforming. Adv. Mater. Res. AMR. **939**(1), 68–75 (2014)
12. Venkateswarlu, K., Saurabh, S., Rajinikanth, V., Sahu, R.K., Ray, A.K.: Synthesis of TiN reinforced Aluminium metal matrix composites through microwave sintering. J. Mater. Eng. Perform. **19**, 231–236 (2010)

Chapter 13
Optimization of Laser Cutting Parameters for Tailored Behaviour of Scrap (Ti6242 + Ti) Based Composites Through Semisolid Sintering

Sonia Ezeddini, Emin Bayraktar, I. Miskioglu, and D. Katundi

Abstract Tailored behaviour of scrap titanium based composites can be done reasonably by laser cutting at the optimum process parameter ranges. Laser cutting is frequently used for manufacturing of composites as an advance machining process. It is suitable for geometrically complex profile cutting in the metals used in manufacturing engineering. In the present work, laser cutting process of scrap titanium matrix composites obtained from scrap (Ti6242 + pure Ti) chips reinforced with aluminium, tin, zirconium, molybdenum were carried out. Actually these materials are being widely used in the aerospace such as aircraft structure. For this reason, it was interesting to develop a new composite from scrap (Ti6242 + Pure Ti) chips reinforced with certain elements and to examine its tailored beahviour of these composites for using in manufacturing engineering. Due to thermal effects of laser cutting process, it should be optimized main operating parameters such as laser power, and cutting speed, gas pressure, etc. on the cutting surface. The evolution of the microhardness underneath the cutting surface due to laser power was examined. The composite used in this study was produced through combined method of powder metallurgical (P/M) and thixoforming (Semi solid). Microstructure of cutting edge and cutting surfaces are investigated in detail by scanning electron microscopy (SEM). Cutting surfaces have been analyzed with 3D optical surface roughness-meter (3D–SurfaScan). Roughness evaluations were taken as optimization criteria as a function of the cutting surface and cutting parameters (power, cutting speed, gas pressure etc.) that have been carried out by Taguchi method. A simple and useful tool was proposed for using in real manufacturing environment.

Keywords Tailored behaviour of recycled composite • Laser cutting • Taguchi method • SEM

13.1 Introduction

The most characteristic behaviour of composites is that, while the individual constituents retain their properties, they synergize to contribute properties which may not be found in any one of them alone [1–4]. Recently, many research studies in the literature have been carried out, and it was revealed that there is remarkable ability of metallic and nonmetallic reinforced metal-matrix composites (MMCs) [5–9]. Several processing techniques have been developed to engineering composites for various fields of applications [4–11] such as aerospace, defense, automobile, and sports sectors, etc. In MMCs, the addition of a small amount of second-phase materials with high shear strength imparts unique properties to the base materials [4–7]. Metal matrix composites (MMCs) have been developed to respond the demand for lighter materials with high specific strength, stiffness and wear resistance [1, 2, 7, 8, 10, 11]. In aircraft engineering, titanium is extensively favored like aluminium as a basic matrix material in MMCs because of its low density, and high performance index. It means that high performance indices due to their light weight, high strength-to-weight ratio, and worthy engineering properties [9–15]. However, particulate reinforced titanium matrix composites (TiMCs) are much more attractive engineering materials due to their strength, ductility and toughness as well as their ability to be processed by non-conventional methods such as sinter + forging and/or sinter + thixoforming; they are relatively low costs regarding to classical manufacturing techniques. Then, for environmental and economic reason, it is very useful to precede scrap titanium and create new composites for engineering applications. Therefore, in recent years, TiMCs can be applied successfully to structural components and the number of applications is expected to increase with the development of low-cost processing methods [10–17].

S. Ezeddini • E. Bayraktar (✉) • D. Katundi
Supmeca/Paris, School of Mechanical and Manufacturing Engineering, St-Ouen, France
e-mail: bayraktar@supmeca.fr

I. Miskioglu (✉)
Michigan Technological University, ME-EM Department, Houghton, MI, USA
e-mail: imiski@mtu.edu

P.R. Thakre et al. (eds.), *Mechanics of Composite and Multi-functional Materials, Volume 6*,
Conference Proceedings of the Society for Experimental Mechanics Series, DOI 10.1007/978-3-319-63408-1_13

In fact, in industrial applications, to improve the structure and properties of particulate reinforced TiMCs various processing techniques have evolved over the last 20 years. Thixoforming is one member of semi-solid forming processes and it possesses characteristics of both casting and forging [4, 7, 8].

For this reason, it was interesting to develop a new composite from scrap titanium matrix composites obtained from mixed scrap of (Ti6242 + Pure Ti) chips reinforced with aluminium, tin, zirconium, molybdenum In the present work, The thermal effects of laser cutting and effects of main operating parameters such as laser power, and cutting speed on the cutting edge and on the cutting surface were examined. The evolution of the microhardness underneath the cutting surface due to laser power is also examined.

The composite used in this study was produced through combined method of powder metallurgical (P/M) and thixoforming (Semi solid). Microstructure of cutting edge and cutting surfaces are investigated in detail by scanning electron microscopy (SEM). Cutting surfaces have been analyzed with 3D optical surface roughness-meter (3D–SurfaScan). Roughness evaluations were taken as optimization criteria as a function of the cutting surface and cutting parameters (power, speed, gas pressure etc.) that have been carried out by Taguchi method. A simple and useful tool was proposed for using in real manufacturing environment.

13.2 Experimental Conditions

13.2.1 Manufacturing of Composites and Microstructural Evaluation

Basically, composition of the specimens designed in this work was arranged as shown in the Table 13.1: mixed scraps of (Ti6242 + Pure Ti) chips were reinforced with aluminium, tin, zirconium, molybdenum and as secondary reinforcements, (Si, and Boron) were used.

Processing route has been carried out through combined process of powder metallurgical (P/M) and thixoforming. Firstly, all of the element of this composition have been mixed and milled in a pulverize device during 4 h by using a protection and successfully ball milled (4000 rpm) during 1 h and compacted (CIP) at a pressure of 250 MPa and after that thixo forming sintering by heating at the temperature of 700 °C for 45 min under argon inert atmosphere.

After manufacturing of the composite, microstructural evaluation has been carried out. The dispersion of reinforcement particles in the matrix and interface at matrix/reinforcements was evaluated. Microhardness tests have been carried out on the polished and etched specimens. Thickness of work pieces is 3 mm. Certain basic mechanical parameters have been given here as an indicative results of these composites.

13.2.2 Laser Cutting Process and Cutting Parameters

All of the laser cutting processes have been carried out on a CO_2 laser machine called "4000-TLF TURBO" Basic parameters of laser cutting process have been given in the Table 13.2. In the present work, three levels of powers (Pw) and three levels of cutting speeds (S), with three different gas pressures (Gp) have been chosen in order to examine their impact on the basic performance parameters. Nine samples were cut for each composition. Totally, 27 different Laser-cutting-samples were used and these samples have been prepared for damage evaluation (Fig. 13.1).

After achievement of the cutting process, the widths of resulting cuts were measured and SEM micro-photography of the cutting surfaces was carried out. To measure the striation depths and widths, the surface roughness measurement was carried out using a 3D optical roughness meter (Surf-Scan). For intensive evaluation, measurements were taken in different zones for each composition (Remelted Zone, Heat Affected Zone (HAZ) and the centre of the specimen) by using SEM.

Optimization process for the experimental matrix with variables was employed to develop for each experimental test. Taguchi method and L9 Orthogonal Array were used to reduce number of the experiments. The design of experiments

Table 13.1 Chemical compositions of composite designed in this work

Specimen Name:	Ti6242 + Pure Ti	Al (wt %)	Zr (wt%)	Sn (wt %)	Mo (wt %)	BN (wt %)	Si (wt %)	B (wt %)
Ti6242 + Ti	B	15	8	4	2	<1	0.5	0.25

Table 13.2 Basic parameters of CO_2 laser cutting process

	Levels		
Cutting parameters	Min	Medium	Max
Power (W)	3000	3500	4000
Cutting speed (m/min)	0.048	0.144	0.192
Gas pressure (nitrogen) (bar)	4	8	14

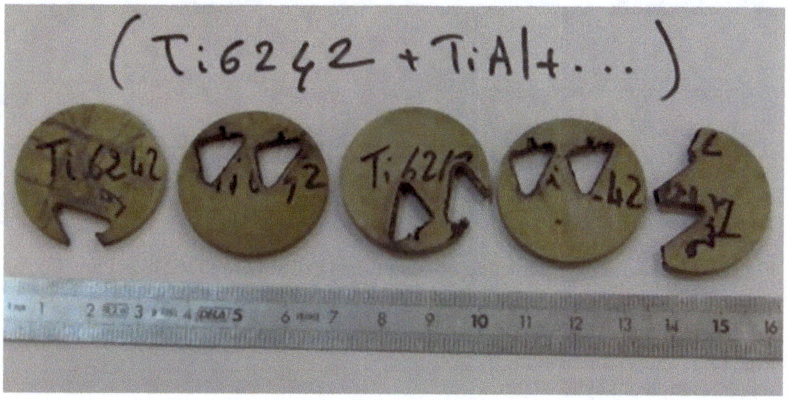

Fig. 13.1 Experimental specimens of (Ti6242 + pure Ti based composites) cut using CO_2 laser cutting process

(DOE) and measurements of the average roughness values "Ra" have been carried out with Minitab 17 software. All of the optimization parameters were evaluated for each composition with these diagrams obtained by software.

13.3 Results and Discussion

13.3.1 Microstructure of ((Ti6242 + Pure Ti) Based Composites

The structures of the composite from mixed scrap of (Ti6242 + Pure Ti) chips reinforced with aluminium, tin, zirconium, molybdenum as secondary reinforcements, Si, and Boron were used. In the Fig. 13.2, general structure of this composite has shown. Actually, Ti6242 is used extensively by French Aeronautical Company (SNECMA-SAFRAN) as a high resistance and high temperature alloy for very good creep behaviour up to 500–600 °C in manufacturing of HP-compressors. Its structure is very complex and generally sensitive to the thermomechanical history. Generally, a mixed α (soft) + β (hard) structure was identified [12–13]. By mixing the scrap of (Ti6242 + Pure Ti) chips reinforced basically with Al, Zr, and Mo, very resistance new composite can be created for industrial application again in that area.

During the ball milling of the mixture with a high mechanical energy, mechano-chemical reactions are created between multiple particles and new characteristics of this composite. This phenomenon is generally called solid state phase reactions start during the treatment of mechano – fusion.

13.3.2 Evaluation of the Results for Laser Cutting Process

Laser gas assisted cutting process of (Ti6242 + Pure Ti) reinforced composite pieces have been carried out. The effect of laser output power, gas pressure and cutting speed on the surface striations and surface roughness evolution of the cutting edge surfaces were analyzed. Profiles of the laser cutting surfaces were also evaluated depending on the process parameters (power, gas pressure and cutting speed, etc.).

In this work, laser beam was conducted with different conditions. As well known, when higher power densities are used, some amount of material is vaporized, forming a cavity called keyhole containing vapour, plasma or both of them. Around this cavity, there is a molten metal. The pressure inside the cavity can give stability to the liquid pressure and keep the cavity

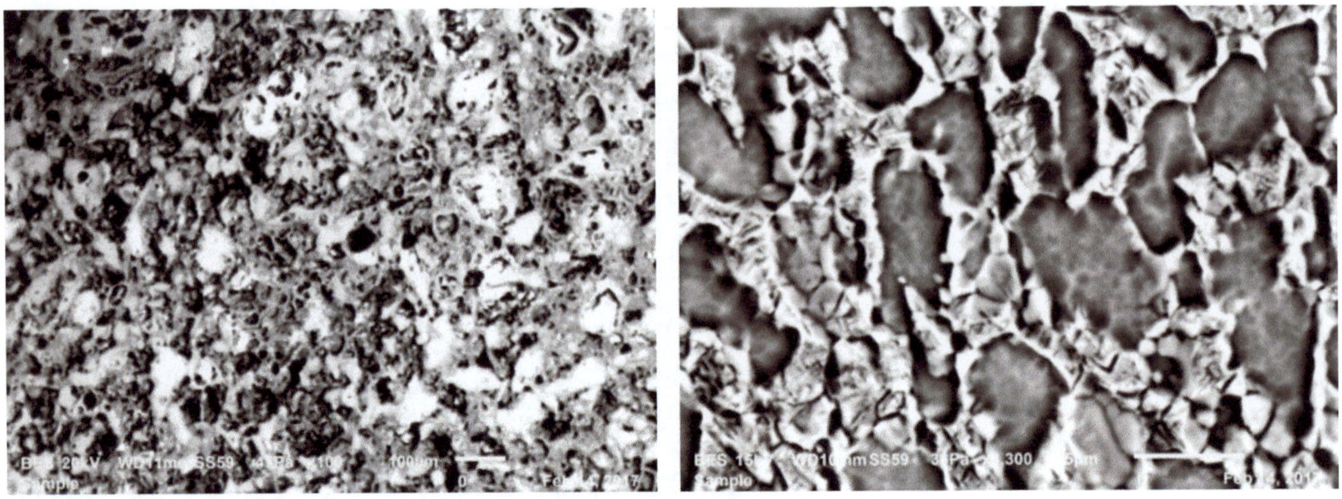

Fig. 13.2 General microstructure of composites called here (Ti6242 + Pure Ti)

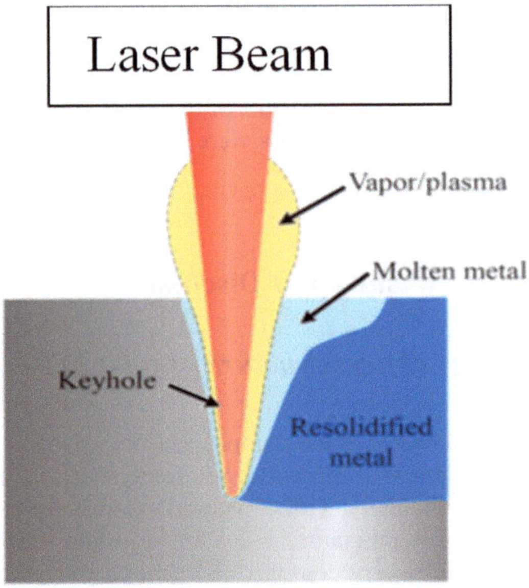

Fig. 13.3 shows a schematic representation of keyhole laser melting mode during the cutting process

opened during the process. Figure 13.3 shows a schematic representation of keyhole laser melting mode during the cutting process.

Normally, the keyhole mode is used for cutting processes of the pieces because laser absorption in keyhole mode is much higher than in the conduction mode. Additionally, the larger depth to width ratio of the melt pool is a benefit for most cutting and very often for welding processes. For the tailored pieces, mainly is very suitable to use this mode in manufacturing engineering. But, if certain parameters could not be optimized, certain instabilities of the cavity can cause in the formation of porosity due to gas bubbles that are imprisoned inside the remelted zone of the material after solidification.

Figures 13.4, 13.5 and 13.6 show surface roughness evolution of the workpieces for two laser output power settings (3 and 3.5 kWs) at three cutting speeds (0.048, 0.144 and 0.192 m/min). Three different gas pressures (4, 8 and 14 bar) were used of these cutting processes. These surface roughness evolutions after cutting were presented at the left column. In the same figures, 3D surface topographies for the cutting edge surfaces were presented at the right column under the same conditions. As can be observed from these SEM pictures, the cutting edge surfaces seem similar at higher cutting speed conditions. The widths of striations fluctuate with flow of liquid metal depending on the cutting parameters used in this work. And also surface topographies justify striations defined mainly by remelted zone.

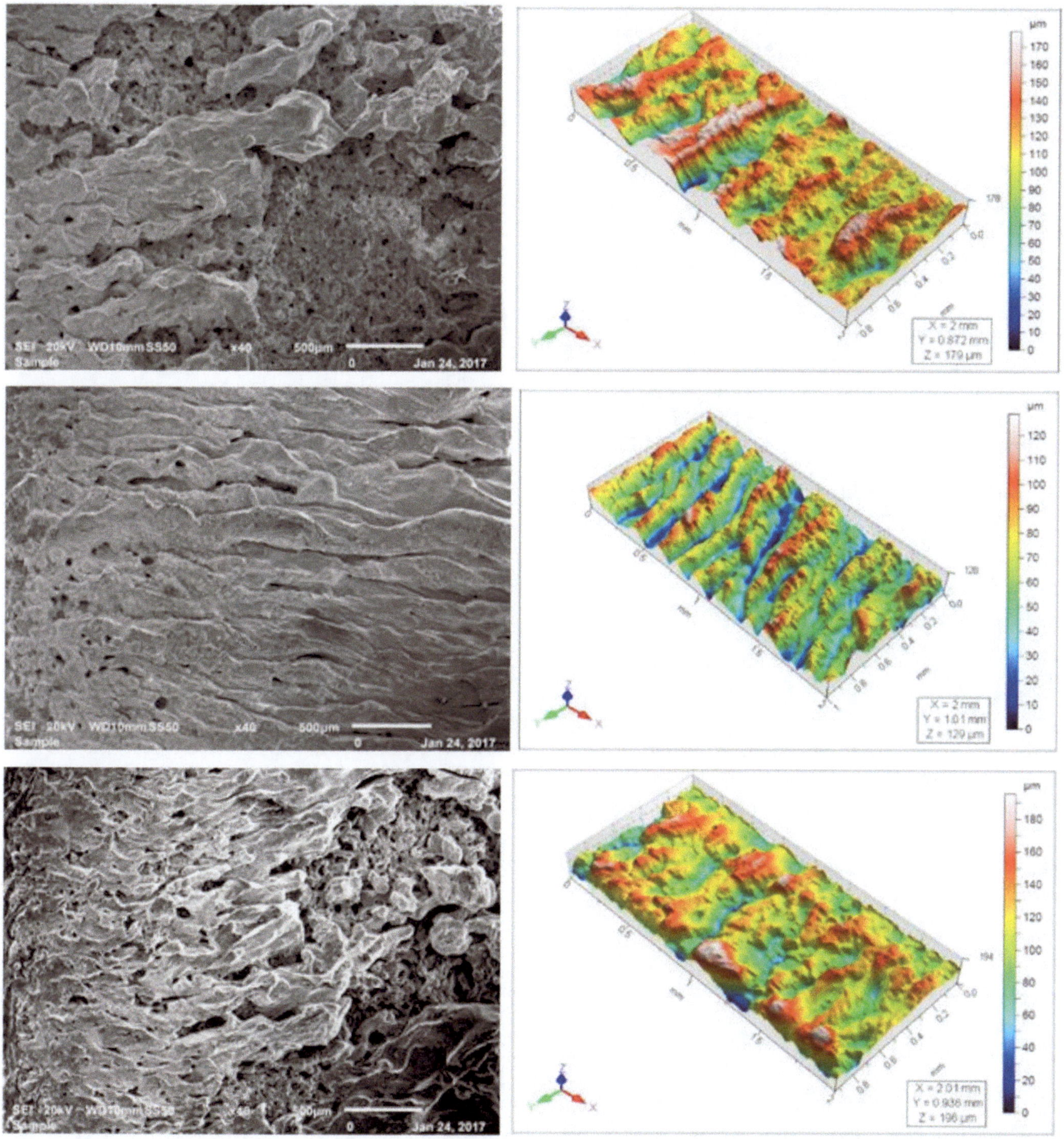

Fig. 13.4 (**a**) SEM image showing the variation of the laser cutting surface morphology of the composite at the left column under the cutting conditions: P = 3 kW, S = 0.048 m/min and pressure = 4 bar. And 3D surface topography of the laser cutting surface of the composite at the right column under the same cutting conditions. (**b**) SEM image showing the variation of the laser cutting surface morphology of the composite at the left column under the cutting conditions: P = 3 kW, S = 0.144 m/min and pressure = 8 bar. And 3D surface topography of the laser cutting surface of the composite at the right column under the same cutting conditions. (**c**) SEM image showing the variation of the laser cutting surface morphology of the composite at the left column under the cutting conditions: P = 3 kW, S = 0.192 m/min and pressure = 14 bar. And 3D surface topography of the laser cutting surface of the composite at the right column under the same cutting conditions

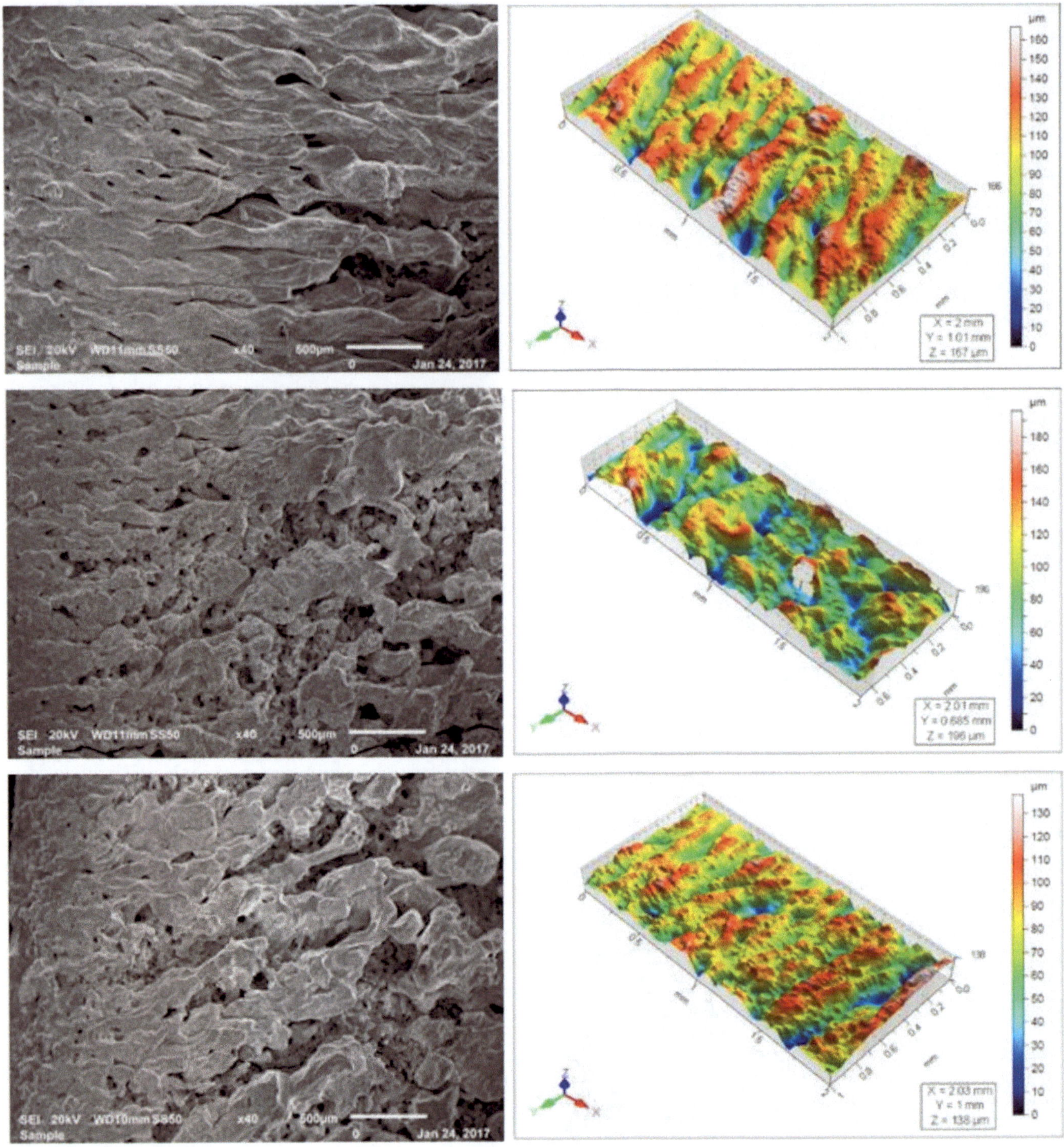

Fig. 13.5 (**a**) SEM image showing the variation of the laser cutting surface morphology of the composite at the left column under the cutting conditions: P = 3.5 kW, speed = 0.048 m/min and pressure = 8 bar. And 3D surface topography of the laser cutting surface of the composite at the right column under the same cutting conditions. (**b**) SEM image showing the variation of the laser cutting surface morphology of the composite at the left column under the cutting conditions: P = 3.5 kW, speed = 0.144 m/min and pressure = 8 bar. And 3D surface topography of the laser cutting surface of the composite at the right column under the same cutting conditions. (**c**) SEM image showing the variation of the laser cutting surface morphology of the composite at the left column under the cutting conditions: P = 3.5 kW, speed = 0.144 m/min and pressure = 14 bar. And 3D surface topography of the laser cutting surface of the composite at the right column under the same cutting conditions

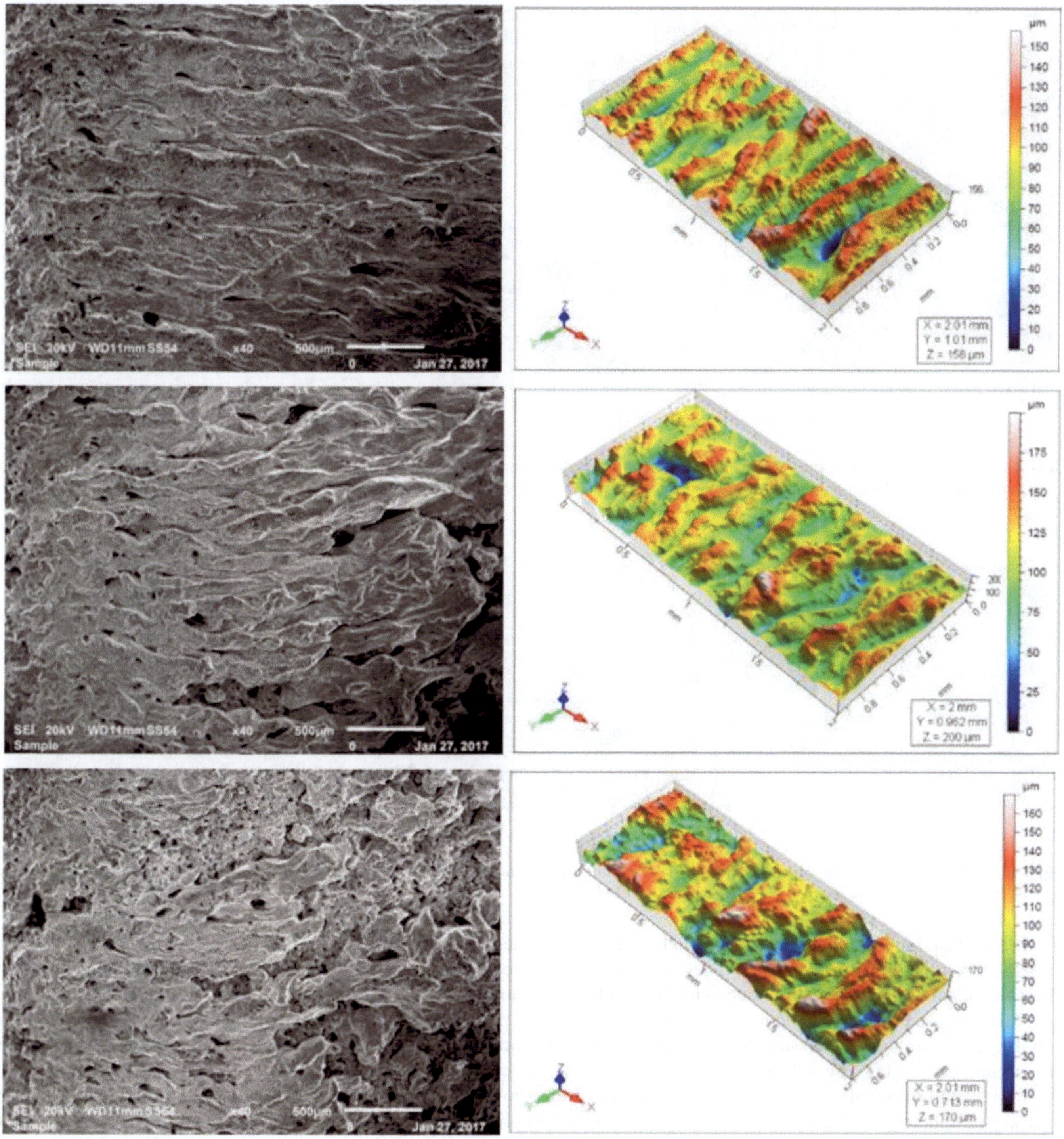

Fig. 13.6 (**a**) SEM image showing the variation of the laser cutting surface morphology of the composite at the left column under the cutting conditions: P = 3 kW, speed = 0.048 m/min and pressure = 14 bar. And 3D surface topography of the laser cutting surface of the composite at the right column under the same cutting conditions. (**b**) SEM image showing the variation of the laser cutting surface morphology of the composite at the left column under the cutting conditions: P = 3.5 kW, speed = 0.144 m/min and pressure = 4 bar. And 3D surface topography of the laser cutting surface of the composite at the right column under the same cutting conditions. (**c**) SEM image showing the variation of the laser cutting surface morphology of the composite at the left column under the cutting conditions: P = 3.5 kW, speed = 0.192 m/min and pressure = 8 bar. And 3D surface topography of the laser cutting surface of the composite at the right column under the same cutting conditions

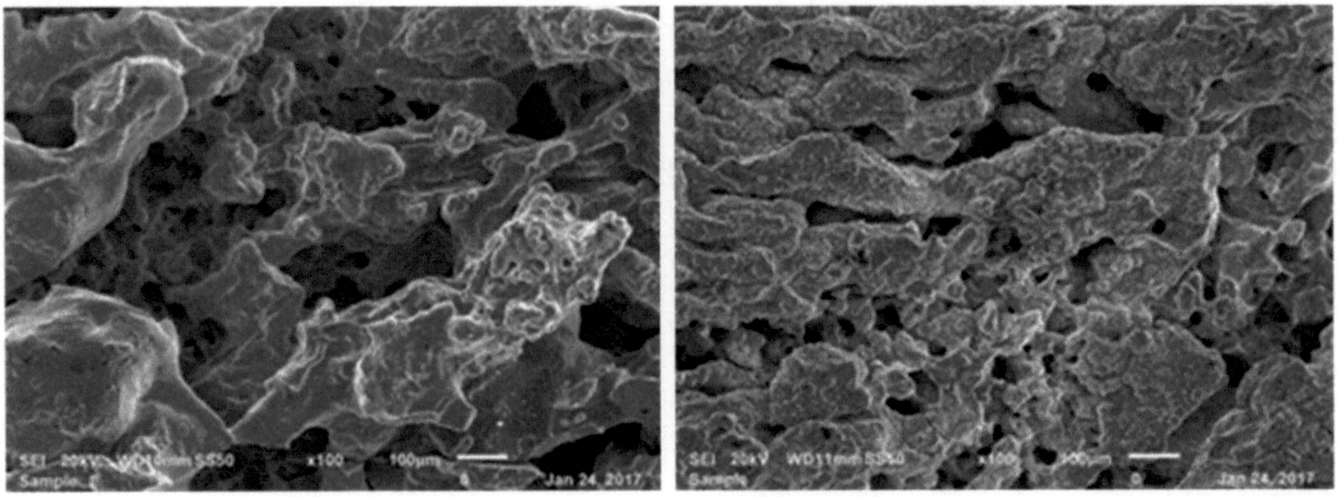

a) Cutting speed=0,048m/min, pressure =4bar b) Cutting speed= 0,144m/min, pressure= 14bar

Fig. 13.7 SEM image showing the variation of the laser cutting surface morphology of the composite at two different laser cutting conditions: (**a**) low cutting speed = 0,048 m/min and low pressure = 4 bar, and (**b**) high cutting speed = 0,144 m/min and high gas pressure = 14 bar

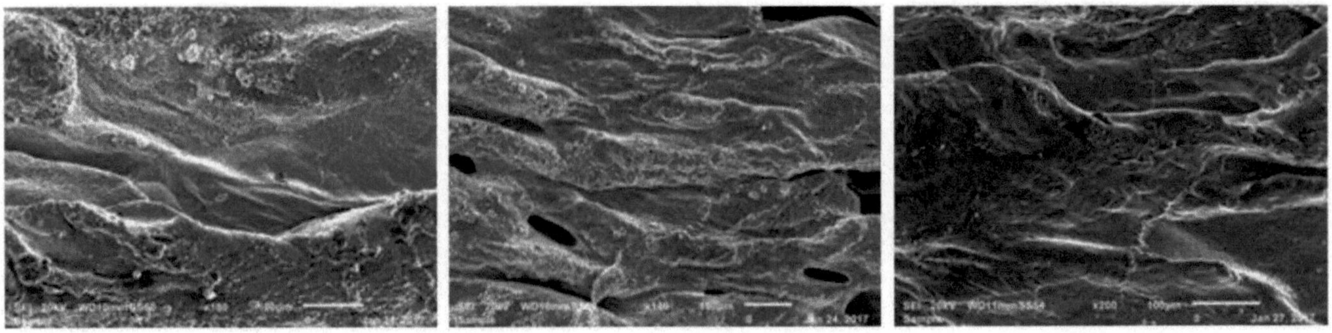

3 kW 3.5 kW 4 kW

Fig. 13.8 SEM image showing the variation of the laser cutting surface morphology depending on the three types of powers used in this work; 3 kW, 3.5 kW and 4 kW, respectively at constant speed (0.144 m/min)

In fact, heat input is bigger close to the melted surface, facilitating diffusion and very often cause in grain growth. As the distance from cutting edge increases, a new area very similar recrystallization zone is observed. This zone can occur after heat treatment at the high cooling rate combined with high heating rate.

This is expected phenomenon met very often in laser cutting processes. Thickness of the remelted zone depends strongly on the gas power and gas pressure that influences strongly surface roughness value much higher than other parameters such as cutting speed. This case will be discussed in the next two sections see later in 13.3.2.1 and 13.3.2.2 respectively.

More detail information on this case, some pictures were shown in the Fig. 13.7. Here, the variation of the laser cutting surface morphology of the composite can be observed depending on two different laser cutting conditions. Figure 13.7a shows a disturbed cutting surface with low cutting speed = 0,048 m/min and low pressure = 4 bar, and Fig. 13.7b shows a smooth cutting surface with high cutting speed = 0.144 m/min and with high gas pressure = 14 bar. One may explain that two different cutting conditions are taken as main parameters that influence cutting surface that is explained with the decrease in surface roughness values.

In another aspect to look at evolution of surface roughness values, effect of only one cutting parameter has been shown in the Figs. 13.8 (only power) and 13.9 only (gas pressure). Here, only a high cutting speed used (0.144 m/min) as constant parameter. Variation of surface roughness is observed very simply with these pictures. It means that high cutting power and high gas pressure can improve surface roughness of the cutting surfaces.

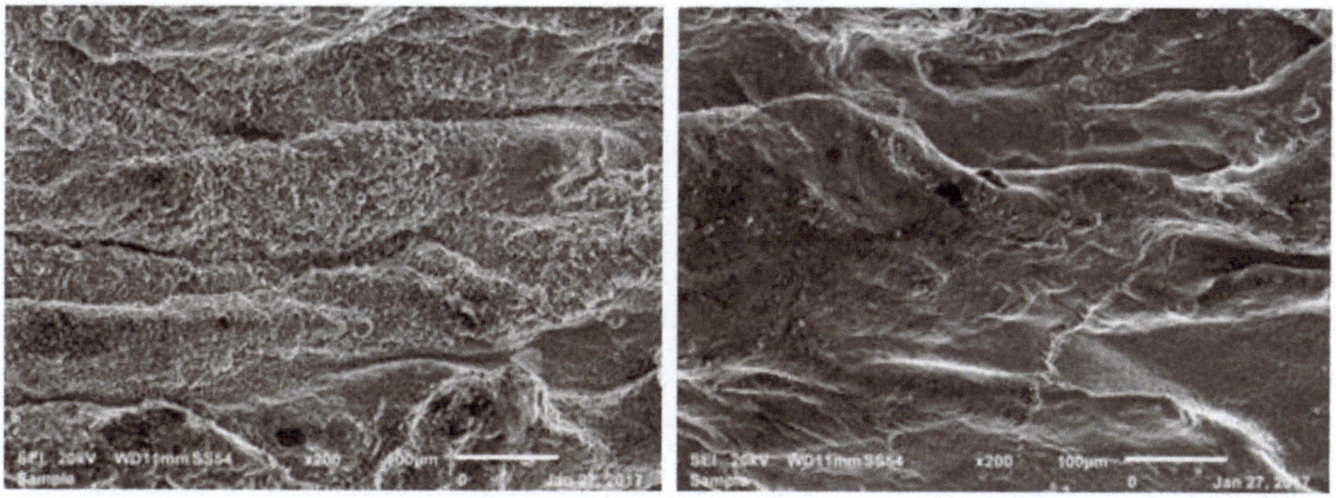

4 bar 14 bar

Fig. 13.9 SEM image showing the variation of the laser cutting surface morphology of the composite at two different laser cutting conditions low and high gas pressures (4 bars and 14 bars in two extremes) at constant speed (0.144 m/min)

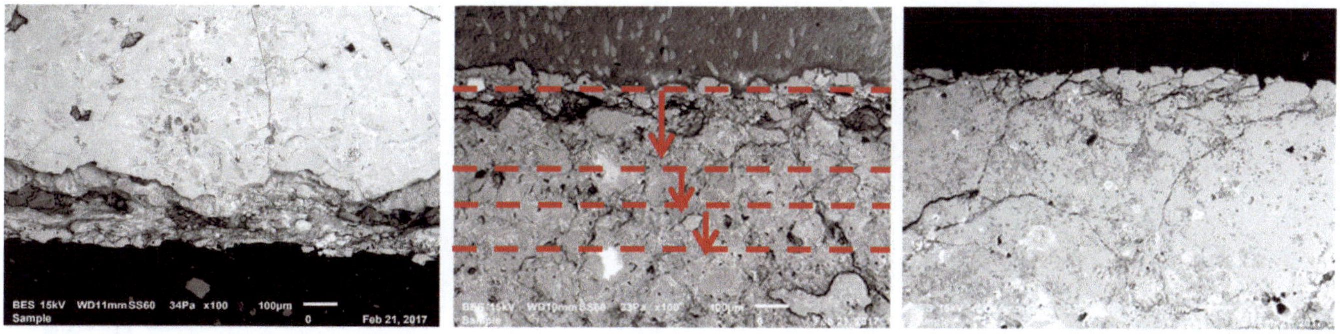

a) b) c)

Fig. 13.10 Microstructural evolution depending on the distance from the remelted zone through HAZ zone to the substrate –or base metal structure; (**a**) Power = 3 kW; Speed = 0.144 m/min; Pressure = 8 bar, (**b**) Power = 4 Kw; Speed = 0,192 m/min; Pressure = 8 bar, and also (**c**) Power = 3, 5 Kw; Speed = 0,048 m/min

Evolution of microstructure as transversal view to the surface under the cutting edge was shown in Fig. 13.10 as a function of the distance from the remelted zone through HAZ zone to substrate –or base metal structure. This zone is not affected from heating. All of the three pictures show transversal view to the surface under the cutting edge.

One may note that the neighbor zone just under or closer to the remelted zone seem as recrystallized zone. This idea should be justified in the next section with microhardness evaluation in the different zones from cutting edge to the base metal.

13.3.2.1 Evolution of Micro-Hardness as a Function of the Laser Cutting Conditions

More detail information on the microstructure evolution from the remelted zone, passing through the heat affected zone until the base metal structure was evaluated by using microhardness ($HV_{0.05}$) measurements. All of the microhardness measurements for three cutting conditions taken as transversal view to the surface under the cutting edge were indicated in Fig. 13.11. One may observe that the values measured in the region closer to the remelted zone seem as recrystallized zone. This zone was identified as distance of 80–90 μm from cutting edge. It is likely as Heat Affected Zone (HAZ) and finally third zone is the base metal that was not affected with heating during cutting process. Microhardness evaluation in these three different zones from the remelted zone, passing through the heat affected zone until the base metal structure show

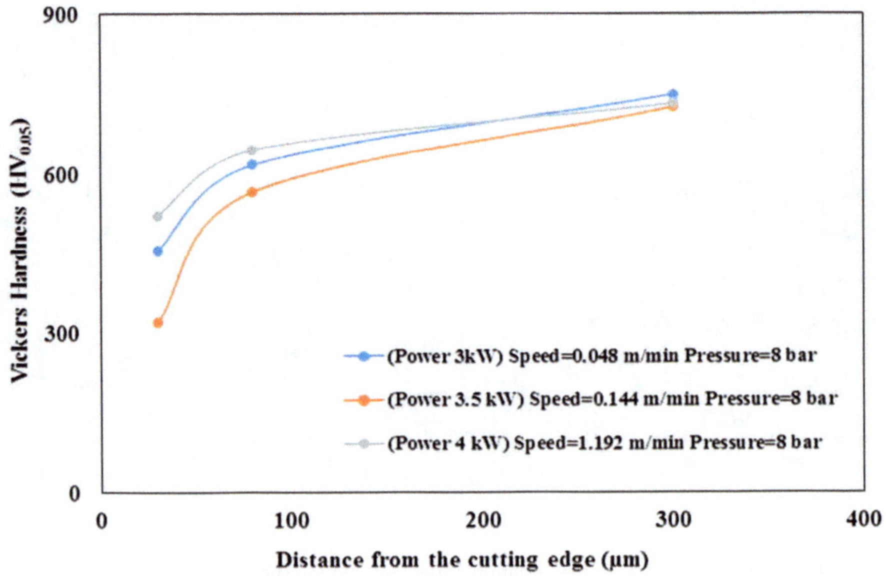

Fig. 13.11 Micro hardness ($HV_{0.05}$) evolution as a function of the distance from the remelted zone through HAZ zone to substrate (unaffected zone) for three different cutting conditions

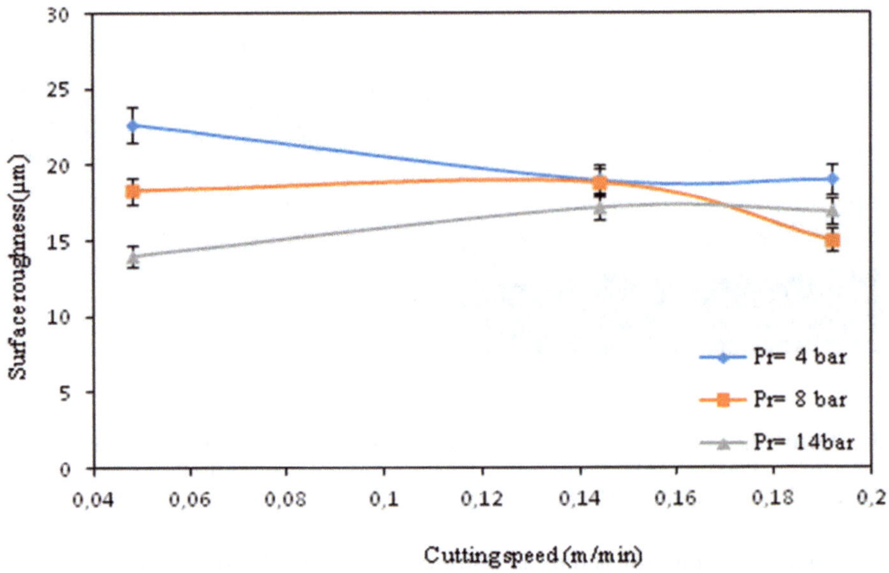

Fig. 13.12 Evolution of surface roughness values depending on the laser cutting speed values in case of three different gas pressures

that the microstructure influences strongly with heat treatment. Evidently, the second zone (HAZ) gives lower microhardness values for all of the three cutting conditions identified as softening phenomenon due to the recrystallization.

13.3.2.2 Evolution and Optimization of Surface Roughness Depending on the three Cutting Conditions

After evaluation of experimental results related to the cutting process, a detail optimization process was carried out. In a simple way, an experimental matrix with variables was employed to develop for each experimental test. Very simply, experimental design (ED) has been carried out with "Minitab 17 Software" by using all of the experimental measurements of the average roughness values "Ra". All of the optimization parameters were evaluated for each cutting condition identified with the optimization diagrams obtained by using this software. First results were presented in the Figs. 13.12. In this figure,

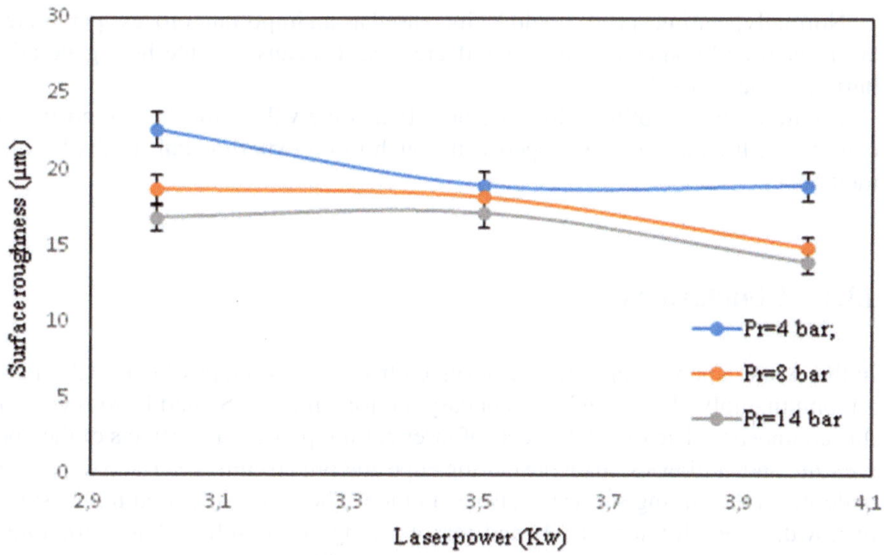

Fig. 13.13 Evolution of surface roughness values depending on the laser cutting power values in case of three different gas pressures

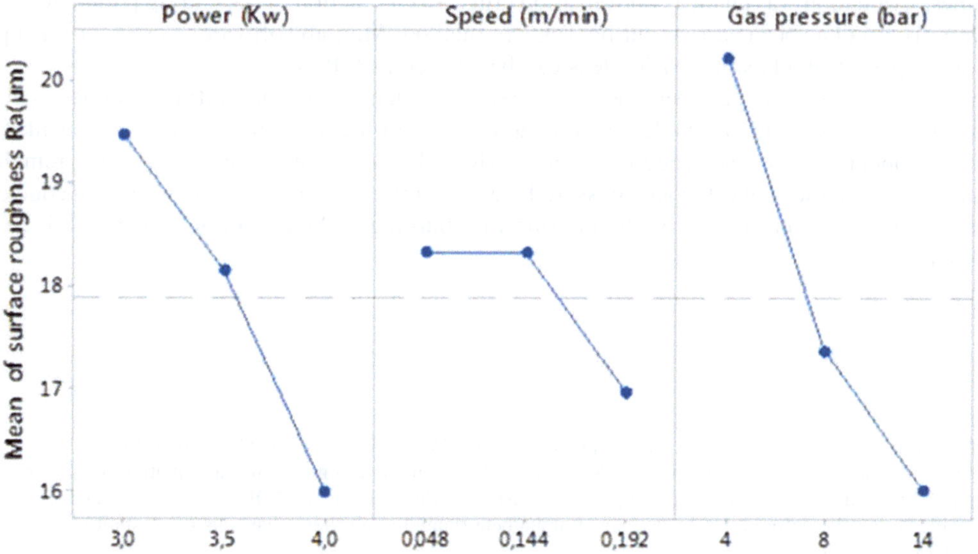

Fig. 13.14 Influence of all of the laser cutting conditions; power, speed and gas pressure. Evolution of surface roughness values (Ra)

evolution of surface roughness values (Ra) was shown by considering only one parameter: it means that evaluation was made as a function of cutting speed for three different gas pressures.

Second evaluation of the surface roughness values (Ra) has been made by considering the variation of cutting power for three different gas pressures (Fig. 13.13). Both of these figures indicate that it is not effective way to look at the optimization conditions by considering only one variable.

Influence of one parameter on the surface roughness values is weak in two figures. However, if one may consider all of three parameters as an interaction each other should be much more effective to see evolution of surface roughness value (Ra) with combination of these parameters.

For simplifying evaluation of surface roughness values (Ra), a new optimization was given as the mean values depending on three parameters as an interaction each other such as cutting power, cutting speed and gas pressure and presented in the Fig. 13.14. It seems that these three parameters at higher values improve well surface roughness values (Ra) of this composite. In a general way, surface roughness values (Ra) can be improved roughly 17% by cutting power and 6–8% with cutting speed and more considerable with gas pressure about 20–25%.

Normally, cutting speed should be increased as an important process parameter because, the interaction time between the laser beam and composite specimen decreases, it means that the heat generation will decrease which leads to minimum burning of lateral side.

Naturally, these optimizations are only indicative values for the composite studied here under the limited experimental conditions. It should be developed with much more experimental results by increasing the reliability of the optimization method.

13.4 Conclusions

In this work, a new composite was designed from mixed scraps of (Ti6242 + Pure Ti) chips reinforced with aluminium, tin, zirconium, molybdenum and as secondary reinforcements, (Si, and Boron) were used and examined its tailored beahviour of this composites. The thermal effects of laser cutting process and effects of the operating parameters such as laser power, gas pressure and cutting speed on the cutting edge and on the cutting surface were examined. The evolution of the microhardness underneath the cutting surface is also examined. The composite used in this study was produced through combined method of powder metallurgical (P/M) and thixoforming (semi solid). Microstructure of cutting edge and cutting surfaces are investigated by scanning electron microscopy (SEM). Cutting surfaces have been analyzed with 3D optical surface roughness-meter (3D–SurfaScan).

Surface roughness evaluation was taken as optimization criteria as a function of the cutting parameters (power, speed, gas pressure etc.). Optimization has been carried out by Taguchi method. Minitab software is used for this optimization.

In the frame of the present work, some basic ideas can be derived as follows:

The values of surface roughness parameter Ra are directly influenced by the cutting parameters It should be make attention the type of chosen assistant gas could generate an increase of the machining cost; if an unsuitable gas pressure is used, it is possible to obtain high surface roughness values (Ra), due to overpowering flow of the remelted zone that can cause in abnormal increase in the supported gas pressure by generating excess of heat on the cutting edge due to exothermic reaction is increased. This case can cause very often in sideways burning of the cut surface and hence increase in the surface roughness (Ra) values.

References

1. Yilbas, B.S.: Study of parameters for CO_2 laser cutting process. Mater. Manuf. Process. **13**, 517–536 (1998)
2. Bayraktar, E., Ayari, F., Tan, M.-J., Tosun Bayraktar, A., Katundi, D.: Manufacturing of aluminum matrix composites reinforced with iron-oxide nanoparticles: Microstructural and mechanical properties. Metall. Mater. Trans. B. **45B**(26), 352–362 (2014)
3. Feaugas, X., Clavel, M.: Cyclic deformation behaviour of alpha/beta titanium, micro mechanisms of plasticity under various loading paths. Acta Mater. **45**, 2685–2701 (1997)
4. Yilbas, B.S., Davies, R., Yilbas, Z.: Study into penetration speed during CO_2 laser cutting of stainless steel. Opt. Lasers Eng. **17**, 69–82 (1992)
5. Lopez, V.H., Scoles, A., Kennedy, A.R.: The thermal stability of TiC particles in an Al7 wt%Si alloy. Mater. Sci. Eng. **A356**, 316–325 (2003)
6. Policarpio Ferreira, L.-M., Robert, M.-H., Bayraktar, E., Zaimova, D.: New design of aluminium based composites through combined method of powder metallurgy and thixoforming. Adv. Mater. Res. AMR. **939**(1), 68–75 (2014)
7. Chen, S.: The effects of the high pressure assistant gas flow on high power CO_2 laser cutting. J. Mater. Process. Technol. **88**, 57–66 (2008)
8. Madic, M., Radovanovic, M.: Surface roughness optimization in CO_2 laser cutting by using taguchi method. U.P.B. Sci. Bull. Ser D. ISSN 1454–2358, **75**(1), 97–108 (2013)
9. Miraoui, I., Boujelbene, M., Bayraktar, E.: Analysis of cut surface quality of sheet metals obtained by laser machining: Thermal effects. AMPT J. Adv. Mater. Process. Technol. Taylor & Francis-USA. **1**(3–4), 633–642 (2015)
10. Senthilkumar, V., Bharath,M., Dhanapal, K., Dhinesh Kumaran, M., Gobinath, R.: Analysis and optimization of laser machining parameters. Int. J. Innov. Res. Sci. Eng. Technol. **5**(Special Issue 8), 33–40 (2016)
11. Yilbas, B.S., Yilbas, Z.: Effects of plasma in CO_2 laser cutting quality. Opt. Lasers Eng. **9**, 1–12 (1998)
12. Kursun, A., Bayraktar, E., Enginsoy, H.M.: Experimental and numerical study of alumina reinforced aluminum matrix composites: Processing, microstructural aspects and properties. Compos. Part B Eng. **90**(1), 302–314 (2016)
13. Kassner, M., Kosaka, Y., Hall, J.: Low-cycle dwell-time fatigue in Ti-6242. Metall. Mater.Trans. **30**, 2383–2389 (1999)
14. Yilbas, B.S.: Effect of process parameters on the kerf width during the laser cutting process. Proc. Inst. Mech. Eng. B J. Eng. Manuf. **215**, 1357–1365 (2001)
15. Katundi, D., Ezeddini, S., Bayraktar, E., Miskioglu, I.: Design of new Al – TiN composites through combined method of powder metallurgy and thixoforming. In: AMPT-2016, Kuala Lumpur, 8–11 November 2016
16. Policarpio Ferreira, L.-M., Robert, M.-H., Bayraktar, E.: Production of aluminum / SiC / $NiAl_2O_4$ MMCs by Thixoforming of recycled chips. Int. J. Solid State Phenom. **217–218**(1), 286–293 (2015)
17. Saleh, Y., Margolin, H.: Bauschinger effect during cyclic straining of two ductile phase alloys. Acta Metall. **27**, 533–544 (1979)

Chapter 14
Studying Effect of CO_2 Laser Cutting Parameters of Titanium Alloy on Heat Affected Zone and Kerf Width Using the Taguchi Method

B. El Aoud, M. Boujelbene, E. Bayraktar, S. Ben Salem, and I. Miskioglu

Abstract In laser cutting process, the cut quality is of great importance. The Heat Affected Zone (HAZ), the kerf width (K_w) of laser cut and quality of the cut edges are affected by cutting parameters as well as the work-piece material. In this paper CO_2 laser cutting of 3 mm thick of Titanium alloy sheet grade 5 Ti-6Al-4 V is investigated. The main objectives of the present work are firstly focused on evaluating the effect of laser power (*Pu*), cutting speed (*V*) and gas pressure (*p*) on Heat Affected Zone and Kerf width quality. Secondly, the relationship between cutting parameters and laser process output variables are analyzed and optimized using the Taguchi method. Results indicate that the thickness of the Heat affected Zone increases with the evolution of laser power and decreases with the increase of cutting speed, and the optimum cutting condition was found to be 1 kW for the laser power, 2400 mm/min for the cutting speed and 2 bars for gas pressure. Also, it is underlined that the kerf width is mainly influenced by laser power and cutting speed.

Keywords Laser power • Cutting speed • Heat affected zone • Kerf width • Taguchi method

14.1 Introduction

Titanium alloys are being increasingly used in the aerospace industry due to their high specific strength, excellent corrosion resistance and good high-temperature properties [1]. The low thermal conductivity, low elastic modulus and high chemical activity have made conventional machining of titanium alloy a very hard task [2]. Laser cutting process of titanium alloy leads to alteration of the surface properties of the machined zone. One of the main phenomena is the formation of the heat-affected zone (HAZ) in the vicinity of the erosion front due the heat conduction resulting from the high density laser beam on the workpiece. However, the experimental results for Shanjin et al. [3] showed that the HAZ layer is characterized by micro-acicular martensite grains with a hardness of about 10% above that of the matrix material. And, they have proved that the average energy the high frequency of pulses, the high cutting speed and the high pressure of the assistant gas (argon) have made it possible to acquire thin layers of HAZ. Scintilla et al. [4] presented the results of laser cutting of the Ti 6Al-4 V alloy with a 2 kW fiber laser in continuous mode. Koster and Field [5] found that a wide range of fatigue strengths could be achieved in titanium alloy Ti–6Al–4 V, as a result of different machining processes. They claimed that these variations occurred because of alterations in integrity induced in the material surface as a result of the machining operation. Studies by Rajaram et al. [6] showed that, for cutting the 4130 steel, the speed had a major effect on the width and size of HAZ, in fact, increasing the cutting speed can result decrease of HAZ width. An experimental investigation of laser cutting of polymeric materials was done by Choudhury et al. [7]. They have concluded that the heat affected zone depth is proportional to laser

B. El Aoud
Supmeca – Paris, School of Mechanical and Manufacturing Engineering, St-Ouen, France

University of Tunis El Manar, ENIT, Ecole Nationale d'Ingénieurs de Tunis, Tunis, Tunisia

M. Boujelbene
University of Tunis El Manar, ENIT, Ecole Nationale d'Ingénieurs de Tunis, Tunis, Tunisia

University of Hail, College of Engineering, Hail, Kingdom of Saudi Arabia

E. Bayraktar (✉)
Supmeca – Paris, School of Mechanical and Manufacturing Engineering, St-Ouen, France
e-mail: bayraktar@supmeca.fr

S. Ben Salem
University of Tunis El Manar, ENIT, Ecole Nationale d'Ingénieurs de Tunis, Tunis, Tunisia

I. Miskioglu
Michigan Technological University, ME-EM Department, Houghton, MI, USA

P.R. Thakre et al. (eds.), *Mechanics of Composite and Multi-functional Materials, Volume 6*,
Conference Proceedings of the Society for Experimental Mechanics Series, DOI 10.1007/978-3-319-63408-1_14

Table 14.1 Chemical composition of Ti-6Al-4 V (%)

Ti	Al	V	Fe	C	N	H	O
Base	6	4	0.3	0.08	0.05	0.01	0.2

power and inversely proportional to compressed air pressure and cutting speed. The findings revealed that low surface roughness was achieved with the focal position inside the work piece, which was associated with the wider kerf for improved melt ejection in thick-section metal cutting. CO_2 laser cutting quality of aluminum was investigated by Stournaras et al. [8]. They monitored the cut quality through measuring kerf width, edge roughness, and size of the heat affected zone. Yilbas et al. [9] studied laser cutting of 7050 Aluminum alloy with presence of Al_2O_3 and B_4C in the alloy matrix. They showed that aluminum alloy reinforced with 20% Al_2O_3 resulted in relatively large kerf width size as compared to that corresponding to 7050 aluminum alloy reinforced with 20% B_4C. Eltawahni et al. [10] have investigated the effect of laser cutting parameters on the kerf width and the cut surface roughness of MDF composite material. They have shown that the upper and lower kerf widths decrease as cutting speed increases and they increase as laser power increase. Uslan et al. [11] has investigated the influence of laser power and cutting speed variations on the kerf width size. A lump parameter analysis is introduced when predicting the kerf width size and an experiment is conducted to measure the kerf size and its variation during the cutting process. He has found that the workpiece surface influences significantly the kerf width size. He has also shown that the variation in the power intensity results in considerable variation in the kerf size during the cutting, which is more pronounced at lower intensities. Several investigations were per- formed using systematic approaches to optimize the LBC process. Dubey and Yadava [12] have applied the Taguchi method and the Response Surface Methodology (RSM) method to optimize the LBC process of thin sheet of high silicon-alloy steel, taking into account multi-performance characteristics. Huehnlein et al. [13] have used the design of experiment (DOE) approach to optimize the laser cutting of thin ceramic layers.

The aim of this paper is to study the effect of laser power, cutting speed and gas pressure on the surface integrity and to obtain optimal cutting parameters for minimum on HAZ and Kerf width, while CO_2 laser cutting of Ti-6Al-4 V. Taguchi's parameter design approach is used to accomplish this objective.

14.2 Experimental Condition

Titanium alloy grade Ti-6Al-4 V in sheet form was used as work-piece material. The sheet dimensions were 20 mm × 15 mm with thickness of 3 mm. The cutting operations were performed on a CO_2 laser machine type 4000 TLF TURBO. Chemical composition of this titanium alloy structure is given in Table 14.1.

Cutting specimens were investigated experimentally using output laser power (1–3 kW), cutting speed (480–2400 mm/min) and gas pressure (2–14 bars). Table 14.2 shows cutting parameters and experimental design levels used in this study.

The heat affected zone depths and the kerf width were measured by using optical microscopy. In order to examine the microstructure changes resulting from the laser cutting, we have used scanning electron microscopy SEM.

The Taguchi method and L9 Orthogonal Array were used to reduce number of the experiments. The design of experiments (DOE) and measured HAZ and K_w values are done with Minitab 17 program.

14.3 Results and Discussion

14.3.1 HAZ Depth Analysis

Laser cutting is a thermal process. Most of the heat melts or vaporizes the metallic material and is then blown by the assist gas stream. The rest of the heat is transmitted to the matrix and modifies the microstructure which results in the formation of HAZ layers. Observation by optical microscope shows three different zones (see Fig. 14.1).

- The melted zone (MZ): zone of interaction beam-metal, thickness varies from a few micrometers to a few tens of micrometers, depending on the irradiation parameters.
- The Heat Affected Zone (HAZ): zone located on the edge of the melted zone. The HAZ width is one of the main performances of laser machining. To study the influence of laser beam diameter on the HAZ width, we will measure the average HAZ width from several measurements performed at the surface perpendicular to the cut surface.
- The base metal zone (BZ): the microstructure of this zone has undergone no structural modification.

Table 14.2 Process variables and experimental design levels

Cutting parameters	Symbol	Level 1	Level 2	Level 3
Laser power (kW)	Pu	1	2	3
Cutting speed (mm/min)	V	480	1440	2400
Gas pressure (bars)	p	2	8	14

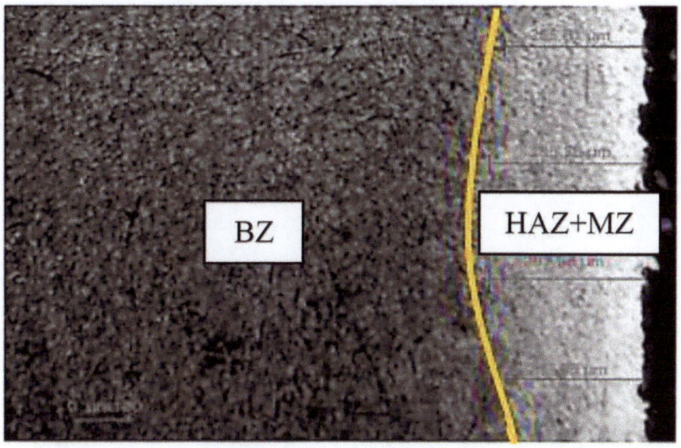

Fig. 14.1 Heat Affected Zone of cut edge at Pu = 2 kW, V = 2400 mm/min and p = 14 bar

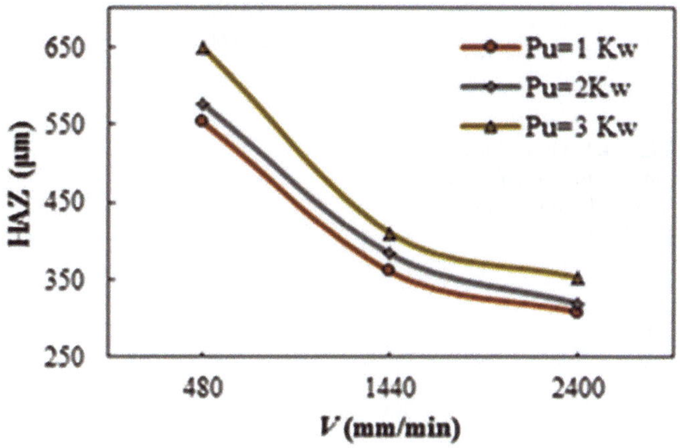

Fig. 14.2 Effect of laser power *Pu* and cutting speed *V* on the HAZ width at *p* = 8 bar

The variation of the heat affected zone width HAZ is investigated as a function of 3 laser machining parameters; Laser power, cutting speed and gas pressure. Figure 14.2 shows the effect of the laser power and the cutting speed on the HAZ width.

The results of the experiment indicate that the laser power and the cutting speed have a very important impact on the HAZ width. As power increases, more energy accumulates on the material, more energy transfers to the matrix, thus the layer of the HAZ becomes thicker. While increasing the cutting speed causes a decrease in energy density and less heat transfer to the matrix, resulting in thin layers of HAZ.

The same relations can be shown with detailed parametric diagrams constructed for the evolution of the HAZ (Fig. 14.3). It indicates that the Heat Affected Zone is proportional to laser power (*Pu*) and inversely proportional to cutting speed (*V*). Table 14.3 indicates that the gas flow under high pressure makes it possible to remove the molten metal from the interaction zone, to obtain a cooling effect on the surface and to reduce the thickness of Heat Affected Zone (HAZ).

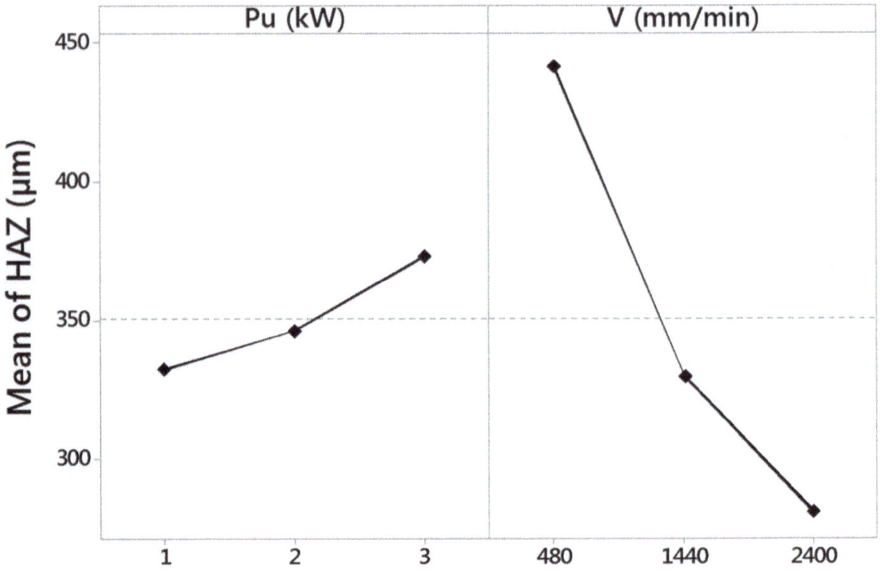

Fig. 14.3 Mean effect of laser power Pu, cutting speed V and gas pressure on the surface roughness

Table 14.3 Effect of the gas pressure on the HAZ for Pu = 2 Kw

	V = 2400 mm/min	*V = 1440 mm/min*	*V = 480 mm/min*
Gas pressure (bar)	HAZ (µm)		
8	320.03	383.45	576.79
14	285.60	328.69	363.63

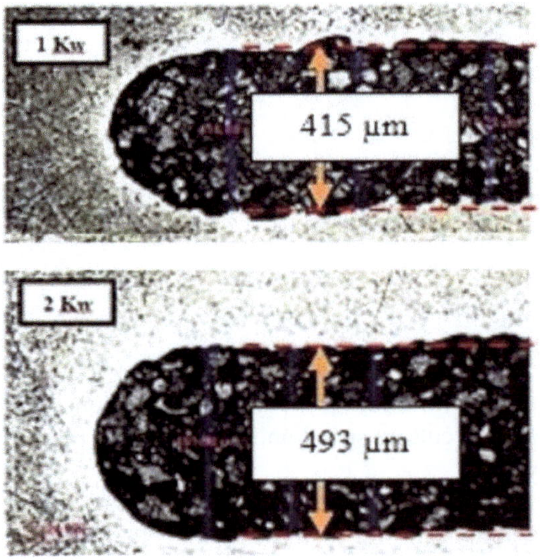

Fig. 14.4 Optical laser cutting micrographs for V = 1440 mm / min and p = 8 bar

14.3.2 Kerf Width Analysis

The cutting width increases with the increasing of laser power (see Fig. 14.4). Indeed, for high power intensities, the molten material remains at high temperature which favors a radial transfer of heat responsible for the increase of the molten zone and consequently the increase in the width of the Kerf. These results confirm the work of B.S. Yilbas et al. [14].

Table 14.4 and Fig. 14.5 show the variation of kerf width with cutting speed. The laser power and assist gas pressure are kept constant.

Table 14.4 Variation of Kerf width with cutting speed at $p = 2$ bar

	$Pu = 1$ kW	$Pu = 2$ kW	$Pu = 3$ kW
Cutting speed in *mm/min*	Kerf width (μm)		
V = *480*	409.29	442.13	453.05
V = *1440*	340.61	424.81	450.81
V = *2400*	333.89	418.05	448.62

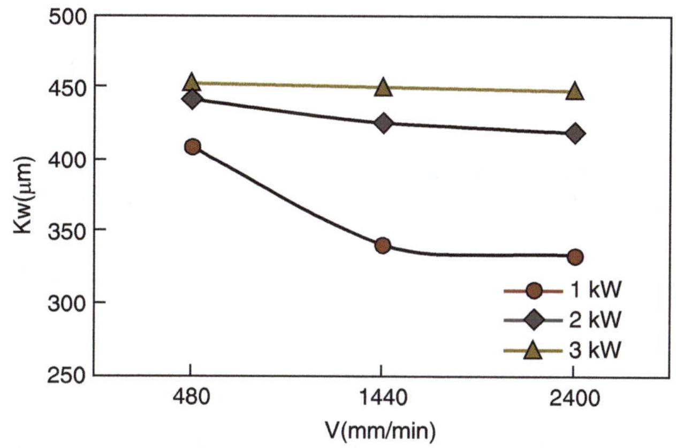

Fig. 14.5 Effect of cutting speed on Kerf width at $p = 2$ bar

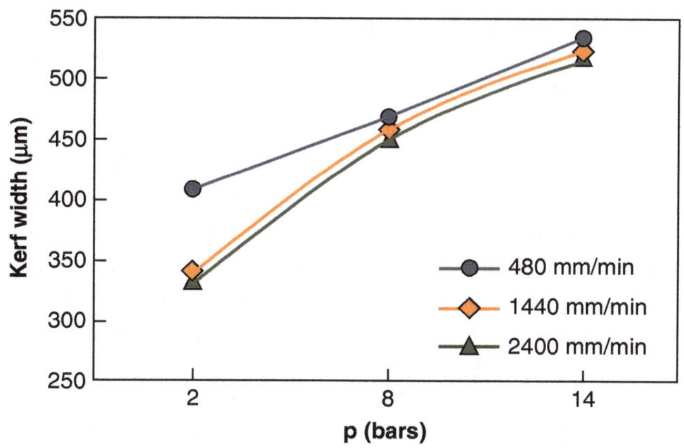

Fig. 14.6 Variation of kerf width with assist gas pressure

The kerf width is found to decrease with the increase in cutting speed. The interaction time decreases between the laser beam and the metal when the cutting speed should be increased. This results in the decrease in the kerf width. However if the cutting speed is made very high the interaction between the laser beam and the metal may not be sufficient to initiate the cut. Also, in laser cutting the cutting speed affects the material removal rate as well as the quality of the cut surface. If the speed is enhanced the laser material interaction becomes less. This may results in the lesser surface damage. The assist gas is used in laser machining primarily to remove the molten metal from the kerf. It impinges on the metal surface co axial to the laser beam. The nozzle shape plays an important role in the effect of assist gas pressure on the kerf width.

Figure 14.6 presents the variation of the kerf width with the assist gas pressure. It is seen that though the kerf width increases with the increase in assist gas pressure, at higher laser powers the kerf is wider but variation is not much. This is because the molten metal expelled from the kerf by the gas jet quickly.

Table 14.5 The results of experiments and S/N ratios values

Experimentalnumber	Pu (Kw)	V (mm/min)	p (bar)	HAZ (μm)	K_w (μm)	HAZ for S/N ratios	K_w for S/N ratios
1	1	480	2	393,01	409.29	−51,88	−52,24
2	1	1440	8	360,88	458,85	−51,15	−53,23
3	1	2400	14	250,75	517,65	−47,98	−54,28
4	2	480	8	576,99	488,02	−55,22	−53,77
5	2	1440	14	328,69	548,09	−50,34	−54,78
6	2	2400	2	236,59	418,05	−47,48	−52,42
7	3	480	14	394,30	580,25	−51,92	−55,27
8	3	1440	2	333,32	450,81	−50,46	−53,08
9	3	2400	8	353,32	489,50	−50,96	−53,79

Table 14.6 S/N response table for HAZ factor

Level	*Pu* (kW)	*V* (mm/min)	*p* (bars)
1	−50.34	−53.01	−49.94
2	−51.01	−50.65	−52.44
3	−51.11	−48.81	−50.08
Delta	0.77	4.20	2.50

14.3.3 Results Analysis Using Taguchi Method

The most essential criterion in the Taguchi method for analyzing experimental data is signal/noise ratio. In this study, the S/N ratio should have a maximum value to obtain optimum cutting conditions, according to the Taguchi method. S/N ratio is used as a measurable value instead of standard deviation because as the mean decreases, the standard deviation also decreases and vice versa. S/N ratios and level values were calculated by using the following equation, "the smaller-the better" in the MINITAB 17 Program.

$$\frac{S}{N} = -10\log\frac{1}{n}\left(\sum y^2\right) \qquad (14.1)$$

where y is the average of observed data, S^2_y the variation of y, n the number of observations, and y the observed data or each type of the characteristics. S/N ratios obtained from this equation are given in Table 14.5.

According to the Taguchi design, the optimum cutting condition is found as −47.48 and −52.24 S/N ratios for HAZ and K_w respectively. The experimental results indicate that the minimum large of HAZ is obtained for laser power of 2 kW, cutting speed of 2400 mm/min and gas pressure of 2 bars. However, the best Kerf width K_w is obtained for laser power of 1 kW, cutting speed of 480 mm/min and gas pressure of 2 bars.

The interpretations can be made according to the level values of *Pu*, *V*, and *p* factors obtained for HAZ given in Table 14.6. The different values of S/N ratio between maximum and minimum are (main effect) also shown in this table.

Figure 14.7 shows the graphic of the level values given in Table 14.6.

The cutting speed and the gas pressure are two factors that have the highest difference between values 4.20 and 2.50 respectively. Based on the Taguchi prediction that the larger different between value of S/N ratio will have a more significant effect on Heat affected Zone (HAZ).

As it seen in Table 14.6, the first level of laser power, the first level of gas pressure and the third level of cutting speed are higher. Consequently, the optimum cutting conditions determined under the same conditions for the experiments to be conducted will be 1 kW for laser power, 2400 mm/min for cutting speed and 2 bars for gas pressure.

The Kerf width values are also obtained using the Taguchi method. The level values obtained from MINITAB 17 Program based on the Taguchi design are given in Table 14.7. The average S/N ratio for every level of experiment is calculated based on the recorded value as shown in Table 14.7.

The gas pressure and the laser power are two factors that have the highest difference between values 2.20 and 0.26 respectively. Thus, it can be concluded that increasing the pressure of assistant gas will increase the K_w significantly.

Figure 14.8 and Table 14.7 show that the first level of laser power, the first level of gas pressure and the third level of cutting speed are higher. Consequently, the optimum cutting conditions determined under the same conditions for the experiments to be conducted will be 1 kW for laser power, 2400 mm/min for cutting speed and 2 bars for gas pressure.

Fig. 14.7 The graphic of mean of S/N ratios for HAZ

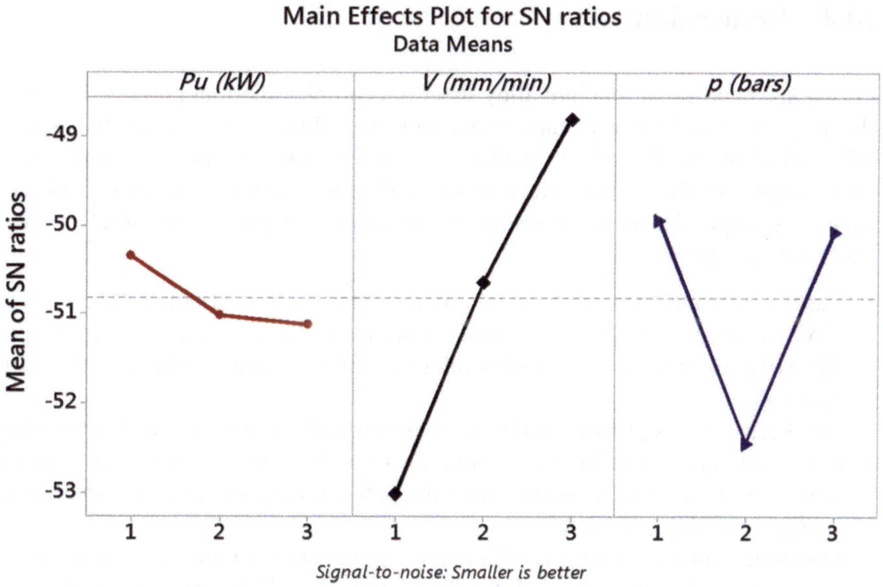

Table 14.7 S/N response table for K_w factor

Level	Pu (kW)	V (mm/min)	p (bars)
1	−53.25	−53.76	−52.58
2	−53.66	−53.70	−53.60
3	−54.05	−53.50	−54.78
Delta	0.80	0.26	2.20

Fig. 14.8 The graphic of mean of S/N ratios for K_w

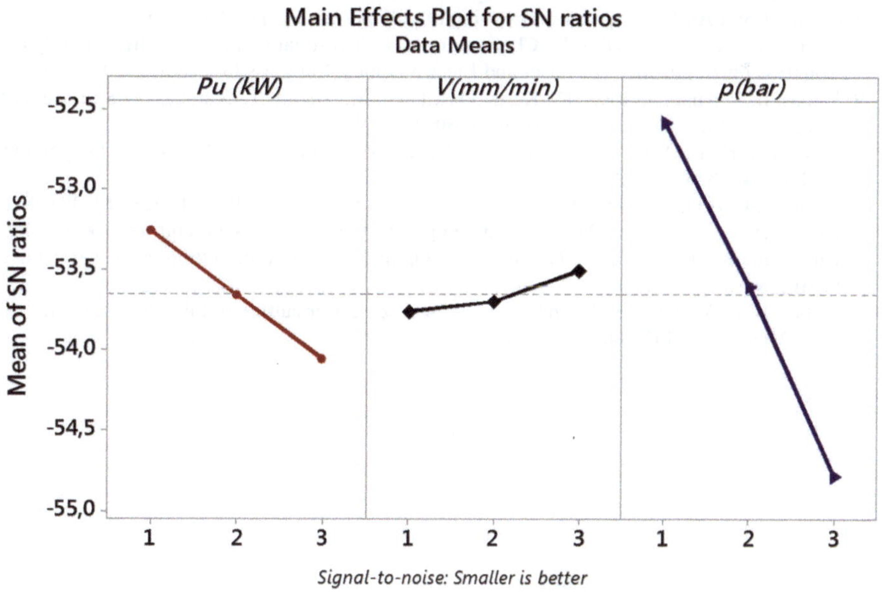

14.4 Conclusion

Pulsed laser cutting of titanium alloy sheet is a complex thermal process. In this paper a CO_2 Laser cutting of Titanium alloy sheets grade Ti-6Al-4 V is carried out and the influence of laser output power, cutting speed and gas pressure on Heat Affected Zone depth and kerf width is analyzed. The resulting cut surfaces are examined using the Scanning electron microscopy (SEM) for heat affected area and optical microscope for kerf of cut edge. Design of experiment approach has been used to predict the relationship between the input process parameters and the responses. The results obtained in this study are given below:

- Laser power and cutting speed had major effect on heat affected zone.
- HAZ depth increases as laser power increases and it decreases as cutting speed increases.
- The HAZ layer is characterized by micro-acicular martensite grains with a hardness of about 22% above that of the matrix material.
- The kerf width is proportional to laser power and gas pressure and inversely proportional to cutting speed.
- HAZ and K_w, S/N ratios were found as a result of experiments conducted according to the L9 orthogonal array. The maximum value was found by using the S/N ratio equation of "the smaller-the better," the maximum S/N ratio yielded optimum cutting parameters.
- Optimum cutting conditions which correspond to maximum S/N value of the smaller surface roughness in CO_2 Laser cutting were found to be 1 kW for the laser power, 2400 mm/min for the cutting speed and 2 bars for gas pressure.

References

1. Hascalɪk, A., Aydas, U.C.: J. Mater. Process. Technol. **190**, 173–180 (2007)
2. Lei, S., Liu, W.: High speed machining of titanium alloys using the driven rotary tool. Int J Mach Tool Manu. **653**, 42–61 (2002)
3. Shanjin, L., Yang, W.: An investigation of pulsed laser cutting of titanium alloy sheet. Opt. Lasers Eng. **44**, 1067–1077 (2006)
4. Scintilla, L.D., Sorgente, D., Tricarico, L.: Experimental investigation on fibre laser cutting of Ti-6Al-4V thin sheet. Adv. Mater. Res. **1281**, 264–265 (2011)
5. Koster, W.P., Field, M.: Effects of machining variables on the surface and structural integrity of titanium. Proceedings of the North American Manufacturing Research Conference, SME, **2**, 67–87 (1973)
6. Rajaram, N., Sheikh-Ahmad, J., Cheraghi, S.H.: CO_2 laser cut quality of 4130 steel. Int J Mach Tool Manu. **43**, 351–358 (2003)
7. Choudhury, I.A., Shirley, S.: Laser cutting of polymeric material: An experimental investigation. Opt. Laser Technol. **42**, 503–508 (2010)
8. Stournaras, A., Stavropoulos, P., Chryssolouris, G.: Investigation of laser cutting quality of aluminum. In: ICALEO 2006 – 25th International Congress on Applications of Laser and Electro-Optics, Congress Proceedings, Art. No. 407 (2006)
9. Yilbas, B.S., Khan, S., Raza, K., Keles, O., Ubeyli, M., Demir, T., Karakas, M.S.: Laser cutting of 7050 alloy reinforced with Al_2O_3 and B_4C composites. Int. J. Adv. Manuf. Technol. **50**, 185–193 (2010)
10. Eltawahni, H.A., Olabi, A.G., Benyounes, K.Y.: Investigating the CO_2 laser cutting parameters of MDF wood composite material. Opt. Laser Technol. **43**, 648–659 (2011)
11. Uslan, I.: CO_2 laser cutting: Kerf width variation during cutting. Proc. I MECH E Part B J. Eng. Manuf. **8**, 571–578 (2005)
12. Dubey, A.K., Yadava, V.: Multi-objective optimization of laser beam cutting process. Opt. Laser Technol. **40**(3), 562–570 (2008)
13. Huehnlein, K., Tschirpke, K., Hellmann, R.: Optimization of laser cutting processes using design of experiments. Phys. Procedia. **5**, 243–252 (2010)
14. Yilbas, B.S., Akhtar, S.S., Bayraktar, E., Gasem, Z.: Laser cutting of thin aluminum and silicon alloy: Influence of laser power on kerf width. Adv. Mater. Res. **445**, 442–447 (2012)

Chapter 15
Fatigue Characterization of In-Situ Self-Healing Dental Composites

D.H. Kafagy, S.S. Khajotia, and M.W. Keller

Abstract The short-term failures of dental composites are a common limitation of these materials. Oral conditions apply fatigue load cycles in the form of chewing and thermal loads, and the damage due to fatigue loads plays a major role in restorative failures in the form of cracks. A lab-based fatigue test is a suitable technique to characterize the crack propagation in dental composites. In this paper, we investigate the ability of a self-healing material to repair or arrest propagating fatigue cracks. In-situ self-healing resin composites were prepared using 15 wt% of activator resin microcapsules and 5 wt% of initiator microcapsules, and equal amounts of 40 wt% each of dental filler and dental resin. Compact Tapered Double Cantilever Beam (cTDCB) specimens of self-healing dental resin composites were prepared by integrating two sets of microcapsules of diameter 45 ± 10 μm containing an acrylate monomer and a polymerization initiator (BPO). Samples were tested in a servo-hydraulic load frame in air at room temperature. Three specimen types are investigated, dental composite without microcapsules, dental composite with non-healing microcapsules, and dental composite with in-situ self-healing microcapsules. The results show that the in-situ self-healing dental composite successfully extended the life of the composite compared to the control samples.

Keywords Fatigue test • Self-healing microcapsules • Dental resin composites • Quasi-static test

15.1 Introduction

The development of fracture mechanics has shown that three key factors control the ability to withstand fatigue loads exposure of a structure to brittle fracture: stress level, crack size, and material toughness. The fatigue characterization is able to provide a good estimate of the material life span. The dental composite is a thermosetting polymer commonly used in the field of restorative dentistry. The lifespan of dental composite restorations depends on, among other factors, articulation and occlusion; these factors cause cyclic loading and wear. These polymers are highly prone to damage in the form of cracks [1].

The exposure of dental composites to cyclic loading is mostly problematic because potential cracks will slowly develop above a threshold of the expected range of stress intensities that is lower than the critical stress intensity. In addition to the cyclic loading, thermal stress has a significant role in lowering the expected lifespan of dental composite restorations [2].

Fatigue in polymers and dental composite have been widely discussed in several studies [2–6]. In most cases, the Paris-Erdogan law was able to precisely describe the progress of fatigue crack rates (da/dN) in polymers. The law is a simplified model of the fatigue crack growth that describes a relationship between the series of the stress intensity factor and the rate of crack growth under a fatigue stress regime. In practice, the Paris law is calibrated to model the linear interval around the midpoint. The law can be used to quantify the residual life in terms of load cycles of a specimen and crack size [7, 8].

The crack propagation behavior can be categorized into three phases: low speed propagation, stable propagation, and high speed propagation. The Paris law can only represent the second phase [5]. According to Azimi et al. (1996); White et al. (2001) and Wetzel et al. (2006), the incorporation of microcapsules, hollow glass spheres or solid particles in brittle thermosetting polymers increased the fracture toughness and improved the fatigue performance [3, 9, 10]. According to Azimi et al. (1996), the addition of various amounts of reactive liquid rubber and glass spheres has improved the resistance of fatigue crack propagation through the whole crack growth regime. This study shows that the amount of shear yielding around the solid spheres was increased and micro-cracking was blocked by particle cavitation.

D.H. Kafagy (✉) • S.S. Khajotia • M.W. Keller
The University of Tulsa, Mechanical Engineering, Tulsa, OK, USA

University of Oklahoma Health Sciences Center, Oklahoma City, OK, USA
e-mail: Dhk542@utulsa.edu

P.R. Thakre et al. (eds.), *Mechanics of Composite and Multi-functional Materials, Volume 6*,
Conference Proceedings of the Society for Experimental Mechanics Series, DOI 10.1007/978-3-319-63408-1_15

Brown et al. (2006), reported that the incorporation of fluid-filled, spherical microcapsules in an epoxy matrix decreased the rate of fatigue crack growth [11].The improved fatigue crack growth resistance was attributed to toughening mechanisms made by the embedded microcapsules that change the morphology of fatigue fracture plane. Also, the incorporation of embedded self-healing microcapsules offered a potential mechanism to further extend the fatigue life of epoxy through self-healing of fatigue damage [11]. Moreover, according to Brown et al. (2004, 2006), the incorporation of self-healing microcapsules induced toughening in the epoxy composite. In Brown's work, both values of the virgin and healed fracture toughness were dependent on the concentration and size of self-healing microcapsules. The average maximum toughness of the epoxy microcapsules composite was higher than the pure epoxy [11, 12]. Wu et al. (2015), reported that the addition of microcapsules up to 7.5 wt% into the dental composite did not harmfully affect the mechanical properties and they achieved significantly high self-healing recovery in dental composite [13].

According to Jones et al. (2007), the fatigue life extension depends on the relative quantity of the chemical kinetics of healing and the mechanical kinetics of crack propagation. When the healing kinetics was accelerated, larger fatigue life extension was realized [1]. According to Brown et al. (2005), when no healing took place at the loads ($K_{max} \geq K_{IC}$) or ($P_{max} \geq P_{IC}$) unless a resting time delay was allowed, above this load, unstable crack growth happened followed by quick failure [5]. In the present work, 50% of the fracture load was considered as a maximum cyclic fatigue load, and no resting period was allowed during the fatigue test.

Mostovoy et al. (1967) introduced the tapered double cantilever beam (TDCB) specimen. A compact version of the TDCB sample was used in the current work (cTDCB). The sample was designed so that the compliance of the sample varies linearly with crack length. The fracture toughness for TDCB specimens is independent of the crack length and depends only on the applied load, which made the TDCB a commonly accepted test for self-healing fracture specimens [11, 14, 15]. Also, side grooves are included to ensure guided crack growth along the centerline of the specimen.

In this work, we present a fatigue characterization study to investigate the impact of incorporating two types of microcapsules in dental composite, these are non-healing microcapsules and in-situ self-healing microcapsules. Three samples have been tested and analyzed using fatigue test at room temperature.

15.2 Methods

15.2.1 Materials and Sample Preparation

Three types of specimens were examined using the fatigue test. These samples were pure dental composite, dental composite with non-healing microcapsules, and in-situ self-healing dental composite. The data was assessed with a cTDCB test, which enabled a crack-length-independent measure of fracture toughness. An experimental resin system was rendered self-healing by the inclusion of a microcapsule system that contained an acrylate resin and a separate capsule system that contained a radical polymerization catalyst. The pure dental composite specimens were made using 40 wt% of dental resin monomer and 60 wt% of an inorganic filler; this composition is commonly used in dental composites [15, 16]. The in-situ self-healing specimens were made by the addition of 10 wt% of healing microcapsules into the composite, an equal amount of filler was removed to keep the dental resin monomer 40 wt% of the composite. Likewise, the dental composite with non-healing microcapsules specimens was made by the addition of 10 wt% of microcapsules into the composite, to examine the impact of fluid-filled microcapsules on the fatigue behavior of dental composite. Urea-formaldehyde microcapsules containing resin monomers or phenol acetate were made with an average diameter of 45 ± 10 μm using the emulsion in-situ polymerization microencapsulation process described by Brown et al. (2003) [17].

Previous research performed by Dailey et al. (2013), on acrylate self-healing, was used as the starting point for developing new healing agent formulations [18]. Four types of monomers that are commonly used in dental composites were considered for the healing chemistry: Tri-methyl Propane Ethoxylate Triacrylate (TMPET), Urethane Di-methacrylate (UDMA), Polyethylene glycol extended UDMA (UDMA PEG), and Bisphenol Ethoxylate Di-acrylate (Bis-GMA). Two other components were also added to the mixture, a solvent phenylacetate (PA) and amine activator 4,4′-methylene-bis (N, N-dimethylaniline) (MBDMA) as a reaction accelerator. BPO was added as a peroxide initiator for free-radical polymerization. A maximum of 9.9 wt% of BPO was dissolved into the solvent phenylacetate to enable encapsulation and delivery of BPO; this rate was constant for all healing agents in the present work.

A 4-part mold was used to fabricate the cTDCB specimens. The mold was fabricated from stainless steel using water-jet cutting to ensure that the mold dimensions were accurate. To fabricate a specimen, the mold was cleaned and lubricated, then the mixture of the dental composite was transferred from the opaque container to the mold inside a light-safe box.

Condensation of each increment of the resin was performed in the light-safe box to avoid premature polymerization of the dental composite. The mold was then removed from the light-safe box, and the specimen was exposed to light from a high-intensity LED dental curing unit (VALO V34596) for 20 s sequentially on each side of the mold. After curing, the specimens were stored in DI water until testing at room temperature. The initial crack was made in the last step before the fatigue test. The initial crack was made using a diamond saw (Buehler IsoMet 1000 Precision Saw) with a 1 mm thick blade. Then the crack tip was sharpened using a new razor blade. After precracking, the specimen was inserted into the pin grips for fatigue testing.

15.2.2 Fatigue Procedure

The experimental fatigue approach was adopted from Brown et al. (2005) and Jones et al. (2007) [1, 5]. Fatigue crack propagation studies were performed using a load frame (MTS 370) with 50 N load cell (Interface SMT1-186). A minimum clearance was allowed between the pin and the sample hole to avoid any noise data during the fatigue test. A sine waveform of frequency 15 Hz was applied with a load ratio ($R = P_{min}/P_{max}$). The test was controlled using a custom LabVIEW virtual instrument. Also, the LabVIEW was used to collect the load and displacement data; these data were analyzed using a statistical analysis software (Origin Pro 2016).

A quasi-static test was performed before each fatigue test using a specimen fabricated from the same batch of material as the fatigue specimen to eliminate errors due to batch-to-batch variation. The fracture value determined from this test was used to set the maximum load for the fatigue test. The maximum fatigue load was 50% of the maximum quasi-static fracture load. The minimum cyclic load was 10% of the maximum fatigue load. The load ratio R was always maintained at 0.1. The starting segment command was at the rate of 1 N/s. Crack lengths were measured optically and calibrated with the specimen compliance. Then, a linear relationship between sample compliance and crack length was used to estimate the crack-tip position and determine the crack length during the test.

15.3 Results and Discussions

Figure 15.1 shows typical load vs displacement curves of the quasi-static test of pure dental composite and dental composite with 10 wt% in-situ microcapsules. The average fracture load was determined by testing five samples of both sets. The results show that most samples broke after approximately 1 mm of displacement. The sample fracture load was represented by the highest load reading in the y-axis. The slope of the curve represents the material stiffness. For this particular set, the average fracture load of the pure dental composite was about 30 N, and for the in-situ dental composite was 36 N. The fracture load of the in-situ self-healing dental composite was about 20% higher than pure dental composite. This result was expected due to the microcapsules toughening mechanism. According to several previous studies that reported toughening mechanism in different types of thermosetting polymers, the incorporation of microcapsules in brittle thermosetting polymers increased the fracture toughness [1, 3, 9]. In addition, insignificant differences of quasi-static fracture load were observed between the non-healing and in-situ self-healing virgin dental composite specimens. However, up to 40% of healing efficiency was achieved in the in-situ self-healing specimens.

Figure 15.2 shows the fatigue test results for three specimens, a dental composite, a dental composite with 10 wt% microcapsules, and a dental composite with 10% in-situ self-healing microcapsules. Both samples that contained microcapsules showed improved fatigue performance in terms of number of fatigue load cycles. By considering the pure dental composite as a reference fatigue curve, the life cycle of the non-healing fatigue curve was about 9.6% larger than the pure dental composite. Also, the life cycle of the in-situ self-healing dental composite sample was about 32% greater than the pure dental composite. These results imply that the microcapsule is toughening the dental composite and improving the fatigue life cycle. Also, these results agreed with our quasi-static results in Fig. 15.1 and with previous studies [1, 3, 9, 11]. The life cycle of the in-situ self-healing dental composite was approximately 21% greater than the non-healing dental composite. This result indicates that self-healing mechanism is involved in addition to the microcapsules toughening mechanism. Otherwise, both fatigue curves of the non-healing composite and the in-situ self-healing composite were anticipated to be equivalent. According to the Paris fatigue model, the fatigue results show a retardation of fatigue crack growth was achieved by improved toughness of the composite system in both samples of non-healing and in-situ healing dental composite. This finding is in agrement with previous investigations by Jones et al. (2007) and Brown et al. (2006) [1, 5].

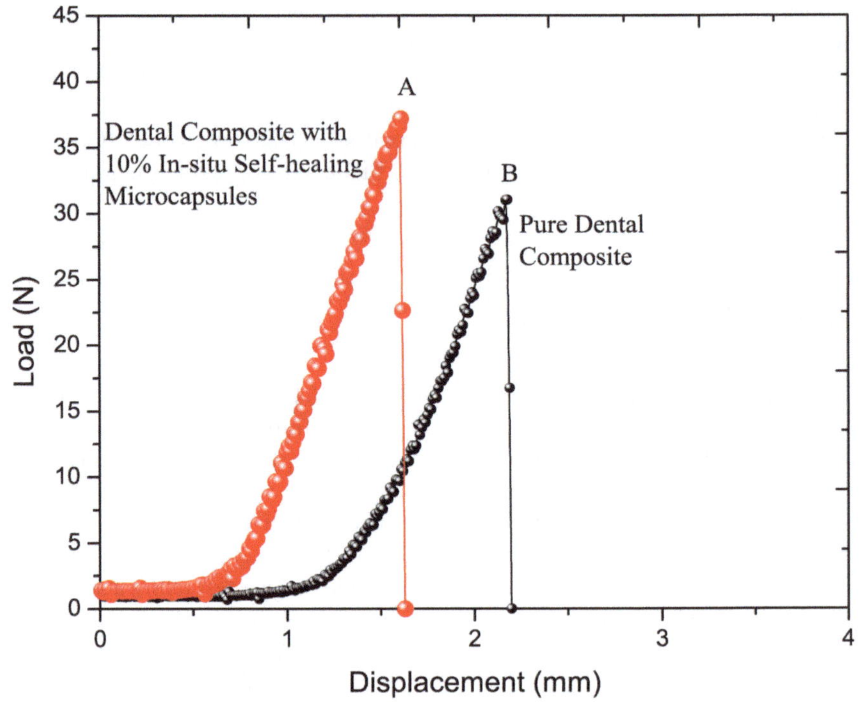

Fig. 15.1 Quasi-static fracture test of a typical cTDCB dental composite. Curve A 10% in-situ self-healing microcapsules. Curve B pure dental composite

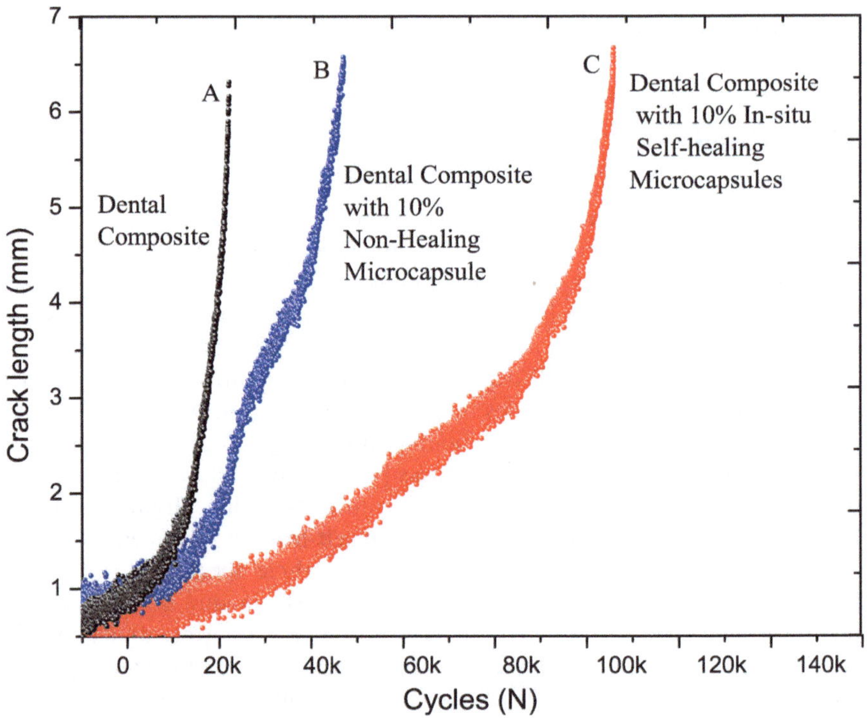

Fig. 15.2 Crack length vs. fatigue cycles at $P_{max} = 15$ N, $P_{min} = 1.5$ N, $R = 0.1$, and Freq. = 15 Hz. Curve A pure dental composite. Curve B dental composite with 10% non-healing microcapsules. Curve C dental composite with 10% in-situ self-healing microcapsules

15.4 Conclusions

Fatigue experiments were successfully performed to study the mechanisms of fatigue crack growth delay and arrest in a microcapsule-containing dental composite. Fatigue life extension was realized by the addition of non-healing and in-situ self-healing microcapsules. A retardation of fatigue crack growth is likely due to both the enhanced toughness of the composite system and the action of the self-healing mechanism. A self-healing system based on encapsulated healing components was able to extend the fatigue life of dental composite.

References

1. Jones, S., Rule, J.D., Moore, J.S., Sottos, N.R., White, S.R.: Life extension of self-healing polymers with rapidly growing fatigue cracks. J. R. Soc. Interface. **4**(13), 395–403 (2007)
2. Studart, A.R., Filser, F., Kocher, P., Lüthy, H., Gauckler, L.J.: Cyclic fatigue in water of veneer-framework composites for all-ceramic dental bridges. Dent. Mater. **23**(2), 177–185 (2007)
3. Azimi, H.R., Pearson, R.A., Hertzberg, R.W.: Fatigue of hybrid epoxy composites: Epoxies containing rubber and hollow glass spheres. Polym. Eng. Sci. **36**(18), 2352–2365 (1996)
4. Bergner, F., Zouhar, G.: New approach to the correlation between the coefficient and the exponent in the power law equation of fatigue crack growth. Int. J. Fatigue. **22**(3), 229–239 (2000)
5. Brown, E.N., White, S.R., Sottos, N.R.: Retardation and repair of fatigue cracks in a microcapsule toughened epoxy composite – part II: In situ self-healing. Compos. Sci. Technol. **65**(15–16 SPEC. ISS.), 2474–2480 (2005)
6. Drummond, J.L.: Degradation, fatigue, and failure of resin dental composite materials. J. Dent. Res. **87**(8), 710–719 (2008)
7. Paris, P.C., Gomez, M.P., Anderson, W.E.: A rational analytic theory of fatiuge. Trend Eng. **13**(4), 9–14 (1961)
8. Janssen, R.P.M., De Kanter, D., Govaert, L.E., Meijer, H.E.H.: Fatigue life predictions for glassy polymers: A constitutive approach. Macromolecules. **41**(7), 2520–2530 (2008)
9. White, S.R., Sottos, N.R., Geubelle, P.H., Moore, J.S., Kessler, M.R., Sriram, S.R., Brown, E.N., Viswanathan, S.: Autonomic healing of polymer composites. Nature. **409**(6822), 794–797 (2001)
10. Wetzel, B., Rosso, P., Haupert, F., Friedrich, K.: Epoxy nanocomposites – Fracture and toughening mechanisms. Eng. Fract. Mech. **73**(16), 2375–2398 (2006)
11. Brown, E.N., White, S.R., Sottos, N.R.: Fatigue crack propagation in microcapsule-toughened epoxy. J. Mater. Sci. **41**(19), 6266–6273 (2006)
12. Brown, E.N., White, S.R., Sottos, N.R.: Microcapsule induced toughening in a self healing polymer composite. J. Mater. Sci. **39**(2003), 1703–1710 (2004)
13. Wu, J., Weir, M.D., Melo, M.A.S., Xu, H.H.K.: Development of novel self-healing and antibacterial dental composite containing calcium phosphate nanoparticles. J. Dent. **43**(3), 317–326 (2015)
14. Jo Ripling, E., Mostovoy, S., Crosley, P.B.: Use of crack-line-loaded specimens for measuring plane-strain fracture toughness. J. Mater. **2**, 664–681 (1967)
15. Beres, R.: A tapered double-cantilever-beam specimen designed for constant-K testing at elevated temperatures. J. Test. Eval. **25**, 536–542 (2009)
16. Braem, M., Finger, W., Van Doren, V.E., Lambrechts, P., Vanherle, G.: Mechanical properties and filler fraction of dental composites. Dent. Mater. **5**(5), 346–349 (1989)
17. Brown, E.N., Kessler, M.R., Sottos, N.R., White, S.R.: In situ poly(urea-formaldehyde) microencapsulation of dicyclopentadiene. J. Microencapsul. **20**(6), 719–730 (2003)
18. Dailey, M.M.C., Silvia, A.W., Mcintire, P.J., Wilson, G.O., Moore, J.S., White, S.R.: A self-healing biomaterial based on free-radical polymerization, **1**, 1–9 (2013)

Chapter 16
Effect of Process Induced Stresses on Measurement of FRP Strain Energy Release Rates

Brian T. Werner, Stacy M. Nelson, and Timothy M. Briggs

Abstract Fiber reinforced polymer composites are frequently used in hybrid structures where they are co-cured or co-bonded to dissimilar materials. For autoclave cured composites, this interface typically forms at an elevated temperature that can be quite different from the part's service temperature. As a result, matrix shrinkage and CTE mismatch can produce significant residual stresses at this bi-material interface. This study shows that the measured critical strain energy release rate, G_c, can be quite sensitive to the residual stress state of this interface. If designers do not properly account for the effect of these process induced stresses, there is danger of a nonconservative design. Tests including double cantilever beam (DCB) and end notched flexure (ENF) were conducted on a co-cured GFRP-CFRP composite panel across a wide range of temperatures. These results are compared to tests performed on monolithic GFRP and CFRP panels.

Keywords Composites • Fracture • Residual stress • Cohesive zone • Analysis

16.1 Introduction

Fiber reinforced polymer composite structures are well suited for the tailoring of material properties. Options available to the designer include various fibers, matrices, orientations, and fabric weaves. Often the most efficient structure, both economically and structurally, combines glass fiber reinforced polymer (GFRP) with carbon fiber reinforced polymer (CFRP) into a hybrid structure. While the same matrix may be used, the fibers themselves have greatly different properties. In this study, the focus is placed on both stiffness and coefficient of thermal expansion. Due to these differences, significant residual stresses can develop either when cured at an elevated temperature, such as in an autoclave, or when the structure's service temperature differs from the material's stress free temperature.

Over the past few years, the cohesive zone element and associated traction-separation law have gained prominence when designers employ a fracture based failure criterion for delaminations [1–4]. As long as linear elastic fracture mechanics holds (LEFM), the most important parameter in the traction-separation law is the strain energy release rate, G. Determining the strain energy release rate, G, for a monolithic fiber reinforced polymer is relatively straightforward and follows a standard experimental procedure. However, once significant residual stresses are present along the interface in question, the coupled effect of residual stresses and fracture toughness can produce uncertainty [5, 6]. If an apparent toughening due to the presence of residual stresses is modeled as a higher strain energy release rate, the condition for a nonconservative design arises which can lead to unexpected failure.

By investigating the effect of residual stress in a hybrid specimen across a wide temperature range, the ultimate goal is to decouple the question of the residual stress state and applied loading to develop a more robust cohesive zone or localization element. Perhaps designers can take advantage of an apparent toughening due to residual stress to develop more efficient structures.

All of the testing was performed at three temperatures: subambient (−54 °C), room temperature (+20 °C), and elevated (+71 °C). While all of the testing remained within the glassy regime for the epoxy matrix, the wide temperature range produced significantly different levels of residual stress at the interlaminar interface of interest.

B.T. Werner (✉) • S.M. Nelson • T.M. Briggs
Sandia National Laboratories, 7011 East Avenue, Livermore, CA 94550, USA
e-mail: btwerne@sandia.gov

P.R. Thakre et al. (eds.), *Mechanics of Composite and Multi-functional Materials, Volume 6*,
Conference Proceedings of the Society for Experimental Mechanics Series, DOI 10.1007/978-3-319-63408-1_16

16.2 Material

Three panel types were fabricated from two materials, a carbon/epoxy (CFRP) 8-harness satin weave fabric prepreg and a glass/epoxy (GFRP) 8-harness satin weave prepreg. The panels consisted of GFRP bonded to GFRP, $[(0/90)_{7s}]_s$, CFRP bonded to CFRP, $[(0/90)_{4s}]_s$, and GFRP bonded to CFRP, $[(0/90)_{10s}{}^{glass}/(0/90)_{5s}{}^{carbon}]$. All three laminates were laid up to produce two symmetric beams with an equivalent bending stiffness and a Teflon strip was embedded along one edge at the warp-to-warp interface between the two beams to produce a precrack. The prepreg layup was co-cured using a standard cure schedule that has been developed for this material [7]. This produced a material with V_f of 0.60 and E_{11} of 30 GPa for the GFRP and 60 GPa for the CFRP.

The panels were cut into specimens approximately 25 mm wide and 200 mm long with the Teflon precrack at one end using a water cooled diamond tile saw. The initial length of the precrack was approximately 38 mm long. The total thicknesses of each specimen were 6.82 mm for the GFRP/GFRP and 7.68 for the CFRP/CFRP bulk material specimens, and 8.75 mm for the GFRP/CFRP hybrid specimens. For the DCB and asymmetric-DCB (ADCB) specimens, hinges were bonded to each face of the precracked end using a secondary bond. The surfaces of both the specimen and the hinge were abraded lightly with coarse grit sandpaper to improve adhesion and two different epoxies were used. For the room temperature and subambient testing a high strength epoxy (Hysol 9430) was used but for the elevated temperature testing a high temperature structural adhesive had to be employed (Hysol 9394).

16.3 Experimental Procedure

The experiments were carried out using an Instron 5989 electromechanical load frame using 2kN and 10kN load cells and an environmental chamber. The three temperatures investigated were +71 °C, +25 °C, and −54 °C. These temperatures provide a wide range of ΔT's and, with that, significantly different residual stress states in the bi-material specimens. A type K thermocouple was attached to the surface of the specimen to monitor its temperature. Each specimen was allowed time for the temperature to equilibrate before testing was initiated. Prior to starting the test, the specimen was held in load control at a low force to allow the fixturing to expand or contract with the temperature without applying significant load to the specimen. The load cell was also held outside of the environmental chamber with a fan blowing on it to ensure that there was minimal thermal drift in the load reading.

16.3.1 DCB Testing

DCB tests were performed in accordance with ASTM standard D5528–13 [8]. Initial tests used a laser extensometer to track the crack opening displacement but after comparison with the stroke measurement, little error was found due to frame compliance at the low loads applied. Each specimen was tested at a constant displacement rate of 1 mm/min (Fig. 16.1).

Crack lengths and the applied energy release rate were calculated using elastic beam theory to back out the current crack length from the measured load-displacement data. For the bulk material the equations used were:

$$a = \left[\frac{\delta E b h^3}{8P}\right]^{\frac{1}{3}} \tag{16.1}$$

$$G_{app} = \frac{12P^2a^2}{Ebh^3} \tag{16.2}$$

For the asymmetric DCB, Brown et al., derived the beam theory solution for the crack length, a, and applied energy release rate at the crack tip, G_{app} [9]:

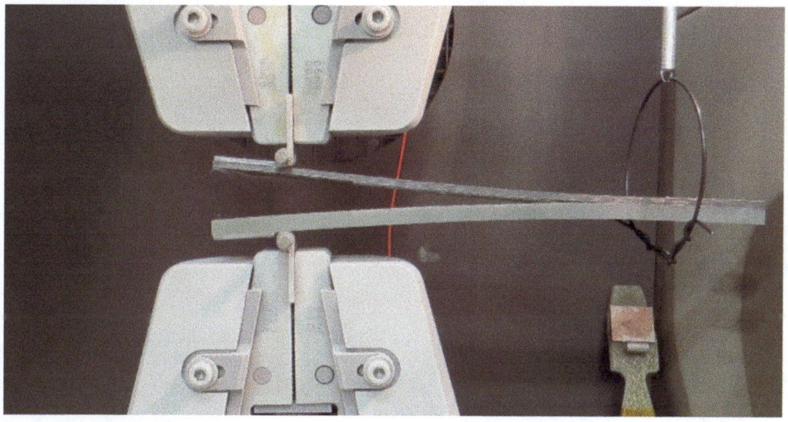

Fig. 16.1 DCB test setup

$$a = \left[\frac{\delta b}{16P}\left(E_1 h_1^3 + E_2 h_2^3\right)\right]^{\frac{1}{3}} \tag{16.3}$$

$$G_{app} = \left[\frac{27\delta P^4}{2b^4}\left(\frac{1}{E_1 h_1^3} + \frac{1}{E_2 h_2^3}\right)\right]^{\frac{1}{3}} \tag{16.4}$$

where P = load
b = width of DCB
d = crack opening displacement
E_i = Young's modulus of material *i*
h_i = thickness of laminate *i*

16.3.2 ENF Testing

For fabric composites there is no ASTM standard for the determination of Mode II energy release rate. For unidirectional composites it is ASTM D7905/D7905M-14 [10] which is the ENF specimen (Fig. 16.2), however when switching to a fabric composite there are additional complications that can make accurate determinations of energy release rate more difficult. The main problem is friction in the pre-cracked section of the specimen. As the beam is loaded, the cracked faces rub against one another which dissipates energy and can lead to error by increasing the apparent energy release rate. This is further complicated by the fact that the ENF test is unstable for initial crack lengths shorter than 0.69 L where the specimen has a total span of 2 L. With such a large length, the precracked area that is susceptible to friction is quite large. Some investigators have balanced this by performing the test with shorter crack lengths which leads to unstable crack growth [11]. This may also produce experimental error as the energy dissipation calculated by the load displacement curve cannot account for the true energy released (Fig. 16.3).

The load displacement curve from the experiment follows the solid line in Fig. 16.3, but the actual energy released follows the dashed line. The experimental error is equivalent to the size of the lobed region to the left of the rapid crack growth line. Some researchers prefer the End Loaded Split (ELS) test that uses a cantilever beam approach with the precracked section at the free end. By pulling up on the free end from both the top and bottom adherend with the same load, the problem of friction is solved. The test is also inherently stable. However, for this study the ENF test was pursued to compare with legacy data. An issue also arises from the overhang on the support roller on the precracked end of the ENF specimen. Despite the beam being unloaded outside of the support rollers, energy may still be dissipated due to friction from contact. For all of these reasons an initial study was performed to determine the ideal test geometry to produce consistent results. Parameters such as temperature (20 °C) and orientation (GFRP up) were held constant while initial crack length and span were varied. With a span of 2 L set at 102 mm, 127 mm, or 152 mm, the initial crack length was varied from 0.3 L, 0.5 L, and 0.69 L. Testing of a minimum of three specimens under each condition was performed to provide some statistical value. Overhang was not considered during this round of testing, but was measured to determine its effect after the tests (Fig. 16.4).

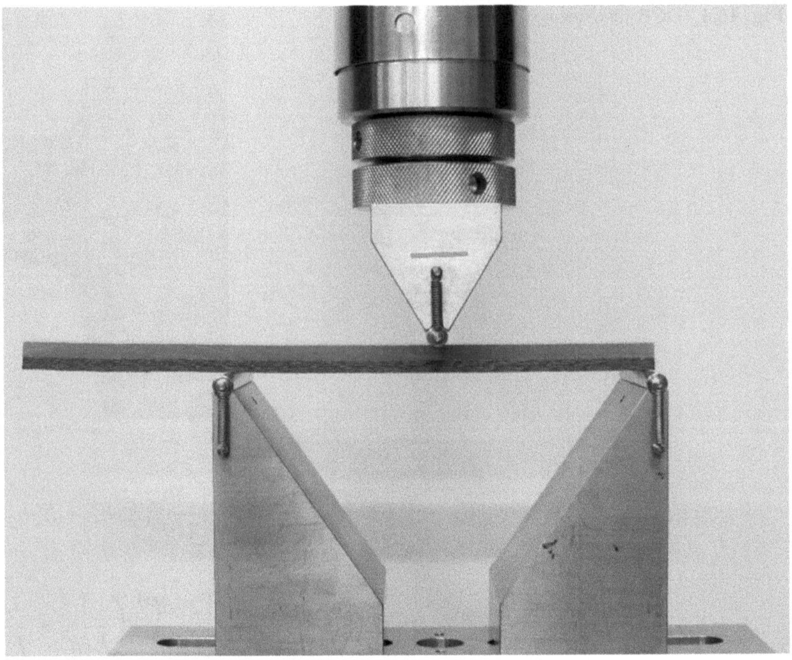

Fig. 16.2 ENF test setup

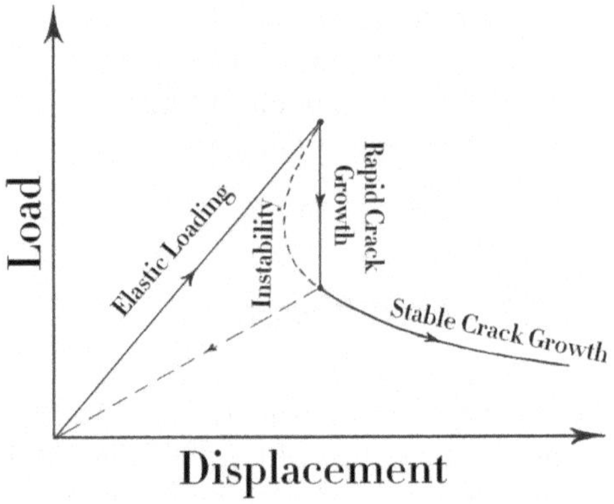

Fig. 16.3 Instability of ENF test [12]

Where the test had similar overhang lengths, no conclusions could be drawn to its effect. However, for tests that had varying overhangs for the same test condition, the effect is quite clear qualitatively. The longer the overhang protrudes from the support roller on the specimen's cracked end, the greater the frictional losses and the more the energy release rate is overestimated. All of the dashed lines in Fig. 16.4 trace back to approximately 1100–1200 J/m^2.

Each test was performed in the Instron 5989 electromechanical load frame. The displacement was measured using a laser extensometer to remove the effect of system compliance. The specimens were loaded at a displacement rate of 1 mm/min and when the crack extended to approximately the midspan they were unloaded at the same rate. The total energy dissipated was obtained by integrating the load displacement curve. The displacement may not have unloaded to 0 as the curvature of the beam changed once residual stress was released. To measure the crack area, the transparency of the GFRP was taken advantage of. The initial crack area could be easily seen and measured before and after testing. A reference line was drawn across the width of the specimen in the precracked region. A Keyence VHX-5000 digital microscope on 30× zoom and image stitching was then used to measure the crack area in the software. This microscope was calibrated using a NIST traceable calibration scale to produce accurate measurements. This area was then compared to the area post-test to determine the total crack amount of crack growth (Fig. 16.5). The quotient of the two values, energy dissipation and crack growth area were then used to determine energy release rate, G.

For the unstable crack growth specimens like that shown in Fig. 16.5, there is a steep load drop once the crack extends. In the stable crack growth specimens the load drops more gradually until it reaches the load roller in which case it begins to pick

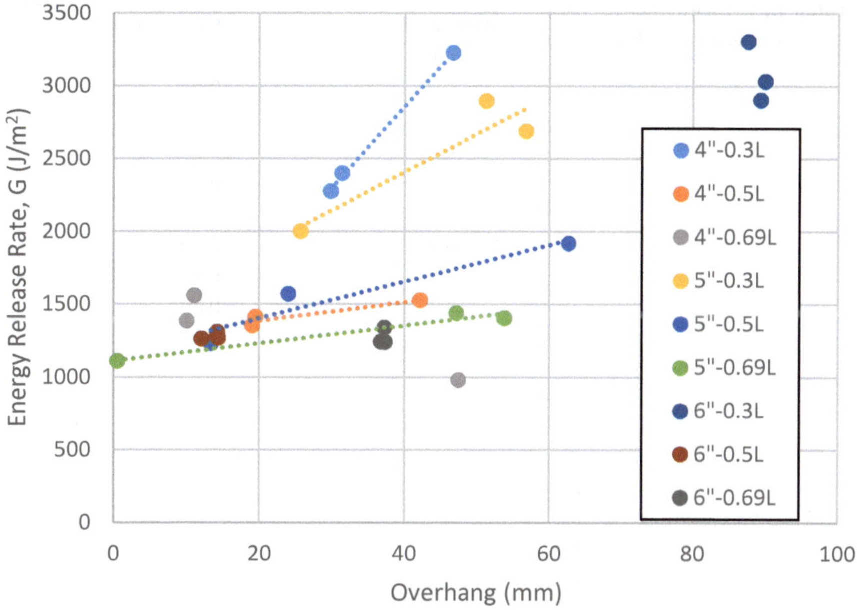

Fig. 16.4 Overhang effect on energy release rate

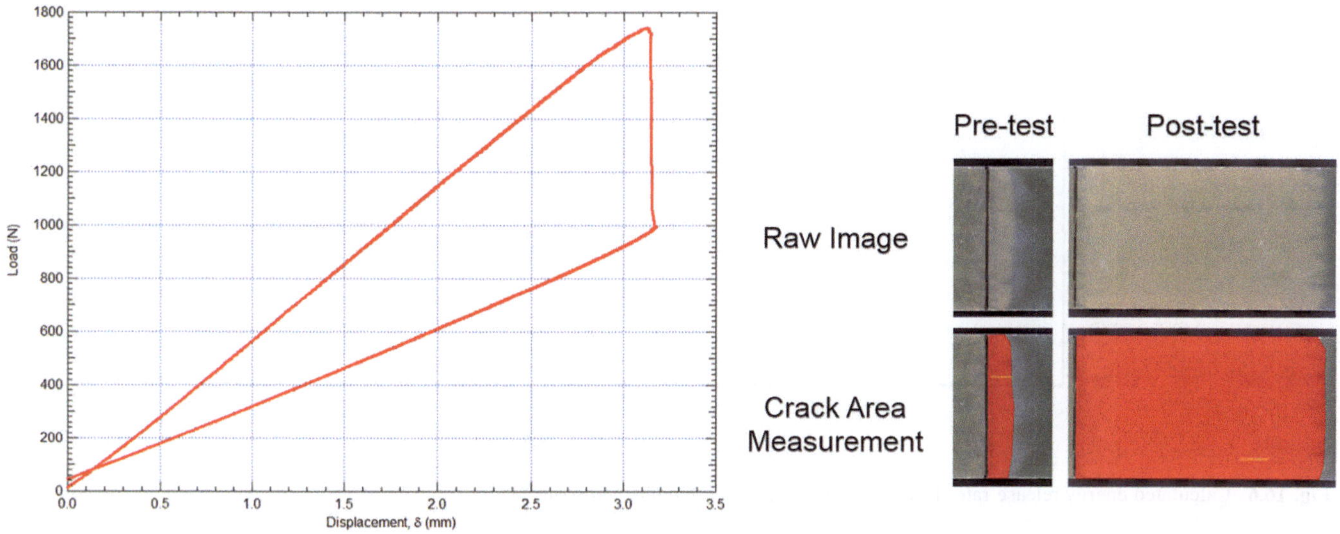

Fig. 16.5 Load-displacement curve of a typical ENF test (*left*) and crack area measurement images (*right*)

back up again. It is at that minimum load at which the specimen is unloaded. The results of the test geometry study are shown in Table 16.1 below.

The short crack length of 0.3 L produced the most energy dissipation and was so unstable the crack tended to grow beyond the load roller. It was also more likely to produce what appeared to be multiple cracks in interlaminar regions away from the interface of interest. The other two crack lengths produced similar results while the 0.5 L crack length was more consistent than the 0.69 L crack length due to its lower coefficient of variation. Either would work fine for a study though. The 0.5 L crack length was chosen for its repeatability, lower friction, larger area of crack growth, and less of an impact to the calculated energy release rate from the variations in the stress field due to Hertzian contact as the crack approached the load roller. The largest span was also chosen because it results in lower applied loads, a larger precracked area to carry those loads, and thus lower frictional losses. A plot tabulating the results in Table 16.1 is shown in Fig. 16.6.

For all of the following testing at the various temperatures and orientations, the span had a length of 152 mm and the specimen had a precrack length of 37 mm. While a Teflon strip was used to produce the precrack during the cure, the crack was manually extended with the use of a razor blade wedging open the precracked side to form a sharp crack tip for the testing. For the bulk material testing, the tests were limited to the GFRP-GFRP specimens as the crack surfaces could be

Table 16.1 Energy release rates for various test geometries

Span (2 L)	Crack length	G Avg (J/m^2)	COV (%)
102 mm	0.3 L	2636	16.1
	0.5 L	1434	5.0
	0.69 L	1308	18.6
127 mm	0.3 L	2532	15.1
	0.5 L	1575	8.8
	0.69 L	1319	25.2
152 mm	0.3 L	3082	5.4
	0.5 L	1282	1.6
	0.69 L	1273	3.7

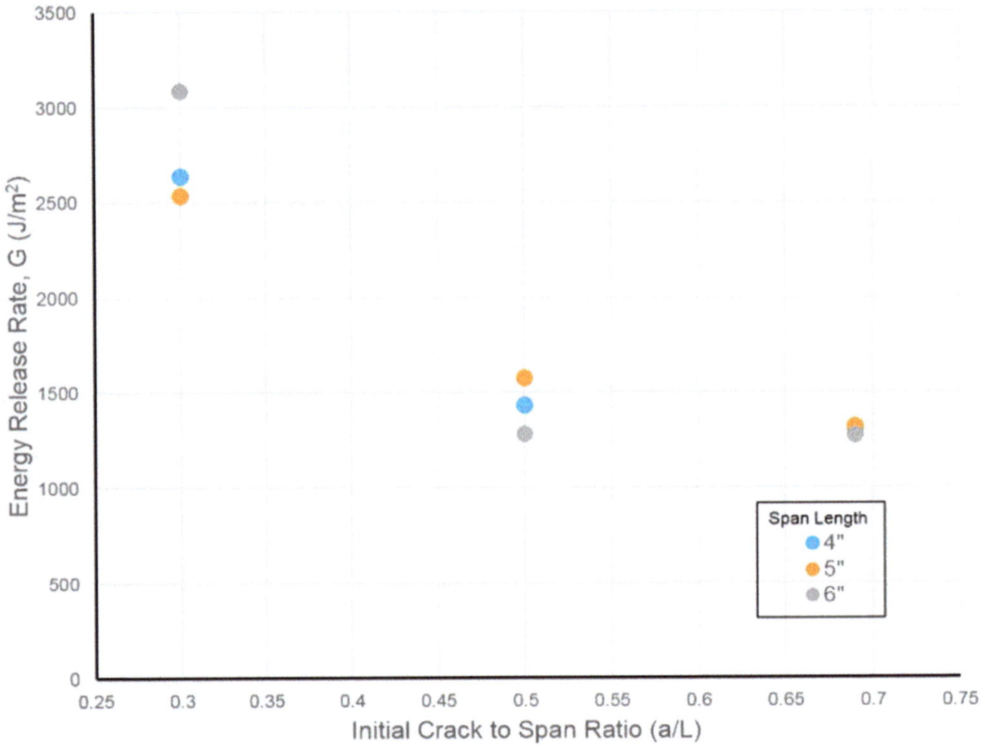

Fig. 16.6 Calculated energy release rates from varying initial crack lengths and spans

inspected visually. The bi-material specimens were tested at all three temperatures in two orientations, with the load roller pressing on the CFRP and with the load roller pressing on the GFRP.

16.4 Results

16.4.1 DCB Results

The bulk material GFRP-GFRP and CFRP-CFRP specimens showed no temperature dependence on the energy release rate as shown in Fig. 16.7. The CFRP-CFRP specimens do display a slight rising resistance curve at the ambient and elevated temperatures. This may appear to be evidence of fiber bridging, like in unidirectional composites, but this phenomenon was not observed during the testing. The GFRP has a critical energy release rate of approximately 280 J/m^2 while the CFRP varied from 280–320 J/m^2. The similarity is not surprising as they both contain the same matrix. The GFRP may be slightly less tough as the fracture surface appeared to fail at the fiber-resin interface as there was dry fiber on one of the fractured

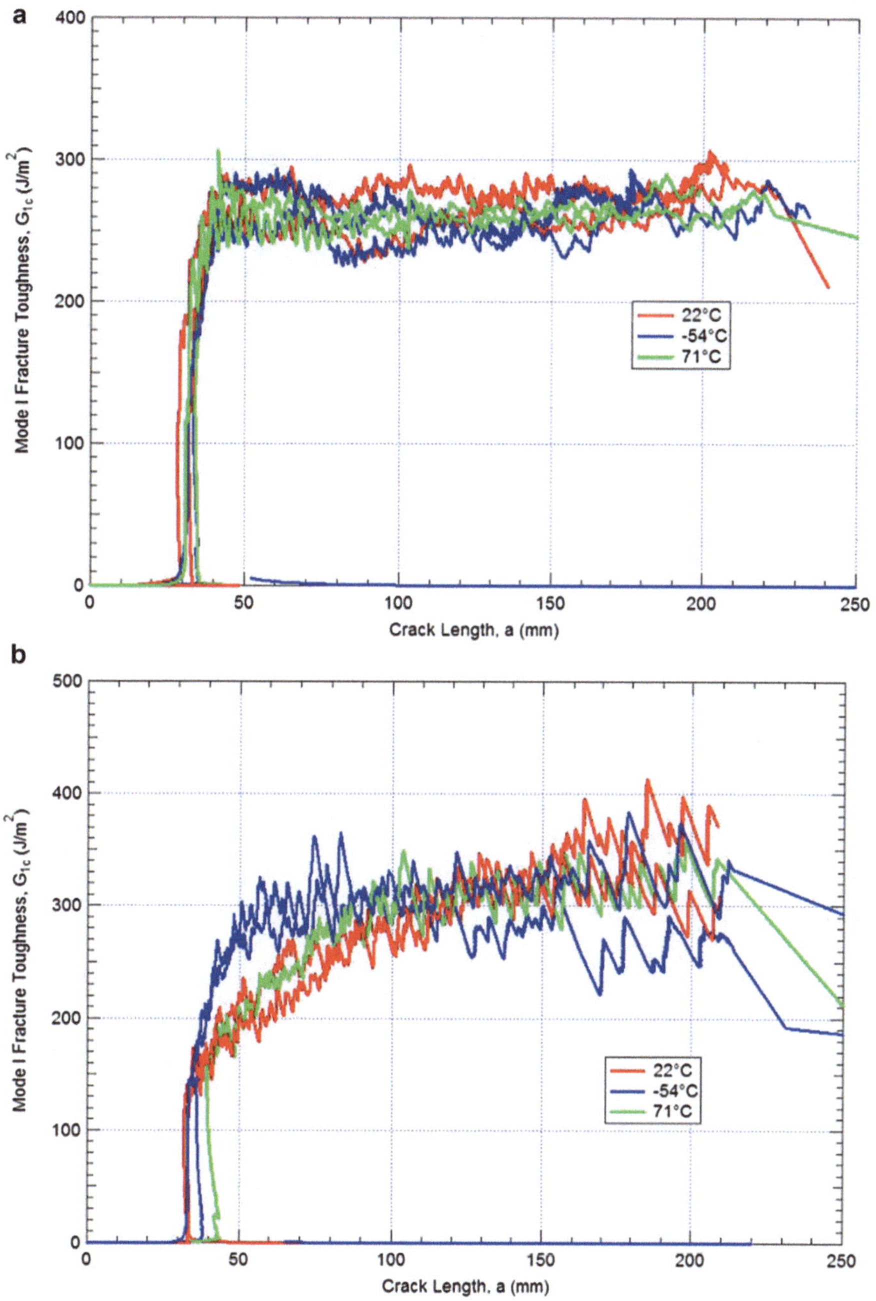

Fig. 16.7 Energy release rate vs crack length for (**a**) GFRP/GFRP and (**b**) CFRP/CFRP laminates

surfaces. The CFRP also presented more of a skip start behavior but this was likely due to larger resin rich regions since it has a coarser weave than the GFRP.

Unlike the bulk material DCB tests, the ADCB shows a significant temperature dependence (Fig. 16.8). As the temperature drops, the ΔT increases which leads to higher residual stresses at the bi-material interface. Since higher residual stresses are correlated with higher apparent strain energy release rate it is assumed that there is a complex stress state at the crack tip. With the help of computational methods, the stress fields at the crack tip can be investigated further to determine whether it is simply the residual stresses producing a stress state that is counter to the applied loading or perhaps the in-plane shear stresses along the interface are producing mixed-mode behavior. It is also worth pointing out that in the bulk material DCB tests, the critical energy release rate for the GFRP was slightly under 300 J/m^2 and the CFRP is between 250 and 350 J/m^2. However, in the ADCB tests, the elevated temperature, least tough experiment showed a critical energy release rate of around 400 J/m^2, higher than that of any of the bulk material tests. This further suggests that the small amount of residual stress even at elevated temperature is beneficial to the crack propagation resistance.

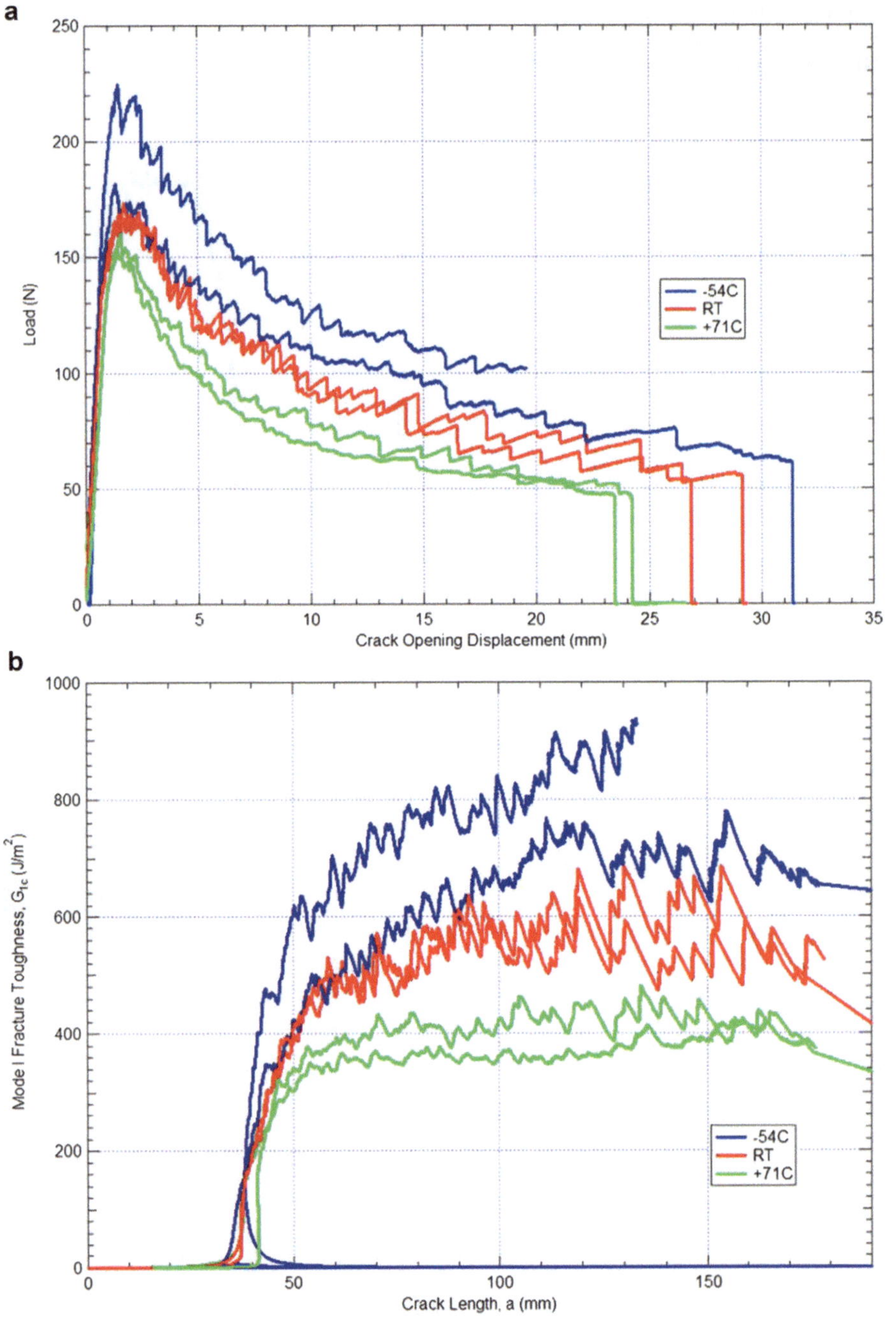

Fig. 16.8 (**a**) Load-displacement and (**b**) energy release rate vs crack length for ADCB laminate

When investigating the fracture surfaces, a change in the mode of crack propagation is seen (Fig. 16.9). At the subambient temperature, all of the resin is pulled off of the CFRP, showing dry fibers. At the elevated temperature, all of the resin is pulled off of the GFRP, showing dry fibers. At room temperature, the failure appears to be more cohesive within the resin layer.

Without accounting for the effect of residual stresses, the bi-material interface can appear significantly tougher. This can lead to a structural design that is nonconservative.

16.4.2 ENF Results

The testing was limited to GFRP-GFRP for the bulk material as ultrasonic inspection would be needed to determine initial and final crack areas. This work is planned and ongoing. Typical load displacement curves for the hybrid specimens can be seen in Fig. 16.10.

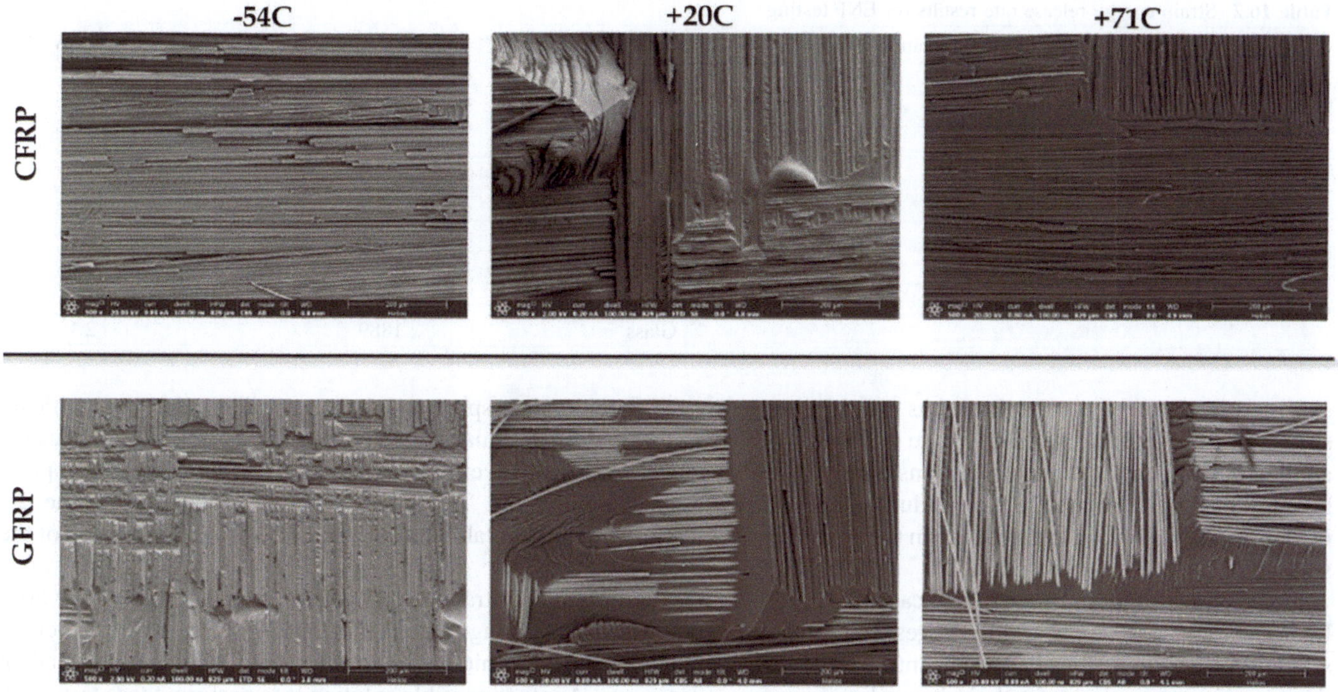

Fig. 16.9 SEM images of the fracture surfaces in the bi-material specimen at each temperature (500× magnification)

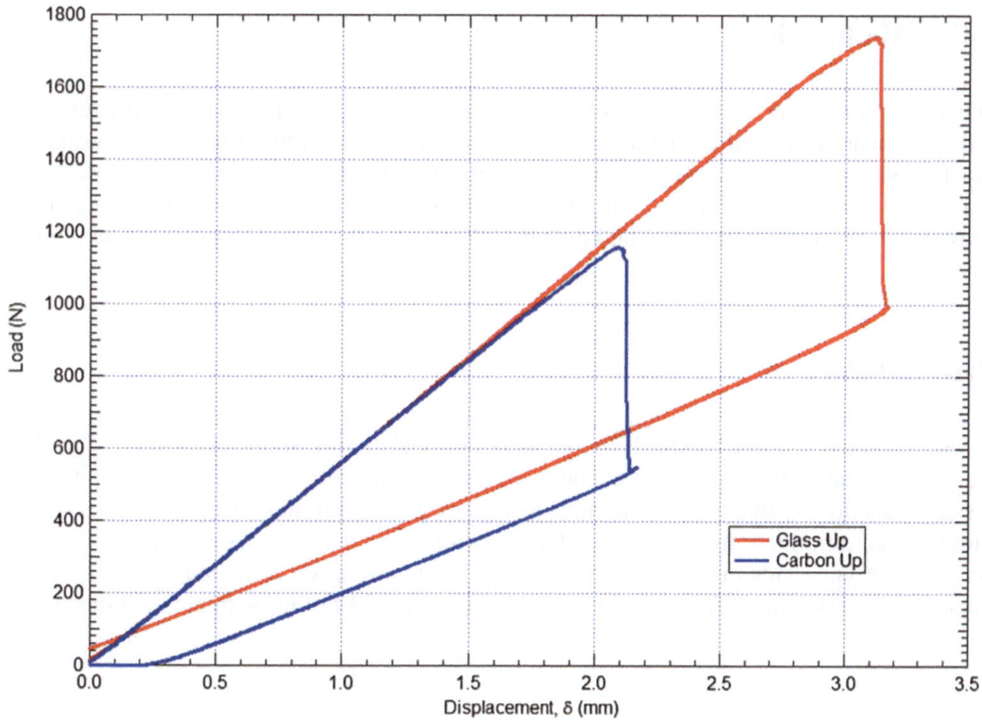

Fig. 16.10 Load-displacement curves for hybrid ENF specimens

Table 16.2 Strain energy release rate results for ENF testing

Specimen type	Temperature	Face up	G_{Avg} (J/m^2)	COV (%)
GFRP	71 °C	N/A	1230	2.1
	20 °C		1172	0.6
	−54 °C		1856	5.6
Bi-material	71 °C	Carbon	869	5.7
		Glass	1224	5.6
	20 °C	Carbon	733	2.9
		Glass	1281	1.6
	−54 °C	Carbon	882	4.9
		Glass	1889	2.1

The orientation of the specimen has a strong impact on the specimen's response. This is most likely due to significant shear at the bi-material interface. In one orientation, CFRP facing up, this residual shear is acting in the same direction as the applied bending load so they work constructively to extend the crack. However, when the GFRP is facing up, the applied bending load must overcome the residual stress before the crack is extended. This trend is further pronounced as the ΔT increases. Such a response is not seen in the bulk material GFRP specimens. A table listing the results from this portion of the study is shown below (Table 16.2).

For each material under each load case, the subambient testing produced strain energy release rates approximately 50% higher than what the trend would suggest from the elevated and ambient testing. This is most likely due to additional crack surfaces being formed in adjacent ply interfaces due to crack jumping. Ultrasonic inspection is currently being performed on these specimens to correct for the enhanced strain energy release rate. Another possible solution is to perform these tests with a longer initial crack length as in the test geometry study, the more unstable crack growth also produced multiple cracks.

16.5 Simulation Methodology

As presented in the previous section, the formation of residual stresses improved the measured fracture toughness only in the carbon-glass DCB specimens. This implies that the observed trend is related to the formation of residual stresses, rather than some temperature dependency in the material properties. Therefore, computational methods were developed to simulate the effect of residual stress formation within the carbon-glass DCB specimens to better understand the bondline's stress evolution, the stress fields around the crack tip, and potential drivers for the increasing fracture toughness.

16.5.1 Analysis Software

All simulations were completed using Sandia National Laboratories' Sierra Solid Mechanics code Adagio. Adagio is a Lagrangian, three-dimensional code for the finite element analysis of solid structures and is suitable for implicit, quasi-static analyses. Also, Sierra Adagio can make use of a multi-level solver, the solution core of which is a nonlinear, conjugate gradient algorithm that can iteratively find a solution that is within some user-defined tolerance of equilibrium. Use of the multi-level solver within Adagio aids in the solution of problems containing such challenging aspects as contact, materials with non-linear responses or extreme differences in stiffness, and material failure with element death [13].

16.5.2 Element Formulation

The modeled components were simulated with eight-noded hexahedral elements. While Sierra Adagio's default element formulation utilizes a computationally efficient, single point integration rule, avoidance of potential numerical inaccuracies common to an under-integrated element (e.g., hourglass modes) was necessary as a failure process would be modeled in these simulations. Therefore, Sierra Adagio's selective deviatoric element formulation was instead applied. This element

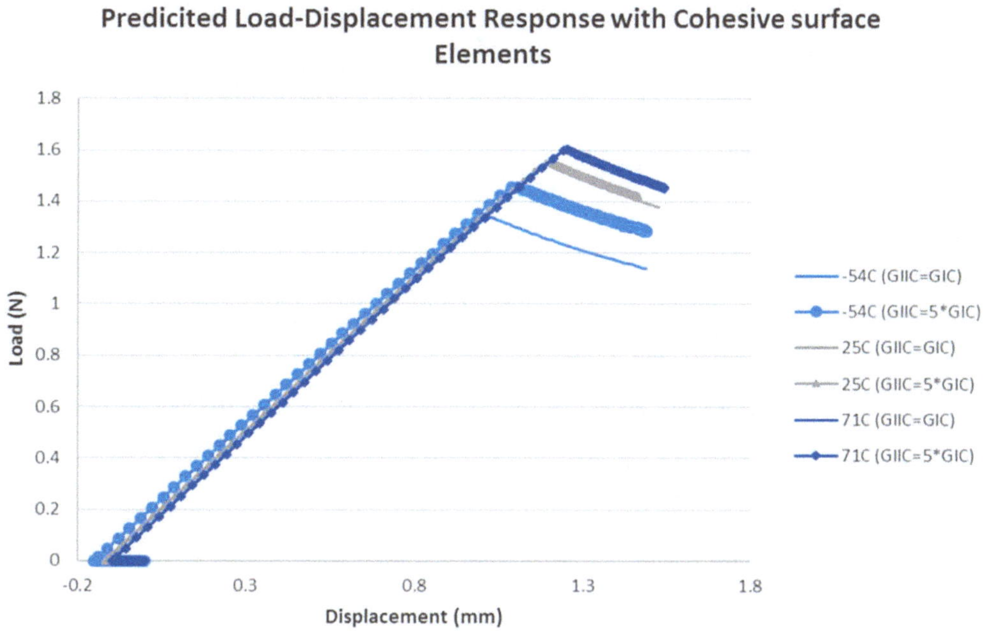

Fig. 16.11 Sample load-displacement response of DCB experiment predicted with cohesive zone methods

type is fully integrated with respect to the deviatoric stress response and under-integrated with respect to the hydrostatic pressure response [13].

Also, note that while most investigations regarding the simulation of crack growth in fiber reinforced polymer composites make use of cohesive surface elements to extrapolate the complexity of a bonded interface to a surface element, the bondline in the current study was modeled with the same hexahedral elements as the composite adherends. As it has been postulated that the fracture toughness increase may in part be due to contraction of the bondline material, it was necessary to model the adhesive layer with an element capable of thermal contractions and surface elements are not able to exhibit such behaviors. To verify the assumption that the correct experimental trend could not be modeled using a traditional cohesive zone modeling approach, several preliminary simulations were completed utilizing a surface element in conjunction with cohesive models with and without a mixed-mode resistance. The results of these preliminary simulations were not able to match the experimentally observed trend. Rather the cohesive zone model results imply that increasing amounts of residual stresses should decrease the experimentally measured fracture toughness and failure load. Figure 16.11 describes the typical results from these cohesive zone model simulations.

While the results shown in the above plot do not match the experimentally observed trend, they do indicate the potential importance of mode mixity. Specifically, as shown in the plot, increasing the mode II fracture toughness effectively increases the predicted peak loads for the 25 °C and −54 °C simulations. Alternatively, increasing the mode II behavior had no noticeable effect on the 71 °C predictions. This observation may indicate that large out-of-plane shear strains, that must be overcome prior to fracture, form along the bondline and increase progressively with greater excursions from the stress-free temperature.

16.5.3 Material Models

The individual components of the simulated DCB geometry were modeled with three materials: the adherends were made of either the carbon or glass composite and the bondline was modeled as neat epoxy. As no yielding or failure was anticipated in either the carbon or glass adherends, these materials were simulated with Sierra Adagio's elastic-orthotropic model. This model requires the definition of the regular nine elastic constants as well as glassy and rubbery coefficients of thermal expansion in each of the materials' orthotropic directions [13]. These properties, which were determined either through experimental testing or micromechanical modeling, are listed in Table 16.3 for the glass and carbon composite materials.

Table 16.3 Material properties used to define the carbon and glass composite materials

	CFRP	GFRP
E_{11} (GPa)	63.9	24.8
E_{22} (GPa)	62.7	23.1
E_{33} (GPa)	8.6	9.7
G_{12} (GPa)	3.44	3.4
G_{13} (GPa)	3.27	2.9
G_{23} (GPa)	3.25	2.9
ν_{12}	0.048	0.13
ν_{13}	0.408	0.36
ν_{23}	0.408	0.36
CTE_{11} (rubbery) (ppm/°C)	1.14	8.31
CTE_{22} (rubbery) (ppm/°C)	1.36	9.88
CTE_{33} (rubbery) (ppm/°C)	282.9	343.5
CTE_{11} (glassy) (ppm/°C)	3.41	17.3
CTE_{22} (glassy) (ppm/°C)	3.42	17.9
CTE_{33} (glassy) (ppm/°C)	72	65.6
T_g (°C)	122.7	104.5
Stress-Free Temperature (°C)	140	140

Table 16.4 Material properties used to define the bondline adhesive

Young's Modulus (GPa)	4.0
Poisson's ratio	0.35
CTE (ppm/°C)	45.0

Next, according to recommendations from literature, the bondline epoxy material was modeled with Sierra Adagio's elastic model, which requires the definition of only the material's Young's modulus, Poisson's ratio, and coefficient of thermal expansion [13]. These material parameters, which were taken from literature, are listed below in Table 16.4 [14–16].

Lastly, to facilitate crack growth within the modeled DCB specimen, it was necessary to define a criterion for element death within the bondline. For simplicity, and according to literature, a maximum strain failure theory was defined and executed for the previously described epoxy material. With this approach, a maximum tensile strain was defined within the adhesive layer, and if this value was exceeded within an element, the offending element was deleted. The maximum failure strain was defined as 2.0% [14]. Note that this process of element deletion was implemented in order to capture both thermal contraction and crack growth within the bondline in a single simulation. However, since finite element methods incorporating element death and deletion are seldom mesh convergent, the methods currently presented in this study cannot be used to make quantitative predictions regarding the material's failure behaviors. Instead the qualitative observations from these current simulations can help prioritize the physical phenomena that should be accounted for during the future development of a cohesive or localization material model suitable for use with a surface element, upon which mesh convergence is generally readily found.

16.5.4 Model Geometry

The simulations were completed with three-dimensional geometries and meshes created using Cubit, which is a robust software toolkit capable of creating both two- and three-dimensional geometries and meshes. As described in the previous sections, since contractions within the bondline were of interest, the adhesive layer was modeled with solid elements. However, as the bondline had an average measured thickness of 0.1 mm and the carbon and glass adherends had thicknesses of 4.9 mm and 3.9 mm, respectively, geometry simplifications were necessary. In order to achieve a resolved stress state within the bondline, the adhesive layer required high quality, refined elements. Therefore, to maintain a reasonable computational cost, a plane-strain model was implemented along with a truncated specimen length. Specifically, while the full 38 mm pre-crack length was modeled, the specimen's simulated bonded length was only 15 mm. This simplification was deemed appropriate, as the experimentally measured fracture toughness was observed to rise to a nearly constant value after the crack grew approximately 10–12 mm (Fig. 16.8). Figure 16.12 shows the modeled DCB geometry without meshlines.

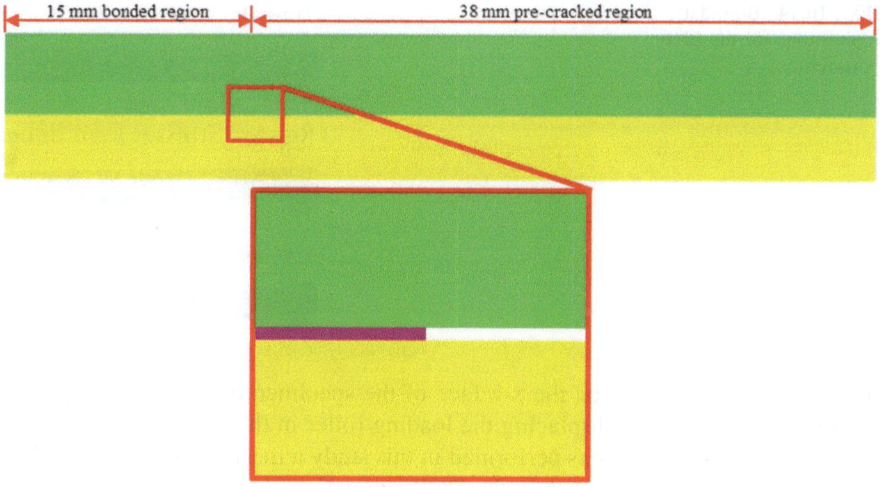

Fig. 16.12 Representative geometry of four-point flexure simulations

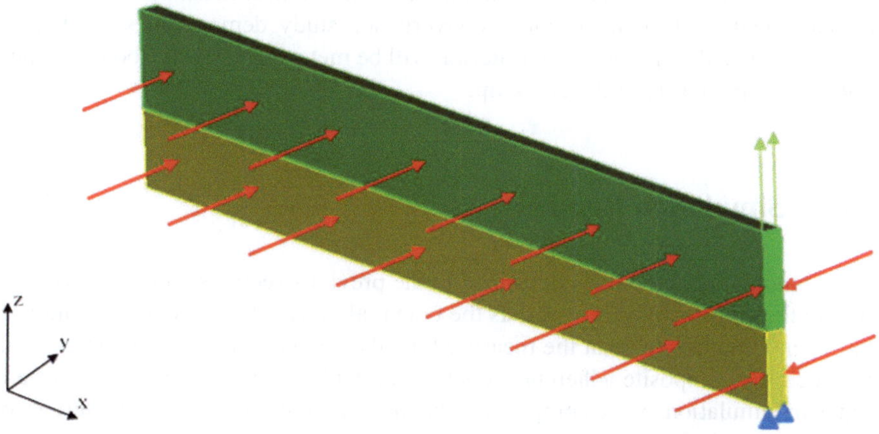

Fig. 16.13 Description of the DCB specimen's applied boundary conditions

Also, note that Sierra Adagio does not include a plane-strain element. Therefore, in order to approximate the plane-strain condition, the geometry shown in Fig. 16.12 was modeled with a depth of only 1 mm and appropriate boundary conditions were applied along the side surfaces of the simulated specimens. These approximations are discussed further in the following section.

16.5.5 Boundary Conditions

Boundary conditions were applied to the DCB experiment simulations to approximate the plane-strain conditions as well as to emulate the formation of the thermal residual stresses and the application of the tensile, mode I loading. First, to approximate the plane-strain boundary conditions, the side surfaces of the specimen were translationally fixed in the depth, or y, direction, as shown by the red arrows in Fig. 16.13. Next, to model the formation of the thermal residual stresses before the application of the mode I loading, the process described in [16] was applied. Specifically, the DCB simulations were initiated will all components at a constant and uniform temperature equal to the experimentally measured stress-free temperature. Then, the modeled specimens were isothermally cooled to the appropriate testing temperature (71 °C, ambient, or −54 °C); and, since the material definitions for the composite and epoxy materials included the appropriate CTE values, residual stresses were formed due to thermal contractions within the bondline and adherends. Lastly, boundary conditions were defined to mimic the associated experimental loading after the simulated cooling process. In the DCB simulation, the bottom-front edge in the pre-cracked region was translationally fixed in the x and z directions (as shown by the blue triangles in Fig. 16.13), while the top-front edge was translationally fixed in the x direction with an applied velocity condition applied in the vertical, or z, direction (as shown by the green arrows in Fig. 16.13). For the ENF simulation, similar boundary

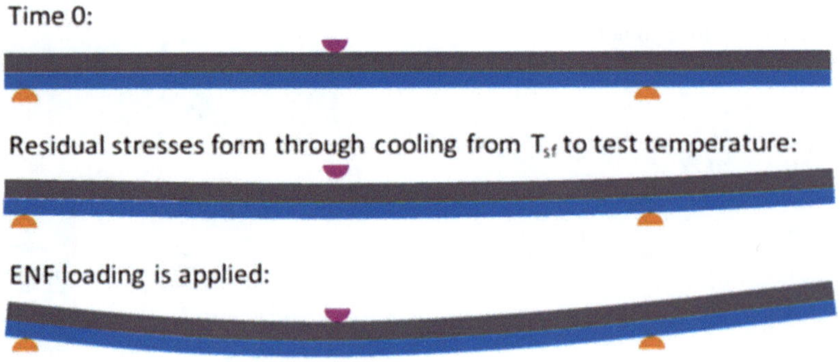

Fig. 16.14 Boundary conditions for the ENF simulations

conditions were applied on the x-z face of the specimen but the two support rollers were fixed in the z direction and the loading was applied by displacing the loading roller in the z direction (Fig. 16.14).

For all of the simulations performed in this study a mesh convergence study was performed in accordance with previous work [18]. The resulting mesh in the DCB simulations was very fine with element lengths of 0.04167 mm which produced discretization errors less than 0.05% when compared to the exact solution obtained through Richardson's Extrapolation at 1 mm of displacement. The ENF simulations used a coarser mesh but the results still yielded errors less than 2% at 1.5 mm of displacement. Note that the mesh convergence study demonstrates verification only prior to the threshold of element deletion, since the critical strain criterion will be met progressively sooner as the element size is made progressively smaller due to the singularity at the crack tip.

16.6 Simulation Results

The computational methods described in the previous sections were developed to aid in understanding the stress evolution within the adhesive layer as well as the potential drivers for the experimentally observed fracture toughness increase. As it has been hypothesized that the measured trends are due to the residual stresses that form as a result of the CTE differential between the composite adherends, contraction within the epoxy bondline, or a combination of these two phenomena, two types of simulations were completed. This first simulation measured the effect of just the residual stresses formed between the carbon and glass composites, while ignoring contractions within the bondline, and the second simulation included the effect of the bondline contractions.

16.6.1 DCB Simulations Without Bondline Contraction

With the modeling approach described in the previous sections and the converged mesh, simulations were first completed to determine the effect of the CTE differential between the carbon and glass adherends on the predicted load-displacement response. Specifically, the temperature boundary condition, described earlier, was applied only to the composite components of the model. The adhesive layer was excluded from the thermal excursion. Figure 16.15 shows the load-displacement response predicted from the three different test temperatures. The simulations were solved only past the point of initial load drop due to instabilities in the implicit solver related to the presence of element death and deletion. Note that Fig. 16.15 is meant to be demonstrate generally the modeled trends. As mesh convergence is not valid for methods involving element death and deletion, the predicted peak loads should not be considered quantitatively.

As shown in the above figure, the trend towards an increasing fracture toughness with increasing levels of residual stress is correctly captured. To better understand the formation of this trend, the out-of-plane shear and normal strains along the length of the plane-strain model's adhesive layer were probed for each of the test temperatures after cooling and before the application of the mechanical load (Fig. 16.16). Note that these strains were chosen to best represent the tangential (mode II) and normal (mode I) deformations within the bondline elements, which would aid in understanding the effect of mode mixity.

As shown in Fig. 16.16, there is a sharp increase in both the shear and normal strains at the crack front as the thermal residual stresses increase. Therefore, it was hypothesized that the large shear strains deform the bondline in such a way as to counteract the mode I, *tensile* loading. When the mechanical loading is applied after the residual stresses are formed, the tangential deformations due to the high shear strains resist the opening of a crack and promote higher failure loads.

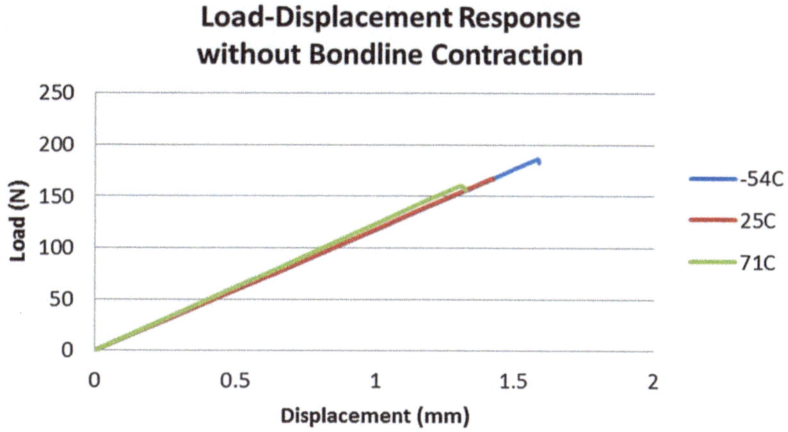

Fig. 16.15 Predicted load-displacement responses with no bondline contraction

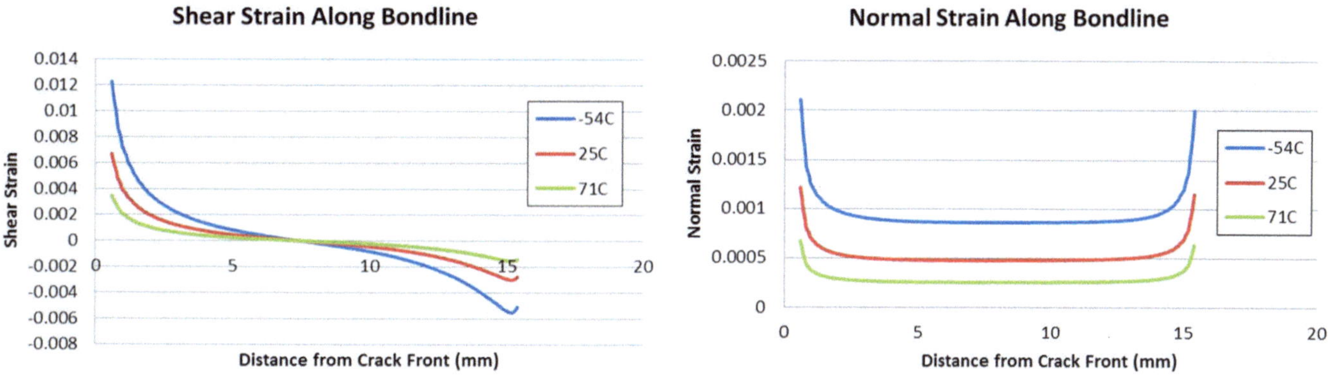

Fig. 16.16 Shear strain (*left*) and normal strain (*right*) along the length of the adhesive layer modeled with no contractions

16.6.2 DCB Simulations with Bondline Contractions

Next, to understand the effect of bondline contractions within the adhesive layer, the simulations described above were repeated with the thermal excursion applied to all model components. The predicted load-displacement responses corresponding to these three simulations are shown in Fig. 16.17.

As shown in Fig. 16.17, the correct trend is again captured, as the predicted peak loads are progressively higher as the residual stress levels increase. Furthermore, when Figs. 16.15 and 16.17 are compared, it is evident that the bondline contractions increase the predicted peak loads slightly. As with the simulations completed with no bondline contractions, the shear and normal strains along the length of the adhesive bondline were examined after the cooling phase of the simulation and, while the shear strains were not affected, the normal strains were significantly decreased (Fig. 16.18).

As shown in Fig. 16.18, it is likely that the thermal contractions of the bondline place the adhesive layer in a state of compression which must be overcome by the mechanical loading, similar to the shear strain effect discussed in the preceding section. Specifically, it is likely that the combined effect of the shear strains formed due to the CTE mismatch of the composite adherends and the compressive strains resulting from the bondline's contraction result is in the higher failure load predictions.

16.6.3 ENF Simulations with Bondline Contractions

The approach taken to modeling the onset of crack growth at various temperatures is similar to that described in the previous section. Prior to element death at the crack tip, the shear stress fields in the bondline are probed approaching the crack tip at each test temperature in both specimen configurations (Fig. 16.19).

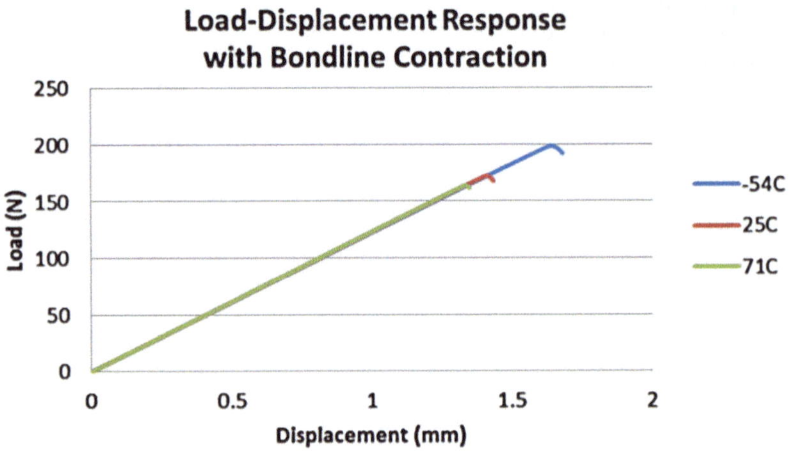

Fig. 16.17 Predicted load-displacement responses including bondline contraction

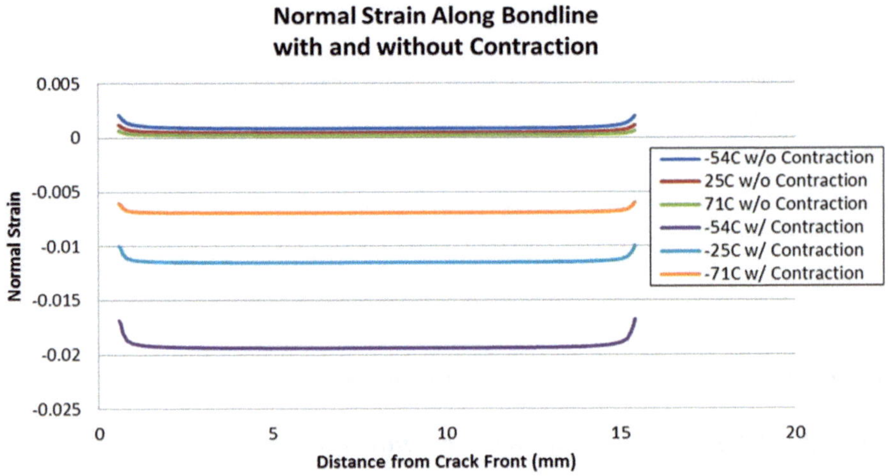

Fig. 16.18 Comparison of the normal strain along the length of the adhesive layer modeled with and without thermal contractions

The trends observed match those of the experiment. When the GFRP is on top, the elevated temperature specimen shows the largest strain value and is closest to failure. However, when the CFRP is on top, the subambient specimen shows the greatest amount of strain. This suggests that the mechanical loading works with the residual strain when the carbon is on top and against it when the glass is on top. Comparison to the experimental results is given in Fig. 16.20.

While the results are heavily mesh dependent, the trend is still accurate. So the failure criteria was tuned to provide good results at room temperature. This resulted in good agreement at elevated temperatures but poor agreement at subambient temperature. This is most likely due to the multiple delaminations seen at cold temperature. If crack growth was restricted to the interface in question, it would most likely have propagated at a lower load.

16.7 Conclusion

Decoupling residual stresses from fracture properties is very important to providing proper material properties to designers of complex structures. Typical zero volume elements such as cohesive zone elements can have trouble accounting for out-of-plane thermal strains that can impact the stress state within a material. It has been shown that these methods capture an improper trend with regards to the effect of increasing ΔT. This may not be a material property as all of the testing in this study was performed well below the glass transition temperature, T_g, of the resin system in the glassy regime. Polymers are known to have significant temperature dependence on their fracture toughness near T_g. By modeling the bondline with a continuum element and using element death, qualitative results match the experimental results quite well. However, this is not a scalable solution as element death is an approach that is significantly mesh dependent and modeling every bondline in a

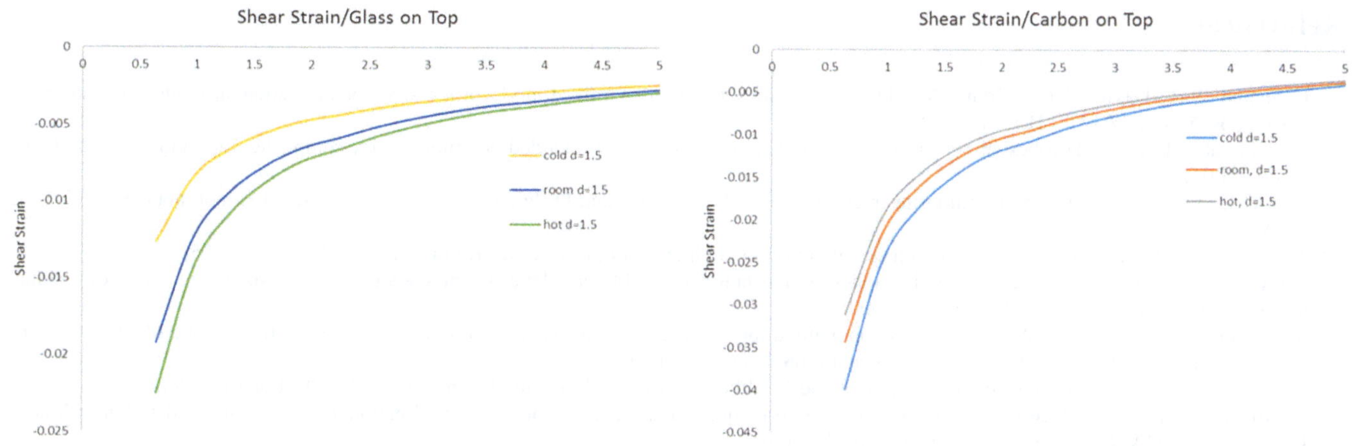

Fig. 16.19 Shear strain profile approaching crack tip for hybrid specimen

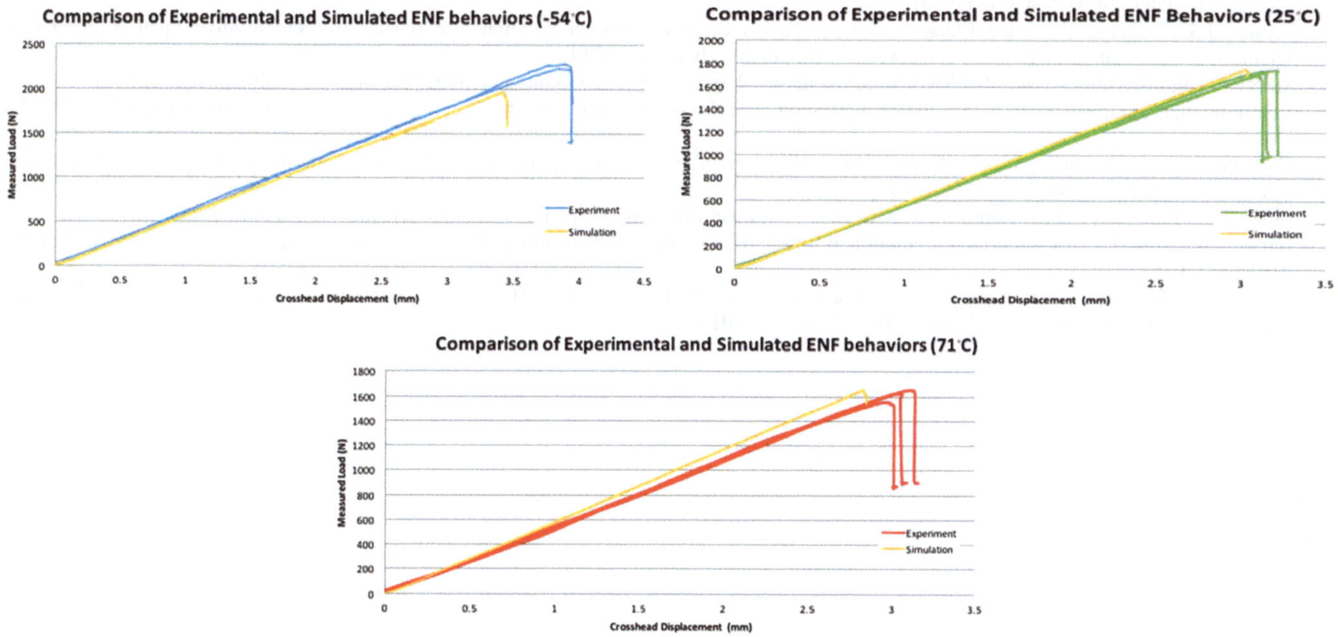

Fig. 16.20 Comparison between experiment and simulated ENF behavior at test temperatures in GFRP on top orientation

laminated structure is not computationally efficient. Further work is needed to characterize whether the residual stress is working to shift the mode-mixity of the stress state at the crack tip from that of the bulk material or if the traction-separation law must be adjusted to account for the residual stress state in the laminate.

Acknowledgement Sandia National Laboratories is a multi-mission laboratory managed and operated by Sandia Corporation, a wholly owned subsidiary of Lockheed Martin Corporation, for the U.S. Department of Energy's National Nuclear Security Administration under contract DE-AC04-94AL85000.

References

1. Camanho, P.P., Davila, C.G., Moura, M.F.D.: Numerical simulation of mixed-mode progressive delamination in composite materials. J. Compos. Mater. **37**(16), 1415–1438 (2003)
2. Li, S., et al.: Mixed-mode cohesive-zone models for fracture of an adhesively bonded polymer–matrix composite. Eng. Fract. Mech. **73**(1), 64–78 (2006)
3. Sun, C.T., Jin, Z.H.: Modeling of composite fracture using cohesive zone and bridging models. Compos. Sci. Technol. **66**(10), 1297–1302 (2006)
4. Yang, Q., Cox, B.: Cohesive models for damage evolution in laminated composites. Int. J. Fract. **133**(2), 107–137 (2005)
5. Nairn, J.A.: Energy release rate analysis for adhesive and laminate double cantilever beam specimens emphasizing the effect of residual stresses. Int. J. Adhes. Adhes. **20**(1), 59–70 (2000)
6. Yokozeki, T., Ogasawara, T., Aoki, T.: Correction method for evaluation of interfacial fracture toughness of DCB, ENF and MMB specimens with residual thermal stresses. Compos. Sci. Technol. **68**(3–4), 760–767 (2008)
7. Warnock, C., Briggs, T.: Cure Cycle Development and Qualification for Thick Section Composites. SAMPE, Long Beach (2016)
8. International, A: Standard Test Method for Mode I Interlaminar Fracture Toughness of Unidirectional Fiber-Reinforced Polymer Matrix Composites. ASTM International, West Conshohocken (2013)
9. Brown, A.A., et al.: Modeling Interfacial Fracture in Sierra. Sandia National Laboratories, Livermore (2013)
10. ASTM: Standard Test Method for Determination of the Mode II Interlaminar Fracture Toughness of Unidirectional Fiber-Reinforced Polymer Matrix Composites. ASTM International, West Conshohocken (2014)
11. Carlsson, L.A., Gillespie, J.W., Pipes, R.B.: On the analysis and design of the End Notched Flexure (ENF) specimen for mode II testing. J. Compos. Mater. **20**(6), 594–604 (1986)
12. Fenner, J.S.: Damage Tolerance Enhancement of Carbon Fiber Reinforced Polymer Composites by Nanoreinforcement of Matrix, in Theoretical and Applied Mechanics. Northwestern University, Evanston (2014)
13. SIERRA SolidMechanics Team: SIERRA/SolidMechanics 4.38 User's Guide. Sandia National Laboratories, Albuquerque (2015)
14. Velmurugan, R., Gurusideswar, S.: Strain rate dependent behavior of glass/nano clay filled epoxy resin composite. Def. Sci. J. **64**(3), 295–302 (2014)
15. Gurusideswar, S., Velmurugan, R., Gupta, N.K.: High strain rate sensitivity of epoxy/clay nanocomposites using non-contact strain measurement. Polymer. **86**, 197–207 (2016)
16. Mallick, P.: Fiber-Reinforced Composites: Materials, Manufacturing, and Design. Marcel Dekker, Inc, New York (1993)
17. Hanson, A., et al.: Experimental measurement and finite element analysis of residual stresses in simple composite structures. CAMX 2016, Anaheim (2016)
18. Nelson, S., Werner, B.: Experimental and computational investigations into the effect of process induced stresses on the mode I fracture toughness of composite materials. CAMX 2016, Anaheim (2016)

Chapter 17
Characterization of UV Degraded Carbon Fiber-Matrix Interphase Using AFM Indentation

Kunal Mishra, Libin K. Babu, and Raman Singh

Abstract In this study, we estimated the effect of fiber constraint on the interphase region using 3D finite element model. Finite element based inverse analysis was employed to optimize the factors influencing fiber bias effect and to quantify the contribution of fiber constraint using an empirical model. We have also investigated the photocatalytic degradation on the interphase region of carbon fiber composites due to ultraviolet (UV) exposure using AFM indentation based force mapping. Samples were kept in ultraviolet (UV) chamber for 0, 0.5, 1, 2, 6, 12, and 24 h at 0.70 W/m^2 irradiance and 50 °C. Force mapping was done to generate "full-field" load-displacement indentation curves, which were then analyzed using the Oliver – Pharr model to determine elastic modulus map near fiber region. For neat CFRP sample, interphase region of thickness was found to be around 100–200 nm. It was observed that the interphase size reduce with the duration of UV exposure. The change in chemical cross--linking of the epoxy resin near interphase region by breaking of C-C bond of the carbon fiber, leads to the formation of new functional groups and chemical bonds. The effect of UV exposure on the carbon fiber was studied by X-ray photoelectron spectroscopy (XPS).

Keywords CFRP • Interphase • UV Degradation • Force Spectroscopy

17.1 Introduction

Interphase region is the area in vicinity of the fiber and predominantly responsible for transfer of load from the matrix to the fibers as shown in Fig. 17.1. Previous studies have indicated that properties of the interphase region evaluated by indentation techniques are different from the bulk matrix [1].

This difference of the properties at interphase region is attributed due to various factors such as fiber constraint, chemical cross linking and indentation depth. The modulus ratio of fiber to matrix is significantly high, which leads to an influence of the stiff fiber on the measurements recorded through indentation experiments in interphase region [2]. There is an observable rise in gradient of the modulus values as indentations are carried out closer to the fiber [3, 4]. Various groups have showed that fiber modification leads to change in the interphase region, which indicates the effect of chemical cross-linking on the interphase [5]. Niu et al. [6] showed that under UV exposure, interphase region of carbon fiber and PEEK decreases suggesting the possible change in chemical cross-linking due to UV exposure. In this study, we have analyzed the combined effect of fiber constraint and chemical cross-linking. We have also evaluated the UV exposed interphase region for different exposure duration using AFM indentation.

K. Mishra (✉) • L.K. Babu
School of Mechanical and Aerospace Engineering, Helmerich Advanced Technology Research Center,
Oklahoma State University, 526 N. Elgin Avenue, Tulsa, OK 74106, USA
e-mail: kunalm@okstate.edu

R. Singh
School of Mechanical and Aerospace Engineering, Helmerich Advanced Technology Research Center,
Oklahoma State University, 526 N. Elgin Avenue, Tulsa, OK 74106, USA

School of Materials Science and Engineering, Helmerich Advanced Technology Research Center,
Oklahoma State University, 526 N. Elgin Avenue, Tulsa, OK 74106, USA

P.R. Thakre et al. (eds.), *Mechanics of Composite and Multi-functional Materials, Volume 6*,
Conference Proceedings of the Society for Experimental Mechanics Series, DOI 10.1007/978-3-319-63408-1_17

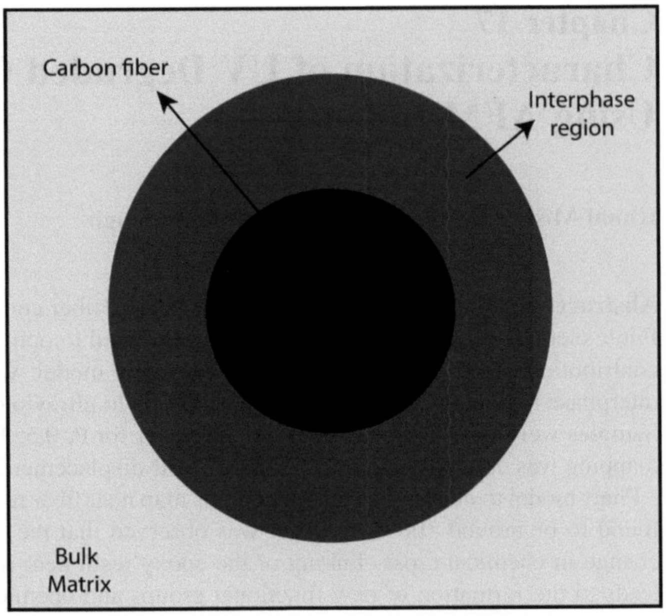

Fig. 17.1 Interphase region in carbon fiber reinforced composites

17.2 Quantifying Fiber: Bias Effect in Indentation Data

3D finite element models representing AFM indentations were developed using ABAQUS to specifically evaluate the impact of fiber bias on the increased modulus in the interphase region. AFM tip AC160TS was modeled as discrete rigid body with tetrahedral shape and back angle of 35°. Indentations were simulated as a function of distance from the fiber, maximum load for different indentation depths and different material properties. The distortion in stress field generated and increased transfer of strain resulting from deeper indentations near fiber region leads to significant constraint effect, resulting in higher modulus values.

Finite element based inverse analysis was used to optimize the parameters contributing to fiber constraint effect. An empirical model to approximate the influence of fiber constraint in the estimation of modulus of interphase region was developed. It is intended to extract the 'true modulus' values of interphase region using indentation data.

17.3 UV Degradation

Figure 17.2 shows the force map obtained by "full field" force displacement curve for UV exposed interphase region at 0 and 6 h respectively. We clearly observe that for non-exposed samples the distinctive interphase area of thickness around 80–100 nm is observed. This interphase region is either due to fiber constraint or chemical cross-linking. While for 6 h exposed CFRP samples, interphase region is almost diminished same as reported by Niu et al. Interestingly, we also observed that the elastic modulus of the epoxy matrix near the fiber region increased, which may be due to the increase in chemical crosslinking from photocatalytic degradation of carbon fiber as indicated in the modulus map.

For a comparison we have exposed the epoxy samples under the same UV conditions. We observed that 6 h exposure of the epoxy resin leads to reduction in the elastic modulus (Fig. 17.3). This clearly indicates that the UV exposure on carbon fiber, changes the chemical cross linking in the vicinity of fiber, leading to the increase in the elastic modulus.

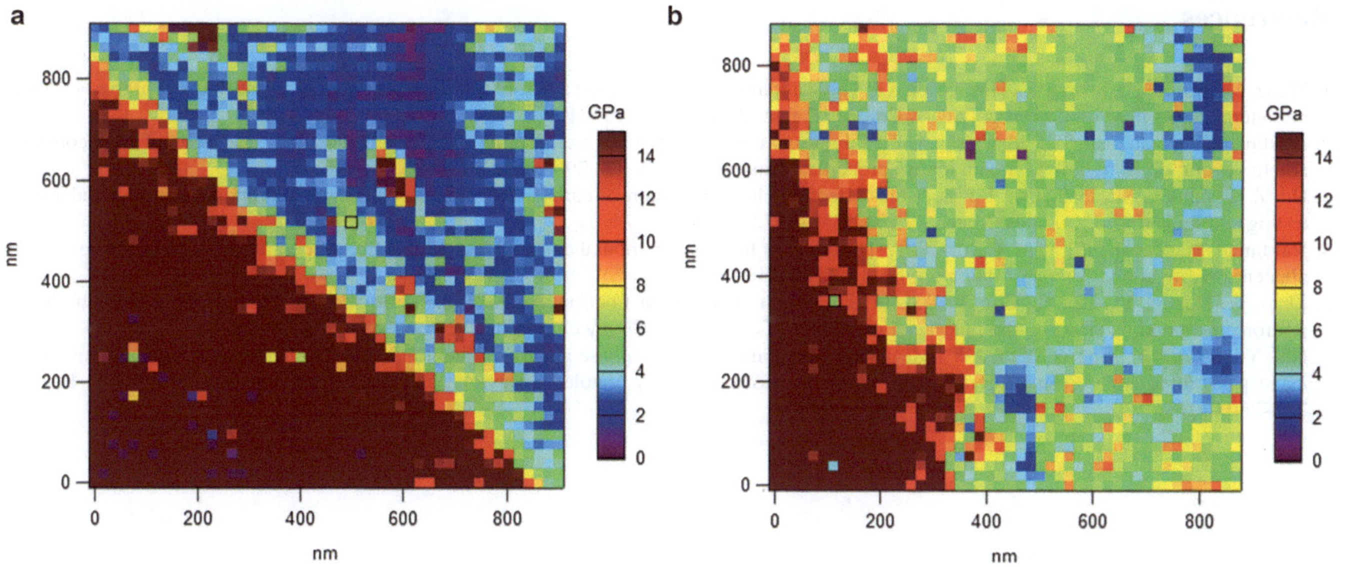

Fig. 17.2 Full field elastic modulus map of (**a**) non-exposed and (**b**) 6 h UV exposed CFRP sample

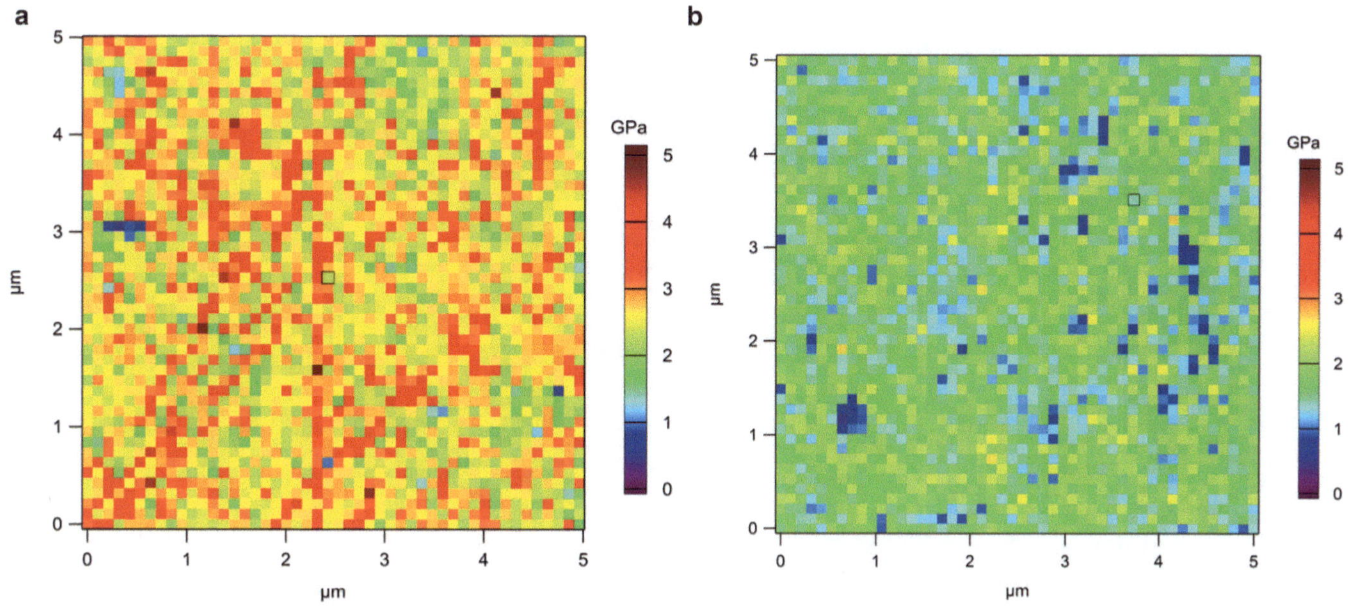

Fig. 17.3 Full field elastic modulus map of (**a**) non-exposed and (**b**) 6 h UV exposed epoxy sample

17.4 Conclusion

From this study, we found that fiber constraint influenced the modulus in the fiber vicinity and also the effect of fiber constraint was governed by the indentation depth. AFM indentation of UV exposed CFRP samples showed that duration of UV exposure leads to the photocatalytic degradation of the carbon fiber. This degradation of the carbon fiber influenced the chemical crosslinking of the interphase region.

References

1. Munz, M., Strum, H., Schulz, E., Hinrichson, G.: The scanning force microscope as a tool for the detection of local mechanical properties within the interphase of fibre reinforced polymers. Compos Part A. **29A**, 1251–1259 (1998)
2. Hardiman, M., Vaughan, T.J., McCarthy, C.T.: Fibrous composite matrix characterisation using nanoindentation: the effect of fibre constraint and the evolution from bulk to in-situ matrix properties. Compos. Part A. **68**, 296–303 (2015)
3. Hu, Z., Farahikia, M., Delfanian, F.: Fiber bias effect on characterization of carbon fiber-reinforced polymer composites by nanoindentation testing and modeling. J. Compos. Mater. **49**(27), 3359–3372 (2015)
4. Hardiman, M., Vaughan, T.J., McCarthy, C.T.: The effect of fibre constraint in the nanoindentation of fibrous composite microstructures: a finite element investigation. Comput. Mater. Sci. **64**, 162–167 (2012)
5. Zhao, F., Huang, Y., Liu, L., Bai, Y., Xu, L.: Formation of a carbon fiber/polyhedral oligomeric silsesquioxane/carbon nanotube hybrid reinforcement and its effect on the interfacial properties of carbon fiber/epoxy composites. Carbon. **49**, 2624–2632 (2011)
6. Niu, Y.F., Yang, Y., Gao, S., Yao, J.W.: Mechanical mapping of the interphase in carbon fiber reinforced poly(ether-ether-ketone) composites using peak force atomic force microscopy: interphase shrinkage under coupled ultraviolet and hydro-thermal exposure. Polym. Test. **55**, 257–260 (2016)

Chapter 18
A Study on Mechanical Properties of Treated Sisal Polyester Composites

G.L. Easwara Prasad, B.S. Keerthi Gowda, and R. Velmurugan

Abstract In the present study an attempt is made to determine the mechanical properties of sisal fiber reinforced polyester composites. Sisal fibers are the natural fibers obtained by processing the leaves of the sisal plants grown in nature. Sisal plant offers hard and strong strands of sisal fibers. The soft tissue of the sisal leaves is removed either physically or by using equipments. The fibers obtained are dried and brushed to remove the dirt left over to get the sisal fibers. In the present study, randomly oriented sisal fiber reinforced polyester matrix composite specimens of thicknesses 2 mm, 3 mm, 4 mm, 5 mm and 6 mm were fabricated by using hot compression moulding technique. 5% NaOH treated sisal fibers of length 10 mm is used as reinforcement for casting the composite specimens. A mixture of polyester resin, methyl ethyl ketone peroxide and cobalt naphthenate of ratio 50:1:1 is used as matrix for the fabrication of composite panels. Composite panels of fiber volume fraction 10%, 15%, 20%, 25% and 30% were casted and the test specimens were cut from the panels and tested for its tensile strength and flexural strength as per ASTM D-3039 and ASTM D-7264 respectively. From the experimental results it is observed that strength of tested specimens was found to show peak values at a fiber volume fraction of 20–25%.

Keywords Sisal • Mechanical properties • Polyester • Natural fibers • Chemical treatment

18.1 Introduction

Continuous studies were reported by researchers in pursuit of new materials which has brought about a large group of new materials and innovations. The need for materials that are stronger, lightweight, corrosion and chemical resistant and permeable to electromagnetic radiations have lead to the use of composites. During the recent 30 years, composite materials, plastics and ceramics have emerged as dominating innovative materials with wide advantages and applications [1]. The constituent material of composites that has higher strength is reinforcement. It can be fiber, fabric particles, particulates or whiskers. In many parts of the world, artificial fibers like steel, carbon or polymeric strands (fibers) are used as reinforcement. Attempts have been made to use naturally available strands obtained from plants and so on as reinforcements in composite materials. A unique aspect of these strands is the low vitality required for their extraction. A noteworthy issue in the utilization of these strands with matrix is that they develop rough surface in the alkaline environment and thus sturdiness of the composite involves concern. Sisal plant offers ascend to hard and strong fibers. The composition of sisal fiber is basically of cellulose, lignin and hemicelluloses. The failure strength and the modulus of elasticity, depend on the amount of cellulose and the orientation of the micro-fibers [2]. As a natural product, these properties have a variation from one plant to another plant. The Sisal fibers are commercially available in different formats viz., fabric, cords, strips, wire and rolls. The sisal fibers are abundantly available in East Africa, Bahamas, Antigua, Kenya, Tanzania, and India. The properties of sisal fibers are very much reported in earlier studies (see Tables 18.1 and 18.2).

The research studies reported in the field of natural fiber and sisal fiber reinforced composites and their respective results are discussed further. Ghosh et al. [4] have reported that banana plant fiber having high specific strength can be used as reinforcement, makes a light weight composite material and can be used to manufacture light weight automobile inner parts. Zaman et al. [5] have reported that the mechanical properties have a strong association with the dynamic characteristics.

G.L. Easwara Prasad
Department of Civil Engineering, MITE-VTU, Moodabidri, Karnataka, India

B.S. Keerthi Gowda (✉)
Department of Structural Engineering, VTU PG Studies, Mysore, Karnataka, India
e-mail: keerthiresearch@yahoo.com

R. Velmurugan
Department of Aerospace Engineering, IIT Madrs, Chennai, Tamilnadu, India

P.R. Thakre et al. (eds.), *Mechanics of Composite and Multi-functional Materials, Volume 6*,
Conference Proceedings of the Society for Experimental Mechanics Series, DOI 10.1007/978-3-319-63408-1_18

Table 18.1 Physical properties of sisal fiber [3]

1	Specific gravity = 1.370
2	Water absorption = 110%
3	Tensile strength = 347 M pa–378 M pa
4	Modulus of elasticity = 15 G pa

Table 18.2 Chemical compositions of sisal fiber [2]

1	Cellulose (%) = 65–78
2	Hemicelluloses (%) = 10–14
3	Pectin (%) = 10
4	Lignin (%) = 9.9
5	Waxes (%) = 2

Both the properties are dependent on the volume of fibers. Dynamic characteristics are predicted by analyzing the mechanical properties. The tensile strength of composite was linearly proportional to natural frequency. Further it is also reported that the damping ratio was increased by incorporation of coir fibers which reduces the high resonance. Gowda et al. [6] have depicted that the tensile strength of coir reinforced polyester composite is relatively high when compared to sisal fiber reinforced polyester composites; and that natural fiber reinforced composites can be regarded as a useful light weight engineering material. Prasad et al. [7] have reported that the feed forward ANN model could be a very better mathematical tool for prognostication of properties of coir reinforced epoxy resin composites.

Further, Verma et al. [8] have reviewed and compiled that many natural fiber embedded composites have shown mechanical properties comparable with glass fiber composites and are being used in furniture and other such related industries. They also mentioned that the surface chemical modification of coir fibers is extensively used in a wide variety of application like packaging, furniture and so on. Karthik et al. [9] have studied the surface roughness value of composites using boiled egg shell, rice husk particulate (15%) and coir fiber (10%). The study has revealed that higher surface roughness values are recorded at fibers of 10 mm length are used when compared with the values obtained using fiber of 30 mm length. Li et al. [10] have presented in their article that natural fibers can be a significant replacement for an artificial fiber in composite materials. Kuriakose et al. [11] have reported that adhesion at fiber – matrix interface can be improved by sodium hydroxide treatment – mercerization of the fibers. This treatment develops porous surface leading to a rough surface texture which allows coir fibers to reinforce strongly with polyester matrix. Idicula et al. [12] have reported the mechanical properties of short arbitrarily (randomly) oriented banana and sisal hybrid fiber reinforced polyester composites, with reference to the relative volume fraction of those two fibers, at a constant total fiber loading of 0.40 volume fractions, keeping banana as the skin material and sisal as the core material, the impact strength of the composites were increased with fiber loading. The tensile and flexural properties of respective composite were found to be better at 40% volume fraction. Girisha et al. [13] have shown that natural fibers (sisal and coconut coir) reinforced epoxy composites were subjected to water immersion tests in order to study the effects of water absorption on mechanical properties. Natural fibers like coir (short fibers) and sisal fibers (long fibers) were used in hybrid combination and the fiber weight fraction of 20%, 30% and 40% for the fabrication of the composite. Water absorption tests were reported by immersing specimens in a water bath at 25 °C and 100 °C for different time durations. The tensile and flexural properties of water immersed specimens subjected to both aging conditions were evaluated and compared with dry composite specimens. The percentage of moisture uptake was found to enhance along with increase in fiber volume fraction due to the high cellulose content of the fiber. The tensile and flexural properties of natural fiber reinforced epoxy composite specimens were found to diminish with increase in percentage moisture uptake. Moisture induced degradation of composite samples was observed at elevated temperature.

Further, Sen et al. [3] have depicted that the materials selected for structural up-gradation must be functionally efficient, qualitatively fine and sustainable. The application of composites in structural facilities is at present mostly concentrated on increasing the strength parameters of the structure with the help of synthetic fibers and does not address the issue of sustainability of the raw materials used. They have also reported that sisal is an abundant source of natural fiber utilized for making ropes, mats, carpets and as reinforcement with cement. In developing countries, sisal fibers are also used as roofing materials in houses. Thomas et al. [14] have reported the mechanical characteristics of hybrid ligno-cellulosic fiber reinforced natural rubber composites. The objective of their study was to explore the effect of the ratios of sisal and oil palm fibers on tensile attributes of natural rubber composites. The mechanical properties of sisal fiber composites are found to be high than oil palm. Sabah et al. [15] depicts that natural fibers, when used as reinforcement, compete with such non-natural fibers as glass fiber. Till date, the most important natural fibers used in composites are Jute, flax, sisal and coir.

Natural Fibers are renewable raw materials and they are recyclable. Joshy et al. [16] have depicted in their article that untreated randomly oriented isora fiber embedded polyester composite exhibited an initial decrease in the tensile and flexural properties at 10% fiber loadings; found to have increased up to 34% fiber loading and then declination with higher fiber loading. 30 mm long randomly oriented isora fiber exhibited as optimal length at which well stress transfer between fiber and the matrix occurs. Authors Inferred that higher fiber lengths lead to fiber-fiber contact, fiber curling results in shortening of effective fiber length below the critical value but less fiber length leads to notching effect results in considerable stress concentration and generation of micro cracks. In evaluation of flexural properties, authors recorded flexural stress-strain behavior at 40 mm fiber length. Authors recorded increase in impact strength with increase in fiber loading and highest impact strength at 48% fiber volume fraction whose further changes with respect to fiber volume fraction is negligible. As overall, authors concluded that in the study, mechanical properties of randomly oriented isora fiber reinforced polyester matrix composite exhibits less fiber matrix interaction at the starting level of fiber loading. At its optimum level of fiber loading shown best fiber matrix interaction but after that authors observed poor dispersion of fibers in the matrix resulted in decrease in the mechanical attributes. Sumaila et al. [17] have reported the influence of length of banana fibers on its mechanical characteristics. Authors used 30% weight randomly oriented non-woven small banana fiber reinforced epoxy composite of 5 mm thickness for their study. With increase in the fiber length here density and impact strength of composites decreases. Tensile strength, tensile modulus and percent elongation of 15 mm long fiber reinforced composite material achieved the highest value. Here fibers were treated with 5% NaOH solution before fabrication, and flexural strength and modulus of composite material increased with enhancement in fiber length up to 25 mm. Silva et al. [18] reported the potential use of long aligned sisal fibers as reinforcement in thin cement based laminates for semi-structural and structural applications. They concluded that the results of this study recognize the high potentiality of the developed material for the use in the construction industry. Naidu et al. [18] showed that the tensile and flexural strength of sisal/glass fiber hybrid component is higher than sisal fiber reinforced composite, but lower than the glass reinforced composite. They also reported that sisal fibre fails first and then the load is transferred to glass fibre, so that the presence of glass fibre in the sisal / glass fibre hybrid composite will influence to improve the tensile and flexural strength of the composite.

From the literature study, it is observed that more detailed study is necessary in the area of study of mechanical properties of chemically (alkaline) treated sisal fiber reinforced polymer matrix composites for different thicknesses and different fiber volume fractions. The specific aim of this study is to identify the influence of thickness of composites and fiber volume fractions on the tensile and flexural strength of composites. The tensile and flexural strength of 5% NaOH treated sisal fiber reinforced polyester matrix composite for different fiber loading between 10% and 30% and different thickness from 2 to 6 mm has been determined and analyzed.

18.2 Materials and Methods

In the present study 5% NaOH treated sisal fibers are used as reinforcement, polyester resin as matrix, cobalt naphthenate as catalyst and methyl ethyl ketone peroxide as curing agent. Casting of specimen is carried out using manually operated, temperature controlled compression molding machine. Such fibers are cut to a length of 10 mm and randomly dispersed while casting the composite panels. Here, the spacers are used for casting composite panels of size 300 mm x 300 mm with varying thicknesses (2 mm, 3 mm, 4 mm, 5 mm, and 6 mm). Resin, catalyst and hardener are mixed in a proportion of 50:1:1 ratio and stirred well. A 10 mm long sisal fiber of required fiber volume fraction (10%, 15%, 20%, 25% and 30%) is weighed and distributed uniformly at the bottom of the mould inside the spacers. Resin is then applied uniformly on fibers. Then the specimen is compressed for 1 h for uniform distribution of matrix and to eliminate the entrapped air bubble, if any. Here the temperature of both base plates is maintained for 80 °C. The composite is removed from the mould and left for curing at room temperature for 24 h after which the desired test specimens were cut from the casted composite panels.

The tensile and flexure strength of each specimen in the present study is evaluated out as per American Society for Testing and Materials (ASTM) standard D-3039 and D-7264 respectively. The specimens subjected to tensile test were of size 250 mm × 25 mm. The thickness of test specimens varied from 2 to 6 mm respectively. All the tension tests were conducted on a computer controlled Universal Testing Machine (UTM) of 50kN capacity and the results were recorded. As per ASTM D7264 standards the specimens of thickness 2 mm, 3 mm, 4 mm, 5 mm and 6 mm of different percentages of fiber content were tested by using the optimal support span-to-thickness ratio of 20:1 respectively.

Table 18.3 Tensile properties of 5% NaOH treated sisal fiber reinforced polyester composites

SI No.	Thickness (mm)	Fvf (%)	Ultimate load (kN)	Tensile strength (MPa)
1	2	10	0.217	04.340
2	2	15	0.314	06.270
3	2	20	0.538	10.760
4	2	25	0.424	08.480
5	2	30	0.342	06.830
6	3	10	0.936	12.480
7	3	15	0.971	12.940
8	3	20	1.076	14.340
9	3	25	0.993	13.240
10	3	30	0.845	11.260
11	4	10	1.750	17.500
12	4	15	1.990	19.900
13	4	20	2.200	22.000
14	4	25	2.420	24.200
15	4	30	1.746	17.460
16	5	10	2.188	17.500
17	5	15	2.440	19.520
18	5	20	2.835	22.680
19	5	25	3.065	24.520
20	5	30	2.585	20.680
21	6	10	2.610	17.400
22	6	15	2.976	19.840
23	6	20	3.426	22.840
24	6	25	3.729	24.860
25	6	30	3.486	23.240

18.3 Results and Discussions

Table 18.3 depicts the tensile strength of 5% NaOH treated sisal fiber reinforced polyester composites. Figure 18.1 shows the variation of the tensile strength of raw sisal fiber reinforced polyester composites with fiber volume fractions. Here, we can observe that 2 mm and 3 mm thick specimens recorded their most noteworthy tensile strength of 10.76 MPa and 14.34 MPa at the 20% fiber volume fractions and with further increase in fiber volume fractions of 25% and 30% the tensile strength is found to decrease. But, tensile strength of 4 mm, 5 mm and 6 mm thick specimens increased up to the 25% fiber volume fractions by recording the high value of tensile strength as 24.2 MPa, 24.52 MPa and 24.86 MPa and decreased at its 30% fiber volume fraction. Table 18.4 shows the flexural strength of 5% NaOH treated sisal fiber reinforced polyester composites. Figure 18.2 shows the variation of the flexural strength of treated sisal-polyester composites with different fiber volume fraction. Here, we can observe that 2 mm, 3 mm, 4 mm, 5 mm and 6 mm thick specimens recorded their most noteworthy flexure strength of 94.354 MPa, 142.307 MPa, 151.38 MPa, 157.692 MPa and 160.00 MPa at their 25% fiber volume fractions respectively.

18.4 Conclusions

Figure 18.1 shows that fiber volume fraction of 0.2–0.25 (20–25%) is the ideal range to get higher tensile properties in 5% NaOH treated sisal-polyester matrix composites. Figure 18.2 unveil that fiber volume fraction of 0.25 (25%) is the optimum amount of fiber volume fraction to obtain highest flexure properties in 5% NaOH treated sisal-polyester composites. In the specimens with low fiber volume fractions upto 20% it can be observed that the fiber content required is less than that required to exhibit higher strengths. On the other hand with higher percentage of fiber volume fractions nearing30%, the fiber content is high and the quantity of matrix required for bonding the fibers is low, it leads to reduction in strength in specimens with higher fiber volume fraction. It can be concluded that 25% fiber volume fraction can be an optimum percentage of fiber volume fraction to get higher tensile and flexural strength.

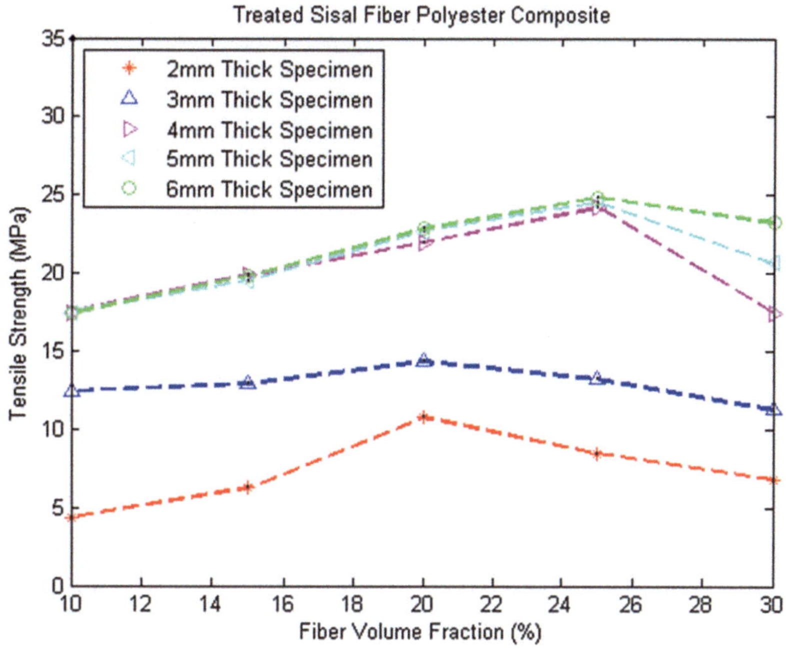

Fig. 18.1 Variation of tensile strength with fiber volume fraction

Table 18.4 Flexure properties of 5% NaOH treated sisal fiber reinforced polyester composites

SI No.	Thickness (mm)	Fvf (%)	Gauge (mm)	Ultimate load (kN)	Flexure strength (MPa)
1	2	10	40	0.071	082.342
2	2	15	40	0.074	085.932
3	2	20	40	0.077	088.761
4	2	25	40	0.082	094.354
5	2	30	40	0.076	087.471
6	3	10	60	0.140	107.692
7	3	15	60	0.149	114.615
8	3	20	60	0.167	128.461
9	3	25	60	0.185	142.307
10	3	30	60	0.163	125.384
11	4	10	80	0.188	108.230
12	4	15	80	0.210	121.153
13	4	20	80	0.225	130.000
14	4	25	80	0.263	151.538
15	4	30	80	0.243	140.000
16	5	10	100	0.238	110.000
17	5	15	100	0.290	133.846
18	5	20	100	0.300	138.462
19	5	25	100	0.342	157.692
20	5	30	100	0.310	143.076
21	6	10	120	0.300	115.384
22	6	15	120	0.362	139.231
23	6	20	120	0.372	143.077
24	6	25	120	0.416	160.000
25	6	30	120	0.380	146.154

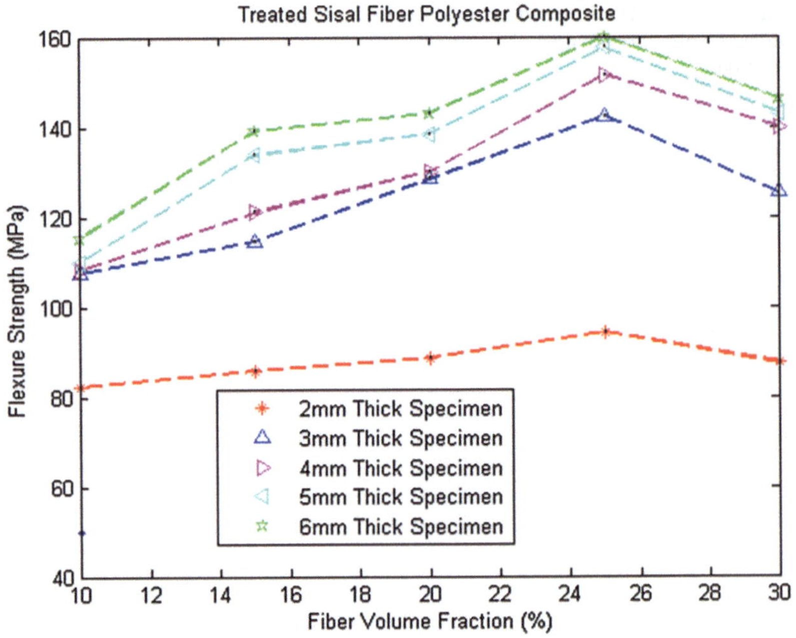

Fig. 18.2 Variation of flexure strength with fiber volume fraction

Good texture and opaqueness of sisal-polyester matrix composite materials supports to use as alternative building material. Usage of this kind of (plant fiber embedded composites) alternative building material adequately decreases the utilization of river sand (now a days, it is one of the eco-social issues) in construction practice and thus, it upgrades the use of agriculture squanders. Usage of this kind of alternative building materials adequately increases the speed of construction practice contrasted with traditional methods. Further, the commercial value of sisal fiber will also increase.

References

1. Strong, A.: History of Composite Materials- Opportunities and Necessities. Brigham Young University, pp. 1–8 (2002)
2. Sreekumar, P., Saiah, R., Saiter, J., Leblanc, N., Joseph, K., Unnikrishnan, G., Thomas, S.: Dynamic mechanical properties of sisal fiber reinforced polyester composites fabricated by resin transfer molding. Polym. Compos. **30**(6), 768–775 (2009)
3. Sen, T., Jagannatha Reddy, H.N.: Application of sisal, bamboo, coir and jute natural composites in structural upgradation. Int. J. Innov. Manag. Technol. **2**(3), 186–191 (2011)
4. Rajesh, G., Rama Krishna, A., Reena, G., Raju,B.L.: Effect of fiber volume fraction on the tensile strength of banana fiber reinforced vinyl ester resin composites. Int. J. Adv. Eng. Sci. Technol. **4**(1), ISSN 2230-7818, 89–91 (2011)
5. zaman, I., Ismail, A.E., Awang, M.K.: Influence of fiber volume fraction on the tensile properties and dynamic characteristics of coconut fiber reinforced composite. J. Sci. Technol. **1**(1), 055–071 (2011)
6. Gowda, K.B.S., Prasad, E.G.L., Velmurugan, R., Akshay, N.K.: Comparative study of tensile strength of coir and sisal fiber reinforced composites. In: Proceedings of the Indo Russian Workshop on Tropical Problems in Solid and Applied Mechanics, Chennai, 12–16 November 2013
7. Prasad, G.L., Easwara, B.S., Gowda, K., Velmurugan, R.: Prediction of properties of coir fiber reinforced composites by ANN. In: Experimental Mechanics of Composites, Hybrid and Multifunctional Materials, vol. 6, pp. 001–007. Springer International Publishing, Cham (2014)
8. Verma, D., Gope, P.C., Shandilya, A., Gupta, A., Maheswari, M.K.: Coir fiber reinforcement and application in polymer composites: a review. J. Mater. Environ. Sci. **4**(2), ISSN: 2028-2508, 263–276 (2013)
9. Karthik, R., Sathiyamurthy, S., Jayabal, S., Chidambaram, K.: Tribological behaviour of rice husk and egg Shell hybrid particulated coir-polyester composites. IOSR J. Mech. Civ. Eng. E- ISSN: 2278-1684, p-ISSN:2320-334X, 75–80 (2014)
10. Xue, L., Tabil, L.G., Panigrahi, S.: Chemical treatments of natural fiber for use in natural fiber – Reinforced composites: A review. J. Polym. Environ. **15**(1), 25–33 (2007)
11. Kuriakose, S., Varma, D., Vaisakh, V.G.: Mechanical behaviour of coir reinforced polyester composites – an experimental investigation. Int. J. Emerg. Technol. Adv. Eng. **2**(12), ISSN 2250-2459, 751–757 (2012)
12. Maries, I., Neelakantan, N.R., Oommen, Z., Joseph, K., Thomas, S.: A study of the mechanical properties of randomly oriented short banana and sisal hybrid fiber reinforced polyester composites. J. Appl. Polym. Sci. **96**(5), 1699–1709 (2005)
13. Girisha C, S., Srinivas, G.R.: Sisal/coconut coir natural fibers – Epoxy composites: Water absorption and mechanical properties. Int. J. Eng. Innov. Technol. **2**(3), 166–170 (2012)

14. Sabu, T., Varughese, K.T., John, M.J.: Green composites from natural fibers and natural rubber: Effect of fiber ratio on mechanical and swelling characteristics. J. Nat. Fibers. **5**(1), 47–60 (2008)
15. Sabah, M., Reem, M.A., Mohannad Saleh, H.: A study on mechanical, thermal and morphological properties of natural fibre/epoxy Comosite. J. Purity Utility React. Environ. **1**(5), 237–266 (2012)
16. Joshy, M., Mathew, L., Joseph, R.: Studies on short Isora fibre-reinforced polyester composites. Compos. Interfaces. **13**(4–6), 377–390 (2006)
17. Sumaila, M., Amber, I., Bawa, M.: Effect of fiber length on the physical and mechanical properties of ramdom oriented, nonwoven short banana (musa balbisiana) fibre/epoxy composite. Asian J. Nat. Appl. Sci. **2**(1), 39–49 (2013)
18. Silva, F., Filho, R., Filho, J., Fairbairn, E.: Physical and mechanical properties of durable sisal fiber–cement composites. Constr. Build. Mater. **24**(5), 777–785 (2010)
19. Naidu, V., Kumar, M., Reddy, G., Khanam, P., Reddy, M., Chakradhar, K.: Tensile & flexural properties of sisal/glass fibre reinforced hybrid composites. Int. J. Macromol. Sci. **1**, 19–22 (2011)

Chapter 19
Strain-Rate-Dependent Failure Criteria for Composite Laminates: Application of the Northwestern Failure Theory to Multiple Material Systems

Joseph D. Schaefer, Brian T. Werner, and Isaac M. Daniel

Abstract The strain-rate-dependent matrix-dominated failure of multiple fiber-reinforced polymer matrix composite systems was evaluated over the range of quasi-static (10^{-4}) to dynamic (10^3 s^{-1}) strain rates using available experimental data from literature. The strain rate dependent parameter, *m*, was found to relate strain-rate dependent lamina behavior linearly to the logarithm of strain rate. The parameter was characterized for a class of laminates comprised of epoxy-based matrices and either carbon or glass fibers, and determined to be approximately 0.055 regardless of fiber type. The strain-rate-dependent Northwestern Failure Criteria were found to fit all data in superior agreement to classical approaches across all strain rates evaluated based on solely lamina-level properties. It was determined that using the determined *m* value with the Northwestern Failure Criteria provided an accurate prediction of material behavior regardless of fiber type for the identified material class, which significantly reduces the material characterization testing required for the typical building block approach used by industry for computational analysis validation.

Keywords Composites • Failure • Strain Rate • Northwestern Failure Theory • Verification and Validation

19.1 Introduction

The Northwestern University Failure Theory for composite lamina has previously been validated using the IM7/8552 material system for lamina and angle-ply laminates [1–6] from quasi-static to dynamic strain rates. This material system was also recently evaluated by Camanho et al. [7] in support of a proposed three-dimensional invariant-based failure criterion for composite lamina. The carbon-epoxy constituent combination provides high strain to failure and intermediate strength for aerospace structural applications. While the experimental results appear to fit the theoretical prediction quite well, they merely serve as preliminary validation (via redundancy) to the underlying theory. Of particular importance is the need to predict the influence of strain rate on composite material properties, the validation of which typically requires extensive experimental data. The objective of the current work is to highlight a simple modeling approach that uses the Northwestern Failure Theory to predict composite strain rate dependence for multiple material classes using reduced and practical data sets.

19.2 Strain-Rate Effects on Matrix-Dependent Properties

To investigate a potential relationship between strain rate and matrix response, the experimental data from [1] is first used as an example for developing the strain-rate-dependence model based on IM7/8552. The strain rate dependent matrix dependent material properties are shown in Table 19.1.

J.D. Schaefer (✉)
The Boeing Company, 6300 James S McDonnell Blvd, Berkeley, MO 63134, USA
e-mail: joseph.d.schaefer@boeing.com

B.T. Werner
Sandia National Laboratories, 7011 East Avenue, Livermore, CA 94550, USA
e-mail: btwerne@sandia.gov

I.M. Daniel
Northwestern University, 2137 Tech Drive, Evanston, IL 60208-3020, USA
e-mail: imdaniel@northwestern.edu

P.R. Thakre et al. (eds.), *Mechanics of Composite and Multi-functional Materials, Volume 6*,
Conference Proceedings of the Society for Experimental Mechanics Series, DOI 10.1007/978-3-319-63408-1_19

Table 19.1 Strain rate dependent properties for IM7-8552 Lamina [1]

Property	Average Strain Rate, $\varepsilon(s^{-1})$		
	10^{-4}	1	800
Transverse Modulus, E_2(GPa)	9	10.6	11.2
Shear Modulus, G_{12}(GPa)	5.6	6.23	6.8
Transverse Tensile Strength, F_{2t}(MPa)	76	89.6	105
Transverse Compressive Strength, F_{2c}(MPa)	288	357	393
Shear Strength, F_6(MPa)	89	109	122.5

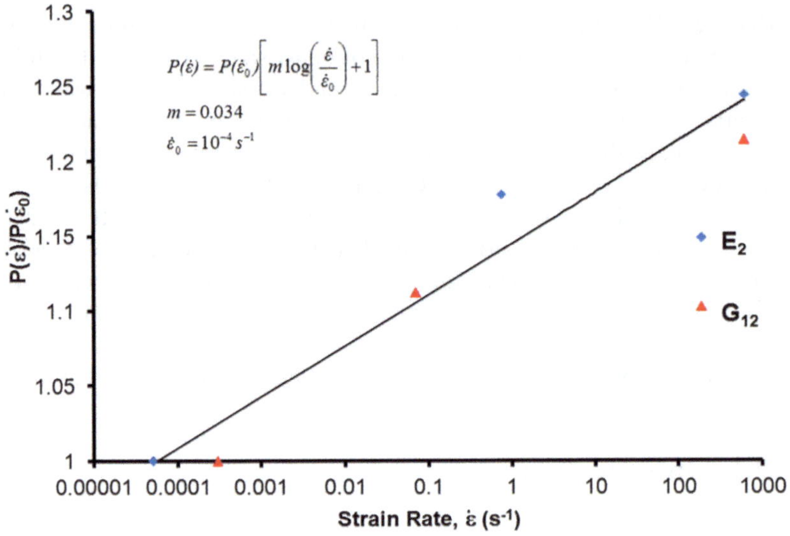

Fig. 19.1 Variation of transverse and shear moduli with strain rate

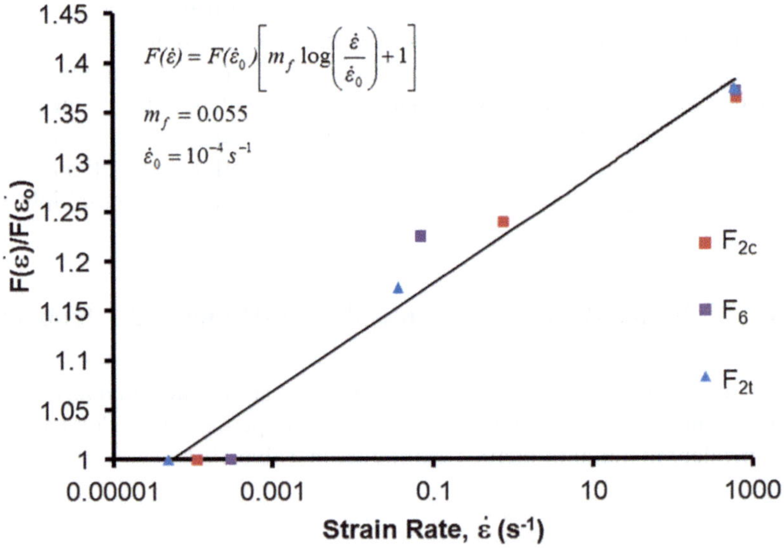

Fig. 19.2 Variation of transverse compressive and shear strength versus strain rate

The modulus and strength data were first normalized by the respective static values. The strain rates imparted span a range of six decades; thus, a logarithmic axis was used as the strain rate scale against which the normalized moduli/strength values were plotted. Important to note in the analysis is that the axial stress and strain are in relation to the global loading axes and not the lamina principal directions. Daniel et al. [7] previously addressed the conversion of these strains to principal coordinates, and noted that the final result was insignificantly different than simply using the loading axis strains as reference. The results are show in Figs. 19.1, 19.2, and 19.3. For the tested strain rates, the moduli in Fig. 19.1 have been fitted with a relationship that is linear with strain rate. The relationship is defined as:

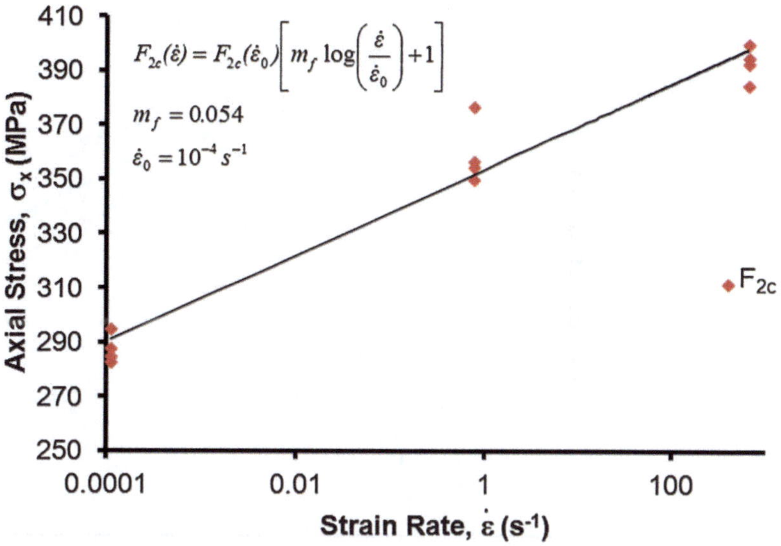

Fig. 19.3 Variation of transverse compressive strength with strain rate

$$P(\dot{\varepsilon}) = P(\dot{\varepsilon}_o)\left[mlog_{10}\left(\frac{\dot{\varepsilon}}{\dot{\varepsilon}_o}\right) + 1\right]$$

where P is the modulus (E_2, G_{12}), m is 0.034, and $\dot{\varepsilon_o}$ is reference strain rate of 10^{-4} s^{-1}.

Fig. 19.2 also shows that the lamina strengths F_{2c}, F_{2t}, and F_6 vary linearly with strain rate. The relationship is modified to reflect strength properties:

$$F(\dot{\varepsilon}) = F(\dot{\varepsilon}_o)\left[mlog_{10}\left(\frac{\dot{\varepsilon}}{\dot{\varepsilon}_o}\right) + 1\right]$$

where F is the strength (F_{2c}, F_{2t}, F_6), m is 0.055, and $\dot{\varepsilon_o}$ is reference strain rate of 10^{-4} s^{-1}.

Figure 66 shows the transverse compressive strength variation with strain rate in relation to the several tests performed at each rate. The relationship and parameters are then:

$$F_{2c}(\dot{\varepsilon}) = F_{2c}(\dot{\varepsilon}_o)\left[mlog_{10}\left(\frac{\dot{\varepsilon}}{\dot{\varepsilon}_o}\right) + 1\right]$$

where m is 0.054 and $\dot{\varepsilon_o}$ is reference strain rate of 10^{-4} s^{-1}.

From Fig. 19.1, the moduli appear to be affected similarly by increasing strain rate. Further, the assumption may be made that since these material stiffnesses behave similarly, the ratio of them may be considered a constant. Therefore, the ratio (defined below) may be obtained by simply performing the static characterization.

$$\frac{E_2}{G_{12}} = \alpha$$

Also important to note is that the shear modulus increases similarly to what is predicted by the theoretical model.

The figures provide an intriguing opportunity for modeling the lamina behavior. For the measured strain range, the strain-rate-sensitive properties have a linear response to the logarithm of strain rate. Using the presented relationship, the properties may be transformed from the current strain rate to that of the reference strain rate, $\dot{\varepsilon_o}$, according to the factor:

$$f(\dot{\varepsilon}) = mlog_{10}\left(\frac{\dot{\varepsilon}}{\dot{\varepsilon}_o}\right) + 1$$

The values for σ_2 and τ_6 may then be determined for these rates through

$$\dot{\sigma_i} = \sigma_i f(\dot{\varepsilon})^{-1}$$

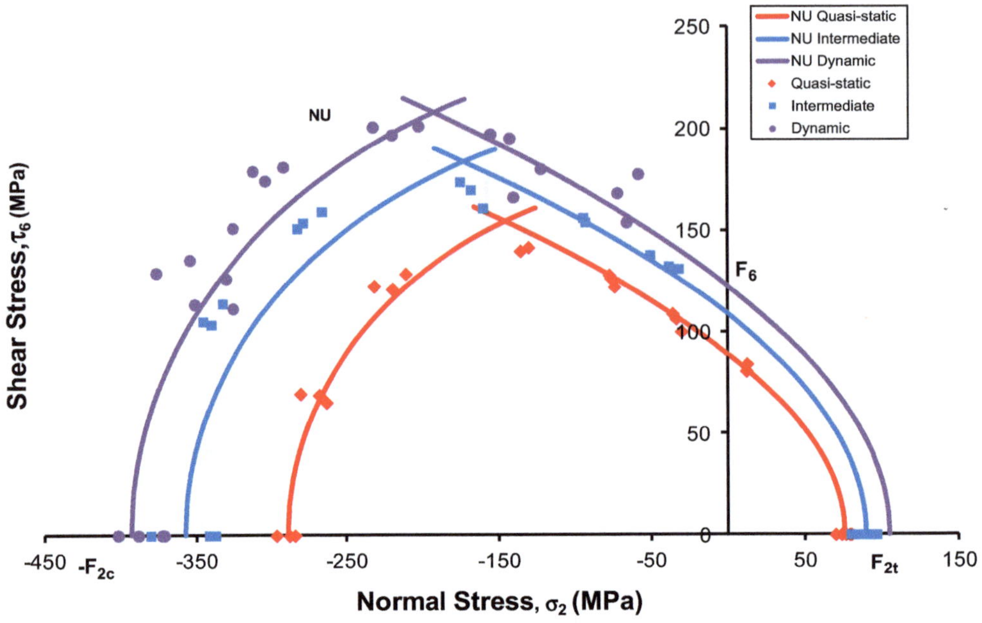

Fig. 19.4 Quasi-static, intermediate, and dynamic strain rate data plotted with Northwestern Failure Theory failure envelopes [1]

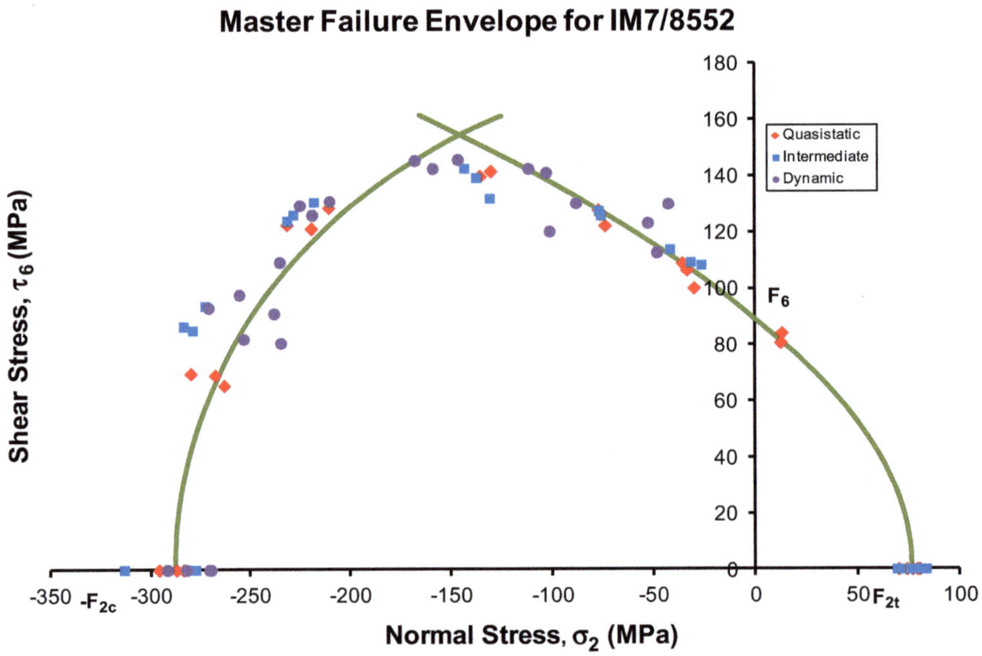

Fig. 19.5 Strain-rate-dependent master failure envelope for IM7/8552 composite lamina [1]

$$\sigma_i = \sigma_2, \tau_6$$

The above relations may then be used to map failure envelope date across the applicable strain rates into a master envelope based on the 'm' parameter, as shown in Figs. 19.4 and 19.5.

The above methodology is next applied to available literature data for a variety of material systems to investigate and evaluate strain rate dependence based on material class.

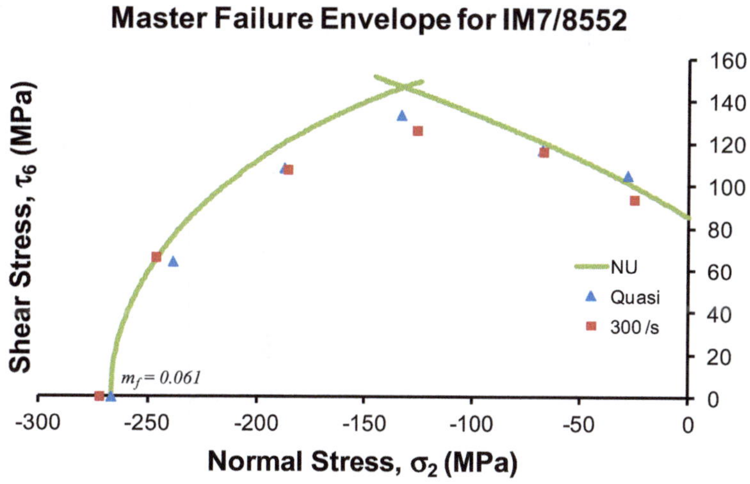

Fig. 19.6 Master NU Lamina Failure Envelope for IM7/8552; data from [8]

Table 19.2 Quasi-static lamina properties from literature

		MPa	MPa	MPa	MPa	MPa	
Fiber	**Matrix**	E_2	G_{12}	F_6	F_{2t}	F_{2c}	V_f
AS4	3501-6	10,300	7	76	57	228	0.63
IM6G	3501-6	10,000	–	–	–	267	–
IM6G	3501-6	10,000	–	–	–	267	–
IM7	8552	–	–	–	–	267	–
IM7	977-2	8570	–	76	73	267	–
IM7	8551-7	8343	5860	72.6	76	215	–
S2	8552	20,000	6900	87.3	–	–	0.56
S2	8552	20,000	6900	87.3	–	–	0.56
S2	8553-40	12,730	4460	60	50	–	–
IM7	8552	9000	5600	89	76.4	288	0.58

19.3 Application to Additional Carbon-Epoxy Material Systems

The behavior of IM7/8552 at varying strain rates was recently investigated by Raimondo et al. [8]. As shown in Fig. 19.6, the data matches well to that obtained in [1], and the NU Theory provides an excellent fit. The data was used to determine a rate dependence parameter (m_f) of 0.061.

The IM7/8552 lamina combines stiff carbon fibers with a toughened epoxy matrix. The fibers are considered to exhibit linear elastic behavior that is independent of strain rate, while the matrix has been shown to undergo significant property variation with strain rate. Therefore, it is generally proposed that similar composite systems-wherein the fiber stiffness greatly exceeds that of the matrix- should be governed by the same behavior. For example, the lamina system IM7/977-2 was investigated by Gilat et al. [10] to determine the biaxial tensile strain rate dependence. 977-2 is a toughened-epoxy formulation that has similar properties (Table 19.2) to those of 8552, except with lower shear strength; thus, this provides a fitting opportunity by which to compare the NU Theory fit. The experimental comparison to the NU Theory prediction is shown in Fig. 19.7.

The matrix dependent properties are clearly influenced by strain rate. The rate dependent parameter, m_f, was determined and the IM7/977-2 data was recast into a master failure envelope, shown in Fig. 19.8.

The resulting fit to the IM7/977-2 system is excellent – indicating that when the fiber type remains the same and the matrix properties (formulation) are similar, the NU Theory aptly predicts failure. This again holds true in another similar composite system, IM7/8551-7, the data for which was obtained by Vinson et al. [11] and is presented graphically in Fig. 19.9.

The quasi-static material properties for IM7/8551-7 are provided in Table 19.2. The rate dependence was determined based on the available data, and the master failure curves are presented in Fig. 19.10.

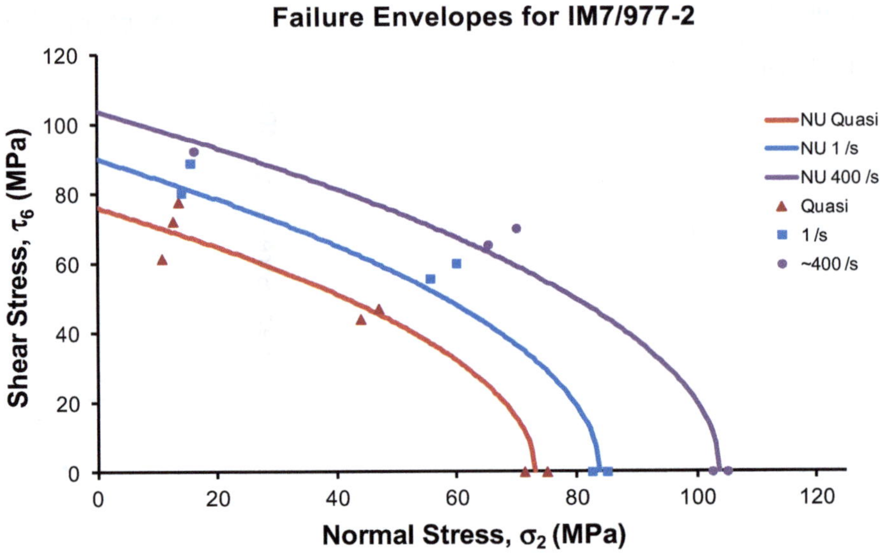

Fig. 19.7 NU Lamina Failure Envelopes for IM7/977-2; data from [10]

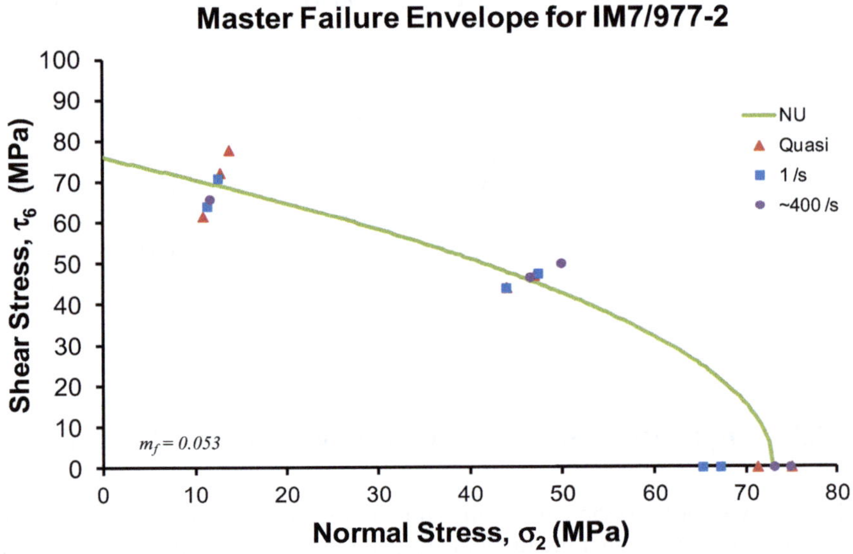

Fig. 19.8 Master NU Lamina Failure Envelope for IM7/977-2; data from [10]

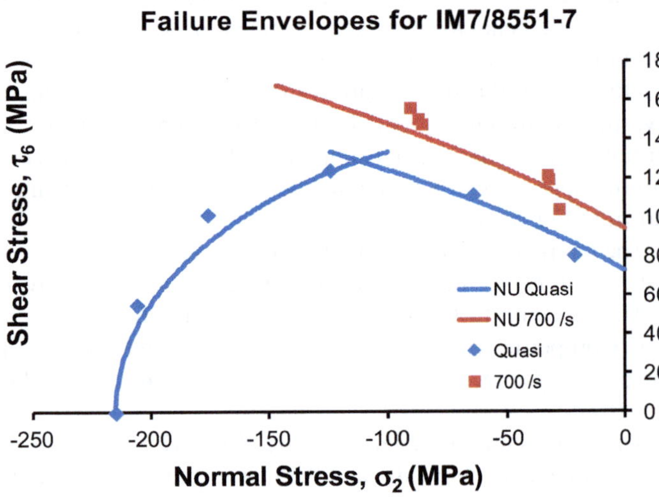

Fig. 19.9 NU Lamina Failure Envelopes for IM7/8551-7; data from [11]

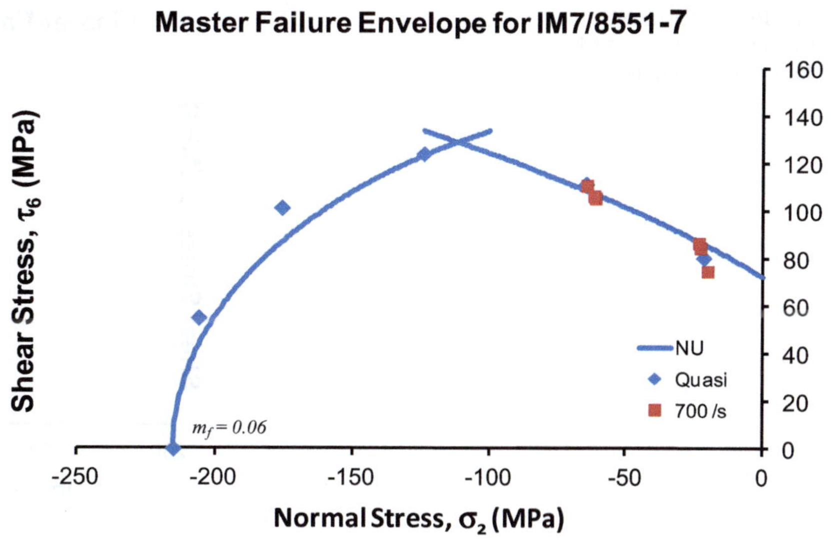

Fig. 19.10 Master NU Lamina Failure Envelope for IM7/8551-7; data from [11]

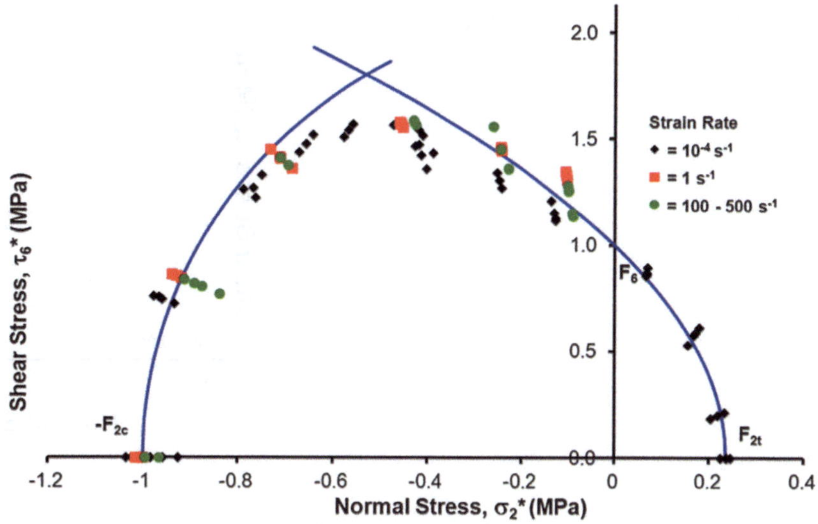

Fig. 19.11 Normalized Master NU Lamina Failure Envelope for AS4/3501-6; data from [7]

The fit to experimental data is again quite good; further, the rate dependence was determined to be similar to that of the IM7/8552 and IM7/977-2 systems. The analysis verifies that the NU Theory is well-suited for predicting failure of toughened epoxy matrix/high stiffness fiber lamina composites. A critical consideration is to investigate the predictive capability of the NU Theory for less ductile matrices partnered with a different carbon fiber. Daniel et al. [7] previously determined the strain rate dependent behavior of AS4/3501-6 composite lamina (Table 19.2), and the results are shown in Fig. 19.11.

The strain rate dependence parameter (*m*) was determined to be 0.057 (~0.06). The AS4/3501-6 lamina has higher stiffness and behaves in a more brittle fashion compared to the 8552, 8551-7, and 977-2 laminae. However, the NU Theory is also shown to predict failure quite well for this system. Bing and Sun [12] also performed an extensive set of tests on the AS4/3501-6 system at varying strain rates. The data from the authors' investigation indicated a strain rate dependence parameter of 0.06, and the fit is shown in Fig. 19.12.

A further comparison is made for the 3501-6 matrix combined with IM6G fibers [13, 14]. The rate dependence was determined to be $m_f = 0.063$. Only data for transverse compression was available in the study; however, the strain-rate-dependence is almost the same as that for the AS4/3501-6 system. This provokes an intriguing question: What role, if any, do the fibers play in matrix-dominated failure behavior? The strain-rate-dependence of glass fiber and epoxy systems is next evaluated.

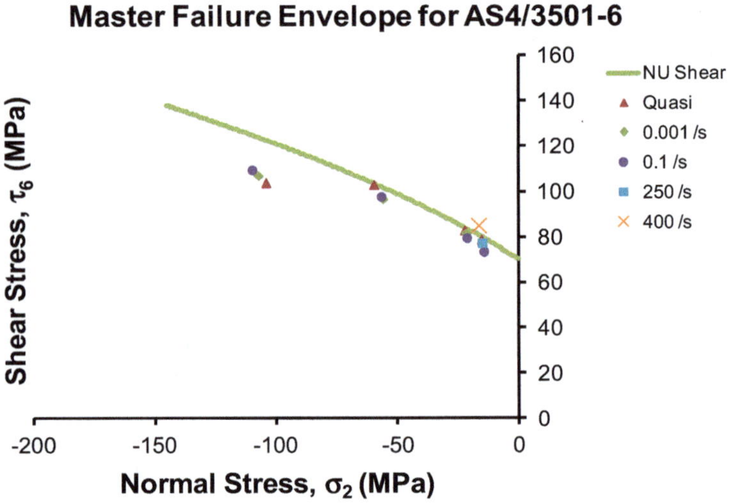

Fig. 19.12 Master NU Failure Envelope for AS4/3501-6; data from [12]

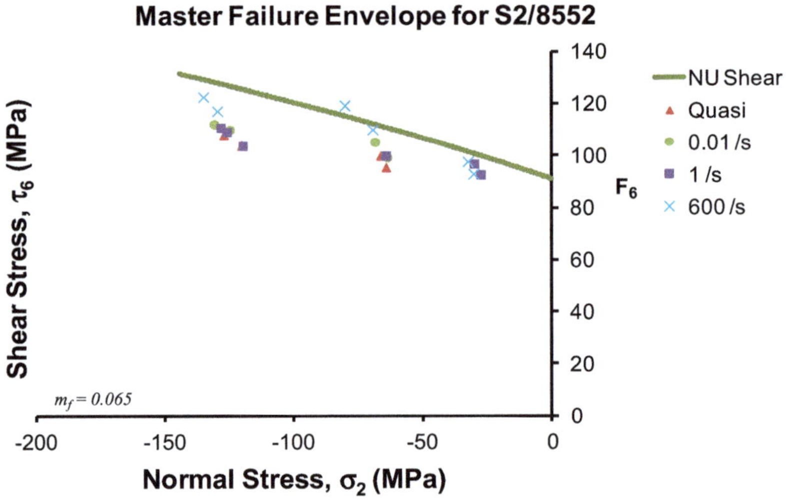

Fig. 19.13 Master NU Failure Envelope for S2/8552; data from [15, 16]

19.4 Application to Glass-Epoxy Systems

Tsai and Sun [15, 16] previously determined the strain rate dependence of an S2/8552 (glass fiber) lamina. S2 glass fibers are considerably less brittle than carbon fibers. The results are plotted with the NU failure envelope in Fig. 19.13, and the properties are listed in Table 19.2.

Interestingly, the rate dependence based on the shear strength (F_6) is similar to the previous values determined for lamina utilizing the 8552 system (although, the value for E_2 appears to be inaccurate). This indicates that for the range of fiber stiffness and fiber volume fractions considered, the strain rate dependence of the matrix-dominated material properties is dependent solely on the matrix constituent. Significantly, the analysis indicates that once the strain rate dependence is determined for a given fiber/matrix combination, in this case IM7/8552, that same rate dependence holds when the matrix is used with other similar fiber types. Thiruppukuzhi and Sun [17] obtained static and intermediate strain rate tensile data for the S2/8553-40 lamina system. The NU Theory envelope is presented in Fig. 19.14.

The rate dependence was determined to be $m_f = 0.06$, which was similar to the other epoxy systems. The master failure envelope is provided in Fig. 19.15. An overall comparison of the evaluated lamina rate dependencies is provided in Table 19.3.

The individual envelope fits appear to be quite good for the experimental data. Thus, the NU Theory is found to be useful for predicting the strain rate dependent failure of a given lamina for the domain of stiff fibers combined with comparatively ductile matrices. For a given matrix, it was determined that the fiber type was insignificant to the material strain rate dependence, which indicates that considerable testing time may be saved in characterizing material performance across

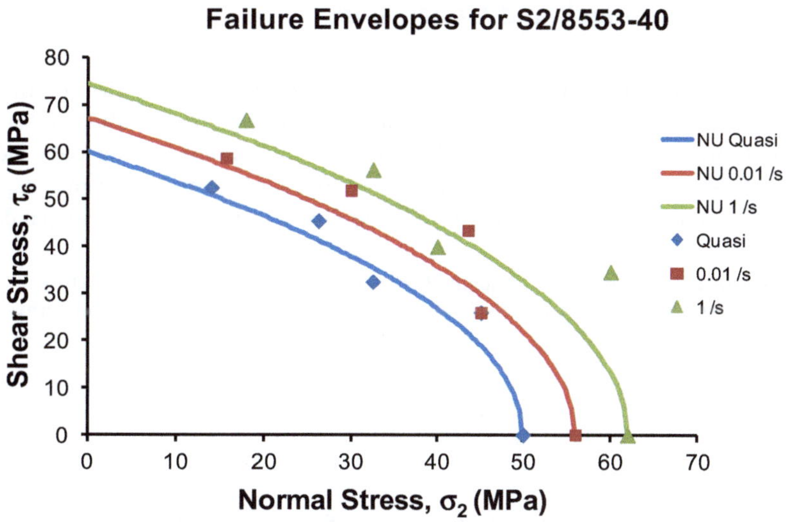

Fig. 19.14 NU Lamina Failure Envelopes for S2/8553-40; data from [17]

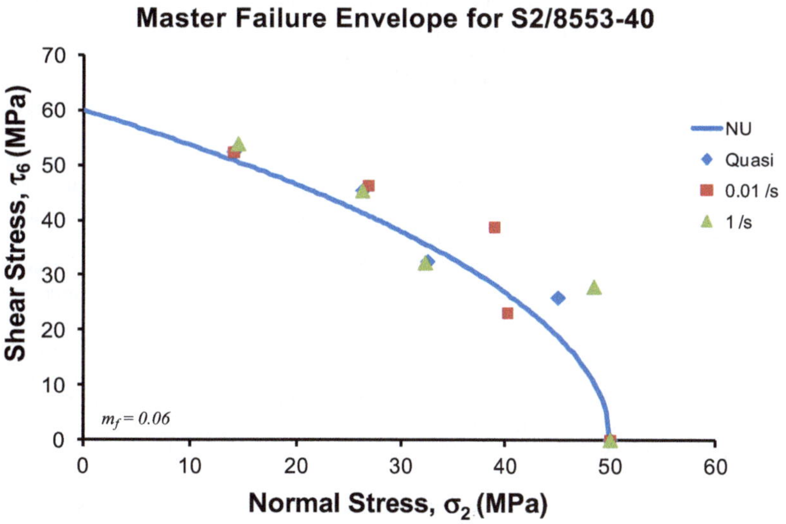

Fig. 19.15 Master NU Lamina Failure Envelope for S2/8553-40; data from [17]

Table 19.3 Lamina rate dependence determine from literature data

Fiber	Matrix	m_f
AS4	3501-6	0.06
AS4	Epoxy	0.05
AS4	Epoxy	0.052
IM6G	3501-6	0.065
IM7	8552	0.061
IM7	977-2	0.053
IM7	8551-7	0.06
S2	8552	0.065
S2	8553-40	0.06

different lamina types utilizing the same matrix but different fibers. Critically, the rate dependence parameter of each system was shown to be ~0.06 regardless of fiber or matrix type. For the general class of laminae investigated, the matrices range quite substantially in relative strength and strain to failure, but this bears little reflection in strain rate dependence. It is proposed that the chemical crosslinking of the epoxy polymer chains during cure are a driver of this similar behavior, but it is outside the scope of the current work. Significantly, the above analysis indicates that for the general combination of stiff fibers with epoxy matrix, the lamina strain rate dependence parameter may potentially be assumed to be 0.06, which greatly reduces the number of characterization tests required.

19.5 Summary and Conclusions

The NU Failure Theory has been verified by an investigation of the IM7/8552 lamina rate dependence, validated by investigation of the AS4/3501-6 lamina system, and further validated for stiff fiber/epoxy lamina systems of varying strength and strain at failure by several additional comparisons using data independently obtained from the current work. The methodology was shown to support the concept of matrix strain-rate-dependence based on material formulation, which may enable reduction of required test data to characterize strain-rate-dependent properties.

Acknowledgement Sandia National Laboratories is a multi-mission laboratory managed and operated by Sandia Corporation, a wholly owned subsidiary of Lockheed Martin Corporation, for the U.S. Department of Energy's National Nuclear Security Administration under contract DE-AC04-94AL85000.

References

1. Schaefer, J.D., Werner, B.T., Daniel, I.M.: Strain-rate-dependent failure of a toughened matrix composite. Exp. Mech. **54**(6), 1111–1120 (2014)
2. Schaefer, J.D., Daniel, I.M.: Strain-Rate-Dependent Yield Criteria for Composite Laminates, Fracture, Fatigue, Failure, and Damage Evolution, vol. 8, pp. 197–208. Springer International Publishing (2016)
3. Daniel, I.M., Schaefer, J.D., Werner, B.: Yield criteria for matrix and composite materials under static and dynamic loading, 20th international conference on composite materials, 19–24 July 2015
4. Schaefer, J.D., Daniel, I.M.: Characterization and modeling of progressive damage of angle-ply composite laminates under varying strain rate loading, 31st ASC technical conference and ASTM D30 meeting 2016
5. Werner, B.T., Schaefer, J.D., Daniel, I.M.: Effect of ply dispersion on failure characteristics of multidirectional laminates; Experimental mechanics of composite, hybrid, and multifunctional materials, Vol. 6, pp. 149–155 (2014)
6. Werner, B.T., Schaefer, J.D., Daniel, I.M.: Deformation and failure of angle-ply composite laminates; Experimental mechanics of composite, hybrid, and multifunctional materials, Vol. 6, pp. 167–171 (2014)
7. Daniel, I.M., Werner, B.T., Fenner, J.S.: Strain-rate-dependent failure criteria for composites. Compos. Sci. Technol. **71**(3), 357–364 (2011)
8. Camanho, P.P., Arteiro, A., Melro, A.R., Catalanotti, G., Vogler, M.: Three-dimensional invariant-based failure criteria for fibre-reinforced composites. Int. J. Solids Struct. **55**, 92. (0)
9. Raimondo, L., Iannucci, L., Robinson, P., Curtis, P.T.: Modelling of strain rate effects on matrix dominated elastic and failure properties of unidirectional fibre-reinforced polymer–matrix composites. Compos. Sci. Technol. **72**(7), 819–827 (2012)
10. Gilat, A., Goldberg, R.K., Roberts, G.D.: Experimental study of strain-rate-dependent behavior of carbon/epoxy composite. Compos. Sci. Technol. **62**(10–11), 1469–1476 (2002)
11. Vinson, J.R., Woldesenbet, E.: Fiber orientation effects on high strain rate properties of graphite/epoxy composites. J. Compos. Mater. **35**(6), 509–521 (2001)
12. Bing, Q., Sun, C.T.: Modeling and testing strain rate-dependent compressive strength of carbon/epoxy composites. Compos. Sci. Technol. **65** (15–16), 2481–2491 (2005)
13. Hsiao, H.M., Daniel, I.M.: Strain rate behavior of composite materials. Compos. Part B Eng. **29**(5), 521–533 (1998)
14. Hsiao, H.M., Daniel, I.M., Cordes, R.D.: Strain rate effects on the transverse compressive and shear behavior of unidirectional composites. J. Compos. Mater. **33**(17), 1620–1642 (1999)
15. Tsai, J.-L., Sun, C.T.: Strain rate effect on in-plane shear strength of unidirectional polymeric composites. Compos. Sci. Technol. **65**(13), 1941–1947 (2005)
16. Tsai, J., Sun, C.T.: Dynamic compressive strengths of polymeric composites. Int. J. Solids Struct. **41**(11–12), 3211–3224 (2004)
17. Thiruppukuzhi, S.V., Sun, C.T.: Models for the strain-rate-dependent behavior of polymer composites. Compos. Sci. Technol. **61**(1), 1–12 (2001)

Chapter 20
Progressive Failure Analysis of Multi-Directional Composite Laminates Based on the Strain-Rate-Dependent Northwestern Failure Theory

Joseph D. Schaefer, Brian T. Werner, and Isaac M. Daniel

Abstract The failure progression of a fiber-reinforced toughened-matrix composite (IM7/8552) was experimentally characterized at quasi-static (10^{-4} s^{-1}) strain rate using crossply and quasi-isotropic laminate specimens. A progressive failure framework was proposed to benchmark the initiation and progression of damage within composite laminates based on the matrix-dominated failure modes. The Northwestern Failure Theory (NU Theory) was used to provide a set of physics-based failure criteria for predicting the matrix-dominated failure of embedded plies using the lamina-based transverse tension, transverse compression, and shear failure strengths. The NU Theory was used to predict the first-ply-failure (FPF) of embedded plies in $[0/90_4]_s$ and $[0_2/45_2/-45_2/90_2]_s$ laminates for the embedded 90° and 45° plies. The Northwestern Criteria were found to provide superior prediction of the matrix-dominated embedded ply failure for all evaluated cases compared to the classical approaches. The results indicate the potential to use the Northwestern Criteria to provide the predictive baseline for damage propagation in composite laminates based on experimentally identified damage response on a length scale-relevant basis.

Keywords Composites • Failure • Strain Rate • Northwestern Failure Theory • Verification and Validation

20.1 Introduction

In multidirectional tape composite laminates, the failure process begins with the initiation and propagation of distributed microcracks in the matrix-dominated modes transverse to the loading direction. In most cases, the formation of such cracks within the multilayer laminate does not initiate catastrophic failure; instead, the lamina, or layer, undergoes 'progressive failure'. In this state, the number and size of the microcracks increases and coalesce within the various laminae constituting the laminate. The growing damage state causes stress redistribution in and between the layers which reduces the overall laminate stiffness.

A symmetric crossply laminate offers a clear example of progressive damage in a lamina within the laminate. Daniel and Ishai [1] analyzed the elastic behavior of such specially orthotropic laminates to develop load-deformation relations. The $[0/90_4]_s$ laminate is particularly suited for analyzing progressive failure. As the axial tensile stress increases, the weaker 90° layer begins to fail due to cracks that initiate within the lamina. The 0° plies continue to support the load as the 90° layers continue to crack. This occurs because the failure of the 90° layers is governed by the lamina's transverse tensile strength, F_{2t}, while the failure of the 0° layers is due to the lamina's axial tensile strength, F_{1t}, which is significantly larger. The process continues until a minimum crack spacing within the 90° plies is reached. This indicates that the 90° layer is unable to support a stress that exceeds F_{2t} because the area available for interlaminar shear (stress transfer) is no longer large enough to exceed the force balance. This is the characteristic damage state (CDS) of the crossply laminate [2–9]. Typically, it has been observed that the minimum crack spacing is on the order of magnitude of the associated layer thickness. Considerable work has been done to evaluate progressive failure in terms of transverse cracking [4–8, 10–15]. Kashtalyan and Soutis [4] and

J.D. Schaefer (✉)
The Boeing Company, 6300 James S McDonnell Blvd, Berkeley, MO 63134, USA
e-mail: joseph.d.schaefer@boeing.com

B.T. Werner
Sandia National Laboratories, 7011 East Avenue, Livermore, CA 94550, USA
e-mail: btwerne@sandia.gov

I.M. Daniel
Northwestern University, 2137 Tech Drive, Evanston, IL 60208-3020, USA
e-mail: imdaniel@northwestern.edu

P.R. Thakre et al. (eds.), *Mechanics of Composite and Multi-functional Materials, Volume 6*,
Conference Proceedings of the Society for Experimental Mechanics Series, DOI 10.1007/978-3-319-63408-1_20

Daniel et al. [5, 16] analyzed the transverse cracking behavior of crossply laminates to show a significant stiffness decrease occurs as the embedded 90° layer approaches the characteristic damage state. A phenomenological approach was taken by Henaff-Gardin and Lafarie-Frenot to show that a CDS variable and shear lag analysis may be used to accurately predict transverse cracking of such laminates in fatigue [3].

In the current work, crossply and quasi-isotropic laminates provide a unique opportunity for further validation of the Northwestern Failure Theory. In the current work, the intent is to demonstrate the point at which FPF occurs in a crossply laminate (90° transverse tension failure), validate the methodology for using the Northwestern Failure Theory, and then provide further validation using quasi-isotropic laminate FPF for 90° and 45° embedded layers. To date, the Northwestern Failure Theory has been shown to predict lamina yielding, lamina failure, angle-ply yielding, angle-ply failure, and total angle-ply failure more accurately than the classical failure criteria.

20.2 Specimen Fabrication and Testing

The IM7/8552 prepreg was cut into 20 cm × 20 cm (8 in. × 8 in.) layers and a $[0/90_4]_s$ was manufactured according to the recommended cure cycle by Hexcel. A water-lubricated diamond grit grinder was used to remove the 6.4 mm (¼ in.) from all in-plane edges to remove inconsistencies due to manufacture, and the panels were fitted with 2.5 cm (1 in.) tabs. The diamond saw was used to cut specimens measuring 19 mm wide × 17.8 cm long (0.75 in. wide × 7 in. long), which were then sanded into parallel alignment using 600 grit sand paper and a wet-sanding approach. The width tolerance was ±25 μm (±0.001 in.) These samples were fitted with a 5 cm (2 in.) extensometer that was connected to the servo-hydraulic testing system (Instron 8500) and calibrated prior to testing. The strain data was recorded using a LabView program and correlated directly to applied load. The specimens were aligned in the Instron grips and loaded in tension at a constant displacement rate to produce a quasi-static strain rate of 10^{-4} s^{-1}. Five tests were performed to determine the FPF.

Quasi-isotropic laminates were manufactured according to the previously described process for crossply laminates. Specimens were cut from 25 cm × 25 cm (10 in. × 10 in.) panels having dimensions of 19 mm wide × 17.8 cm long (0.75 in. wide × 7 in. long), with 5 cm (2 in.) tabs. A $[0_2/45_2/-45_2/90_2]_s$ layup was selected and a 5 cm (2 in.) extensometer was used to measure the strain during loading. The edges of the coupons were polished with 600 grit sandpaper to produce a highly reflective surface on which cracks were easily visible under a microscope. The Instron 8500 testing system was used to load the coupons in tension under constant displacement control to produce a constant quasi-static strain rate of 10^{-4} s^{-1}.

20.3 Crossply Laminates

The thermally-induced residual stresses were determined in order to ascertain the true in-situ lamina stress state [1]. A temperature difference to ambient conditions was determined (REF) ($\Delta T = -150\ K$), and the residual stress for each layer is presented in Table 20.1.

The residual stress in the 0° (exterior) laminae results in the fibers being in compression while the transverse direction is strained in tension. To a lesser degree, this same state exists post-curing for the 90° interior lamina. Importantly, the residual tensile stress within the 90° lamina is approximately half of the transverse tensile strength, F_{2t}, which greatly reduces the available strain to failure. During testing, the laminate is strained in tension and all of the layers deform uniformly. The 90° layers begin to fail when their axial stress reaches F_{2t}. The failure appears as a large transverse macrocrack that is visible along the specimen edge and is notably heard during testing. Once a crack forms, the 90° layer becomes two embedded laminae that are separated by the crack – stress is imparted to these separate laminae via the interlaminar stress between the 0° layer and the 90° segments. Along the loading axis, the tensile stress within the embedded 90° lamina is zero at the crack tip but increases steadily with increasing distance away from the crack. The maximum tensile stress within the 90° lamina occurs approximately at the midpoint between two transverse cracks. A schematic of the loading is shown in Fig. 20.1 (left).

Table 20.1 Residual stresses in crossply laminate layers.

θ°	σ_{1res} (MPa)	σ_{2res} (MPa)	τ_{6res} (MPa)
0	−134.2	38.5	0
90	−9.6	33.5	0

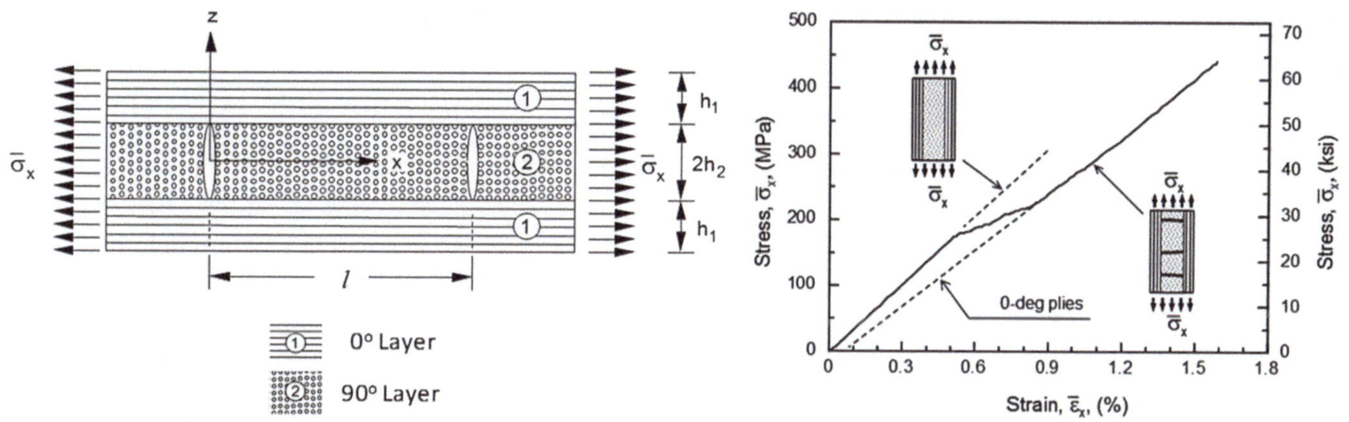

Fig. 20.1 Element of crossply laminate with transverse cracks in 90° layer (*left*), axial stress-strain curve (*right*) [1]

Fig. 20.2 Quasi-static stress-strain behavior for IM7/8552 crossply

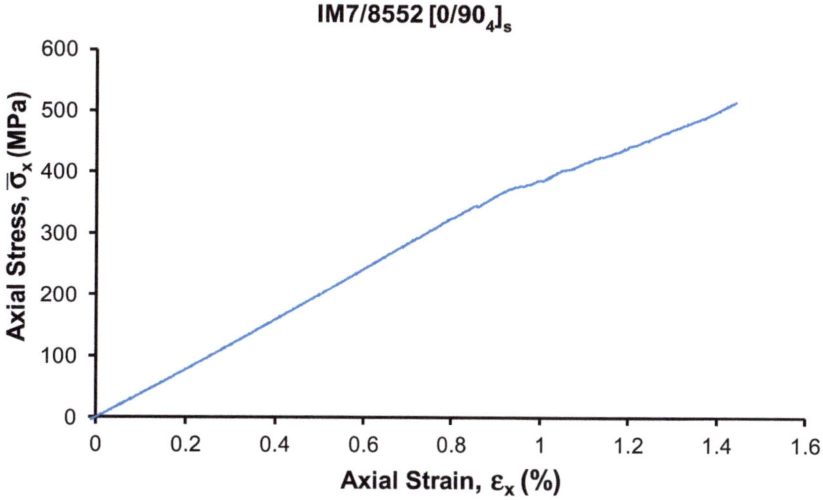

Cracking in the 90° layer continues as tensile strain is applied to the laminate. At the characteristic damage state, the embedded 90° plies are saturated with cracks and no longer provide stiffness or strength to the laminate, and the laminate modulus is governed primarily by that of the 0° plies, as shown in Fig. 20.1 (right).

The transition between laminate behavior to that of the 0° plies is noted in the stress-strain curve by a 'bent' region – this is due to the many local increases in strain that occur when a crack is formed for a given loading. The crossply results for IM7/8552 are discussed in the next section.

20.4 Crossply Results

A representative stress-strain curve for the IM7/8552 crossply laminate is shown in Fig. 20.2. In the figure, the stress-strain behavior clearly deviates from the initial linear-elastic path as the embedded 90° layer fails in tension. A schematic of the failure pattern is illustrated in Fig. 20.3.

In the initial loading stage (left), the stress-strain behavior is linear elastic. As the strain increases, microcracks form around the embedded transverse fibers. Once these layers reach F_{2t}, the microcracks coalesce into large transverse macrocracks (center). The schematic illustrates how the axial stress becomes zero at the crack location and increases to the maximum value halfway between the cracks. At saturation (right) the embedded layer is no longer capable of loading to the F_{2t} threshold. For comparison, all five tests were plotted together in Fig. 20.4.

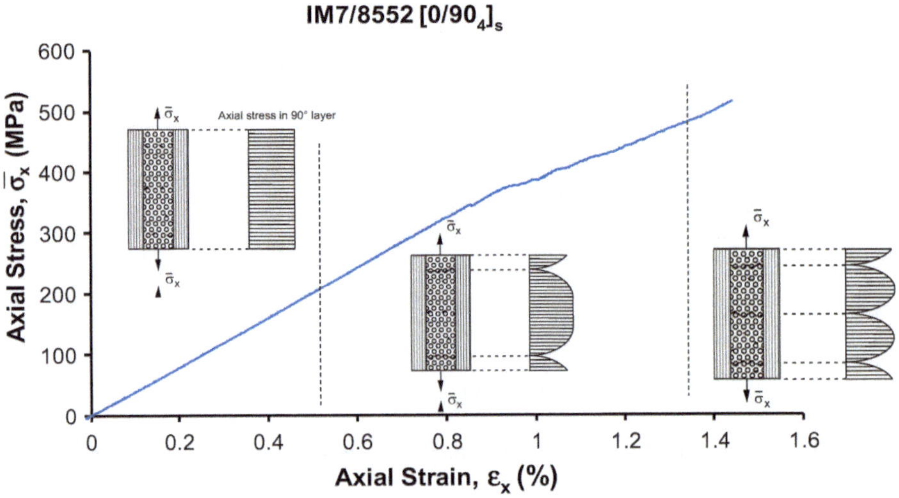

Fig. 20.3 Failure progression of IM7/8552 crossply laminate

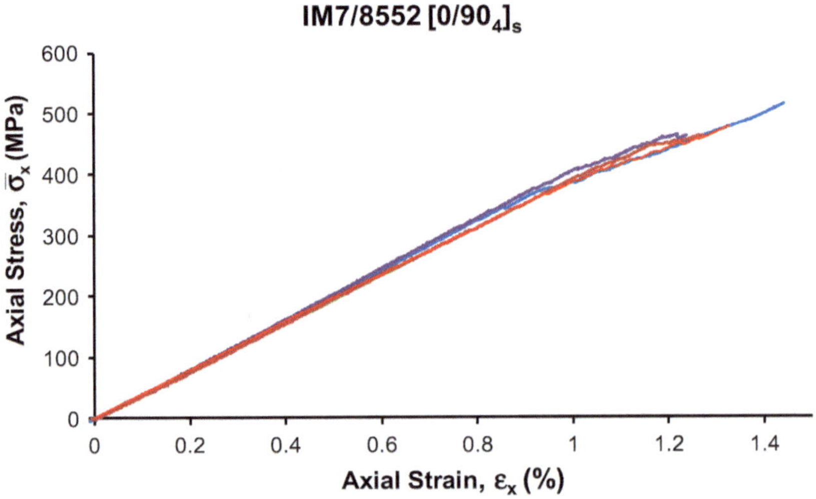

Fig. 20.4 Overlay of the five IM7/8552 crossply tests performed

The figure relates that the linear elastic portion prior to progressive failure is quite consistent for all samples. Furthermore, the non-linear behavior due to extensive embedded lamina cracking occurs at approximately the same strain for each sample. Analyzing and predicting the crossply FPF behavior is discussed in the next section.

20.5 Crossply First Ply Failure

While ultimate failure represents the ultimate design limitation for a given laminate, it is first (and most critically) important to determine the point at which the material begins to fail. Often, progressive failure within laminates is characterized primarily by successive matrix-dominated failure modes. Karthikeyan et al. [17] recently analyzed the ballistic response of various composite plates to ascertain failure sequence in catastrophic events. In such events, the stiff carbon fibers were found to have the highest longitudinal wavespeed compared to other fiber types (Kevlar, S2, E2, etc.) and the lowest specific energy absorption. Epoxy resin, on the other hand, was shown to have a significantly lower wavespeed but possesses comparatively high specific energy absorption. The results showed that matrix damage (in-plane and interlaminar failure) was attributed to the high levels of energy dissipation in the laminate, and the matrix shear and tensile strengths were proposed as the governing material properties. Furthermore, considerable laminate damage occurred prior to final fiber failure (penetration) indicating the need to determine the matrix-dominated failure progression. Pandya et al. previously investigated the attenuation of stress waves within composite panels to arrive at similar results [18]. The use of such panels

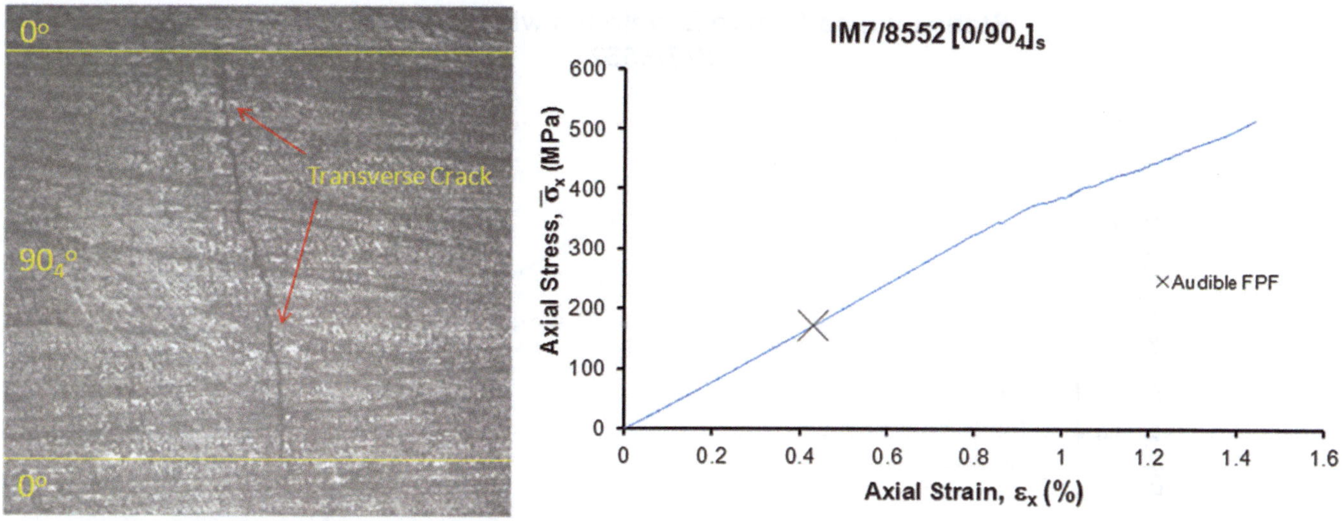

Fig. 20.5 Transverse crack in embedded 90° lamina at FPF

Table 20.2 Audible FPF stress and strain values for embedded 90° lamina in IM7/8552 crossply laminate

#	$\overline{\sigma_x}$ (MPa)	ε_x(%)	σ_2 Total (MPa)
1	201	0.51	79.6
2	180	0.46	74.9
3	167	0.41	70.4
4	175	0.46	75
5	172	0.44	73.1
Avg	**179**	**0.46**	**74.6**

in critical structural and armor applications is an area of rapidly increasing interest [19–25]. Crucial to these research investigations is the ability to understand the entire failure process so that energy modes may be isolated and analyzed.

Historically, non-destructive approaches (Non-Destructive Evaluation, NDE) for determining the damage state within composites are preferable to mechanically isolating the failure surface for analysis; thus, destroying the part [26, 27]. In this way, techniques such as acoustic emission [28–37], thermography [38–41], ultrasonics [42, 43], fiber optics [44], micro computed tomography (μCT) [45], and shearography [46] have been developed to isolate and record failure events. Recently, Schaefer et al. [47] applied in-situ computed tomography (in-situ CT) to identify matrix dominated failure events only previously inferred by acoustic emission. Each of these approaches requires the implementation of specialized expensive equipment and potentially extensive data reduction. Based on the inherent nature of the [0/90$_4$]$_s$ laminate, a starkly audible 'snap' occurs when a transverse macrocrack forms in the 90° layers.

In light of the above NDE techniques, a hybrid acoustic emission – or auditory – approach was employed for determining the occurrence of FPF. An individual test coupon was slowly loaded in tension at constant displacement rate until an audible 'snap' was heard at a distance of 30 cm (12 in.). The test was then terminated and the coupon was slowly unloaded for edge analysis on the microscope. The audible FPF load was recorded, and the entire specimen edge (both sides) was subsequently analyzed for the presence of transverse cracks. A single large macrocrack was visible in the approximate center of the coupon, which validated the approach. An example of the transverse crack at FPF in an unloaded specimen is shown in Fig. 20.5 (left) and the axial stress-strain location is shown (right).

As expected, the FPF strain is considerably lower than that of the lamina F_{2t} $(\varepsilon_{2t}{}^u = 0.9\%)$ and this is attributed to the tensile residual strain imparted by the cure process. Laminate mechanics were used to determine the transverse tensile stress, and the residual stress from Table 20.1 was added to obtain the total transverse stress. A listing of the audible FPF stress and strain values is provided in Table 20.2.

The average values for axial stress, $\overline{\sigma}_x$, axial strain, ε_x, and transverse stress (including residual stress), σ_2, were calculated and the variation was seen to be quite low in each case. The determined transverse stress was noted to be near F_{2t} (76.4 MPa).

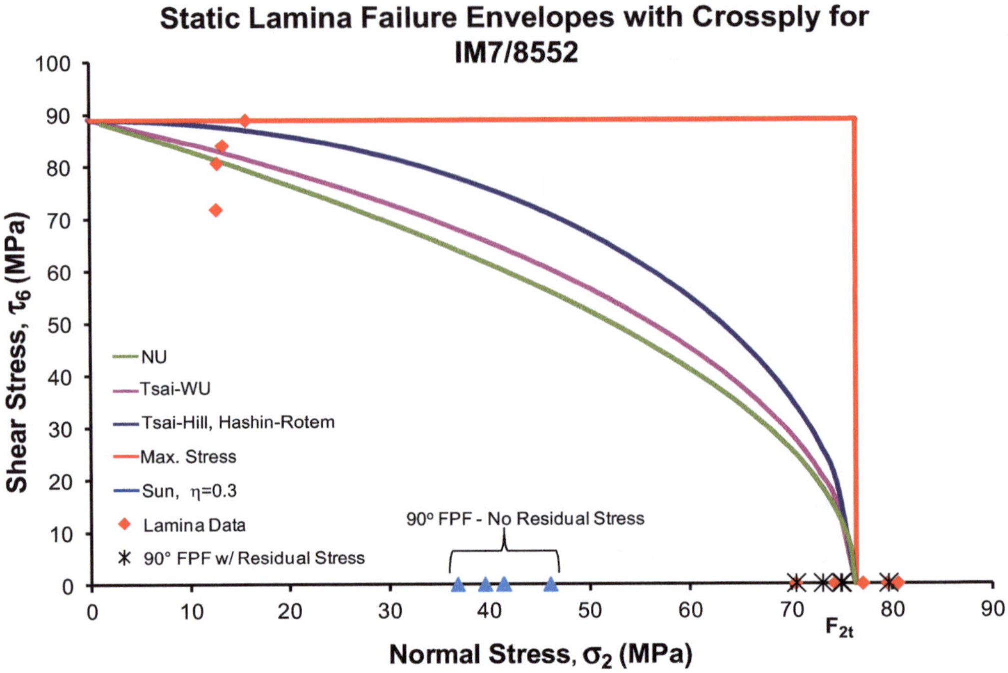

Fig. 20.6 Significance of accurate residual stress determination in FPF analysis

20.6 Application of the Northwestern Failure Theory to FPF of Crossply Laminates

For the case of the $[0/90_4]_s$ laminate, the associated NU Theory relation simplifies directly to the following:

$$\left(\frac{S_f\sigma_2 + \sigma_{2res}}{F_{2t}}\right) = 1$$

where S_f is safety factor on the mechanical loading and $\sigma_{2res}$is the calculated residual stress from Table 20.1. It follows that the NU Theory (as well as that of the classic theories) predicts the embedded 90° lamina will fail when the stress reaches F_{2t}. The determined experimental values were plotted on the quasi-static lamina failure envelope for IM7/8552 along with the classical failure theory predictions and the NU Theory. Plotting the experimental values without the residual stress reiterates the significance of residual thermal stresses on matrix-dominated first ply failure (Fig. 20.6).

Not considering the residual stress within the embedded 90° layers would cause a designer to drastically overestimate the lamina tensile strength. The values may be likewise plotted with those of the angle-ply total failure; the fit is similarly as good (Fig. 20.7).

While the above figures do little to separate the NU Theory from the classical criteria, they do provide important validation of the techniques employed to determine residual stress within the laminate. This provides further confidence in the analysis approach as the investigation moves on to more complicated multidirectional laminates with embedded lamina undergoing biaxial stress states.

20.7 Comparison to AS4/3501-6 Material System

An AS4/3501-6 crossply laminate $[0/90_4]_s$ was fabricated according to the same process as the IM7/8552 material. Samples were cut having the same width and length as that of the crossply specimens previously described. The samples were tested on the Instron at a quasi-static strain rate of 10^{-4} s^{-1}. Three tests were performed, and an average curve was created. The results are shown in Fig. 20.8.

AS4/3501–6 is comparatively brittle to the IM7/8552 material ($\varepsilon_{2t}{}^u$ = 0.6% vs. 0.9%, and E_{2t} = 10.3 vs. 9 GPa, respectively). This is especially noted in the sharp contrast between the three damage zones for the AS4/3501–6 coupons.

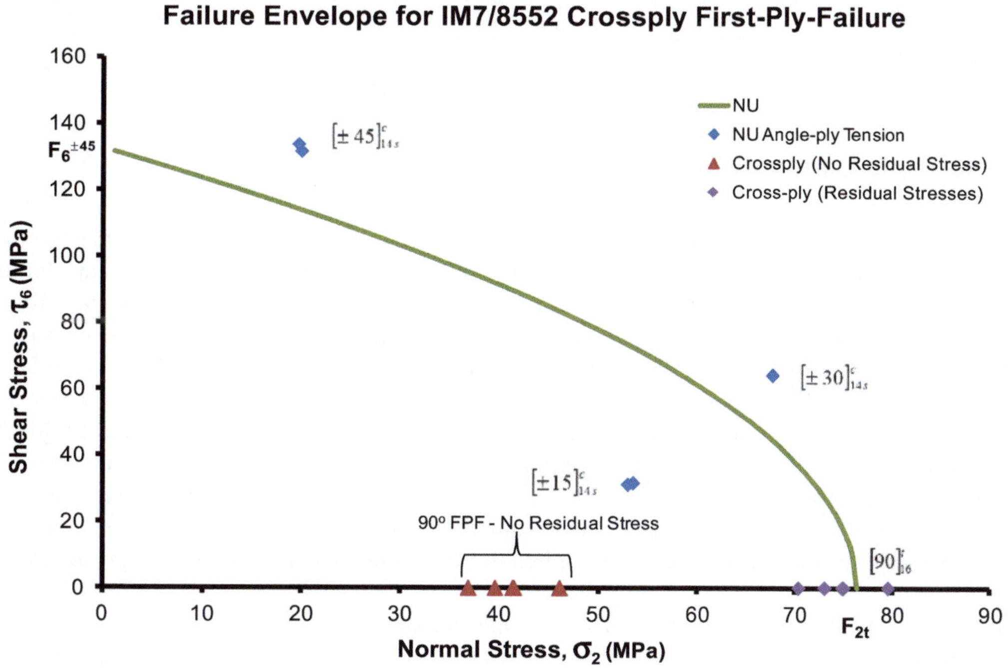

Fig. 20.7 NU Failure Envelope for IM7/8552 Crossply FPF with angle-ply

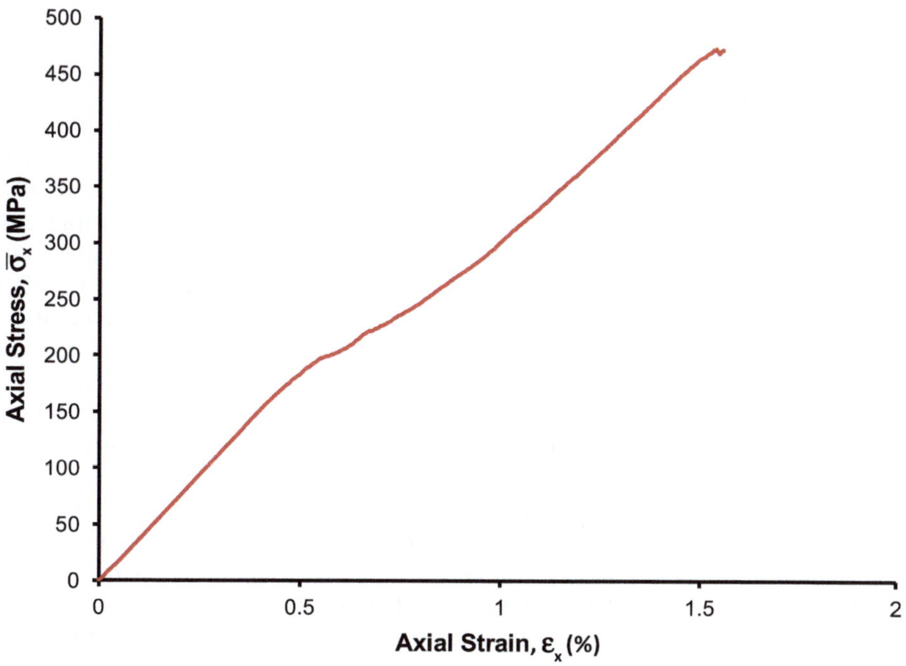

Fig. 20.8 AS4/3501–6 $[0/90_4]_s$ progressive failure

The residual stress was determined from the AS4/3501–6 cure cycle [1], and the temperature difference between the dwell to ambient was set as $\Delta T = -150\ K$. The determined axial residual stress (σ_2) in the embedded 90° lamina was 32.8 MPa. Two tests were performed and the audible FPF was recorded. The strain and transverse stress in the embedded 90° layers is provided in Table 20.3.

The transverse tensile stress for AS4/3501-6 is 57.6 MPa, as previously determined by Daniel et al. [48]. The crossply results correlate well to the lamina value and serve as further validation to the analysis methodology.

Table 20.3 Audible FPF stress and strain values for embedded 90° lamina in IM7/8552 crossply laminate

#	$\varepsilon_X(\%)$	σ_2 Total (MPa)
1	0.27	61
2	0.23	56.3

20.8 Quasi-Isotropic Laminates

While representing an important step towards validating the failure-mode-based approach of the Northwestern Failure Theory, crossply laminates are rarely, if ever, used in practice. In the development of many existing failure theories, it is generally observed that once a particular approach has been shown applicable to a specific test case (typically the exact problem for which the theory was created) the theory is considered 'validated.' Realistically, this is merely validation of the underlying analysis principles and leaves considerable evaluation work yet to be performed. Moreover, a failure theory is not 'validated' until the task for which it has been developed has been addressed by different approaches. For example, a theory that predicts matrix-dominated failure in a lamina should predict matrix-dominated failure in a multidirectional laminate. Thus, the matrix-dominated failure may be analyzed as the same problem inherent to composites in general, and the specific failure modes may be isolated and tested across various laminate configurations. This 'redundant' approach to analysis, noted by Hart-Smith [49–52], provides the robust testing required to effectively signify 'validation'. Therefore, it is highly desirable to evaluate the performance of theory on applicable laminates of considerable design interest: quasi-isotropic laminates.

Quasi-isotropic laminates are a specific class of orthotropic laminates in which the elastic properties are orientation independent; wherein, the in-plane stiffnesses and compliances (including all engineering elastic constants) are identical in all directions. Noted by Daniel and Ishai [1], the simplest type of quasi-isotropic laminate has the configuration $[0/+60/-60]_s$. The general layup of quasi-isotropic laminates has the form:

$$\left[0/\frac{\pi}{n}/\frac{2\pi}{n}/\ldots/\frac{n-1}{n}\pi\right]_s$$

where n is any integer greater than 2.

The $[0/\pm45/90]_s$ layup is commonly employed for general multiaxial loading conditions based on these in-plane properties. The non-linear stress-strain response under load due to embedded ply cracking was analyzed by Deng et al. [53], and Candido et al. [54], while Ogi et al. [55] investigated how moisture and temperature affected quasi-isotropic laminate failure behavior. Quasi-isotropic failure consists primarily of matrix cracking that precedes ultimate laminate failure (final ply failure), the latter being fiber-dominated. Tong et al. [56, 57] recorded the densification of transverse matrix cracks in the embedded 90° and ±45° layers of quasi-isotropic specimens under increasing loading. The results showed that a significant reduction in laminate stiffness occurred with increasing crack density in the 45° plies and the authors noted the need for a model to accurately determine FPF of each embedded layer. This perspective was further reinforced by Swanson and Trask [58] who evaluated quasi-isotropic loading behavior using off-axis tests. Hallett et a. [59], and Chen et al. [60] studied the interlaminar failure and failure progression that coincides with transverse matrix cracking, while Herakovich [61] reported on the effect of free edges on specimen geometry in testing. A through-thickness analysis has previously been performed by Zhou et al. [62] and Paradies investigated effective layup orientations for quasi-isotropic laminates to resist bearing damage [63]. Several attempts have been made to compare available experimental data with an accurate failure model [52, 59–74]; however, Welsh et al. [75] related that the significant paucity of available data for relevant material systems has severely limited the development and validation of useful testing approaches and the understanding of material failure. The authors, working closely with the Air Force Research Laboratory, developed a novel biaxial and triaxial testing system to load laminates in any stress state. While still in development, the work highlights the importance to determine directly practical models for predicting laminate failure.

20.9 Quasi-Isotropic Laminate First Ply Failure

Initially, an incremental loading technique was employed to isolate FPF. A coupon was selected at random and placed in the Instron grips and loaded to ~2.2 kN (500 lbs). The test was then halted and the loading was slowly removed. A microscope was then used to review the entire polished edge of the specimen for cracks in the embedded layers. The number of cracks in

Table 20.4 Embedded layer thermally-induced residual stresses

θ°	σ_{1res} (MPa)	σ_{2res} (MPa)	τ_{6res} (MPa)
0	−38.57	38.57	0
+45	−38.57	38.57	0
−45	−38.57	38.57	0
90	−38.57	38.57	0

each layer was recorded along the gage length, and the coupon was then re-mounted in the testing machine for loading. The sequential loading process continued at increasing intervals of ~2.2 kN (500 lbs) until FPF was audibly detected, at which point the first cracks were seen under the microscope. Further tests were performed continuously and the audible FPF load was recorded. The residual stresses in each embedded layer were determined, and the results are shown in Table 20.4.

The laminate stress-strain behavior under sequential loading is depicted in Fig. 20.9.

The first six loadings indicate linear elastic stress-strain behavior with slight stiffening present by the sixth loading. Audible FPF occurred during the eighth loading. The curves for the first seven loadings were plotted on the same axis to show the linear elastic behavior (Fig. 20.10).

The results show that all of the initial seven loadings fall along the same path – this indicates the laminate stress-strain behavior is linear elastic and that the sequential loading approach did not macroscopically alter the laminate behavior up until this point. No cracks were observed along the specimen edge under the microscope. An example of the coupon edge before FPF is shown in Fig. 20.10 (right) (Fig. 20.11).

The axial laminate modulus (58 GPa) was determined from the stress-strain curve of the seventh iterative loading plot. A value of 59 GPa was predicted using laminate mechanics and this compares well with the experimental result, further verifying the linear elastic laminate behavior. First ply failure was noted during the eighth iterative loading, and the stress-strain behavior is shown in Fig. 20.12.

The sample was unloaded and the polished edge was observed under the microscope. A transverse crack was visible in the embedded 90° and −45° layers near the center of the coupon. The coupon edge is shown in Fig. 20.13.

In the sample, FPF occurred at the same axial stress for both the embedded 90° and 45° layers. Without direct observation of the specimen edge during testing, it is impossible to determine which layers failed first; thus, the potential exists for failure within one layer caused failure in the other. Given this challenge, the axial stress was assumed to be similar for both embedded laminae. The coupon was returned to the Instron and the sequential loading approach proceeded with iterative loading and subsequent crack counts noted until final ply failure occurred. The curves were plotted together for comparison and are presented in Fig. 20.14.

The laminate axial stress-strain behavior deviates from the observed linear elastic trend after the 10th loading due to successive cracking of the embedded layers. Once matrix cracking occurs, an effectively new laminate is 'created' with altered mechanical properties. The curves may be reconstituted into a composite stress-strain curve for the laminate (Fig. 20.15).

For clarity, only the seventh to the final loading segments are shown. The above reconstructed curve was plotted along with the curve for the sequence of uninterrupted tests and is shown in Fig. 20.16.

The reconstructed curve matches relatively-well to the uninterrupted curve. Five additional tests were performed and the axial stress was recorded for the audible FPF. The lamina stresses were determined for each layer using laminate mechanics and accounting for residual stresses.

20.10 Application of Northwestern Failure Theory to Quasi-Isotropic Laminates

The embedded 90° layer audible first ply failure points were plotted in the τ_6 - σ_2 plane. The Northwestern Failure Theory (tensile branch) envelope, along with the classic failure criteria, was included for comparison and the results are shown in Fig. 20.17.

The quasi-isotropic 90° first ply failure points were plotted long with those of the embedded 90° layers in the crossply laminate; further, 10° off-axis lamina tension test data was included for comparison.

When the 90° layers begin to fail, their contribution to axial laminate stiffness and strength diminishes with increasing crack density. Total ply-discounting; wherein the cracked layer is assumed to provide zero support to the laminate, is a conservative analysis that provides a lower threshold for predicting the failure of the surround layers in the quasi-isotropic laminate [76, 77]. Whitney [76] presented an analysis on the reduced stiffnesses determined for various laminates once transverse cracking had initiated and compared ply reduction factors as a function of crack density. Sun et al. [72, 79]

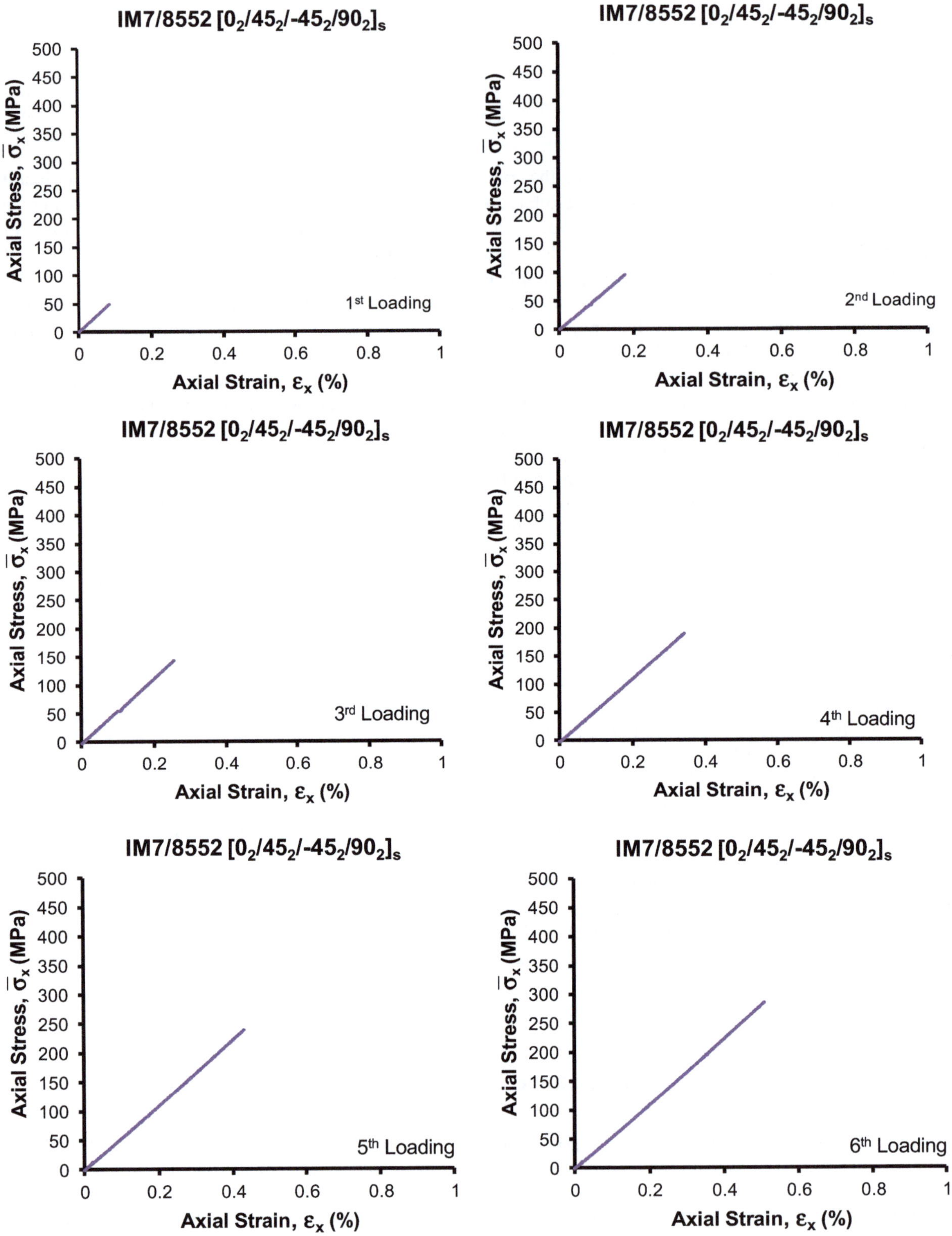

Fig. 20.9 Sequential loading curves for IM7/8552 quasi-isotropic laminate

developed a parallel spring stiffness reduction model and shear-lag analysis to predict the progressive failure several multi-layered laminates. Because the transverse cracks in the 90° layer are quite small and isolated, it requires considerable cracking to occur before the layer becomes so damaged that the overall laminate stiffness is adversely affected. To

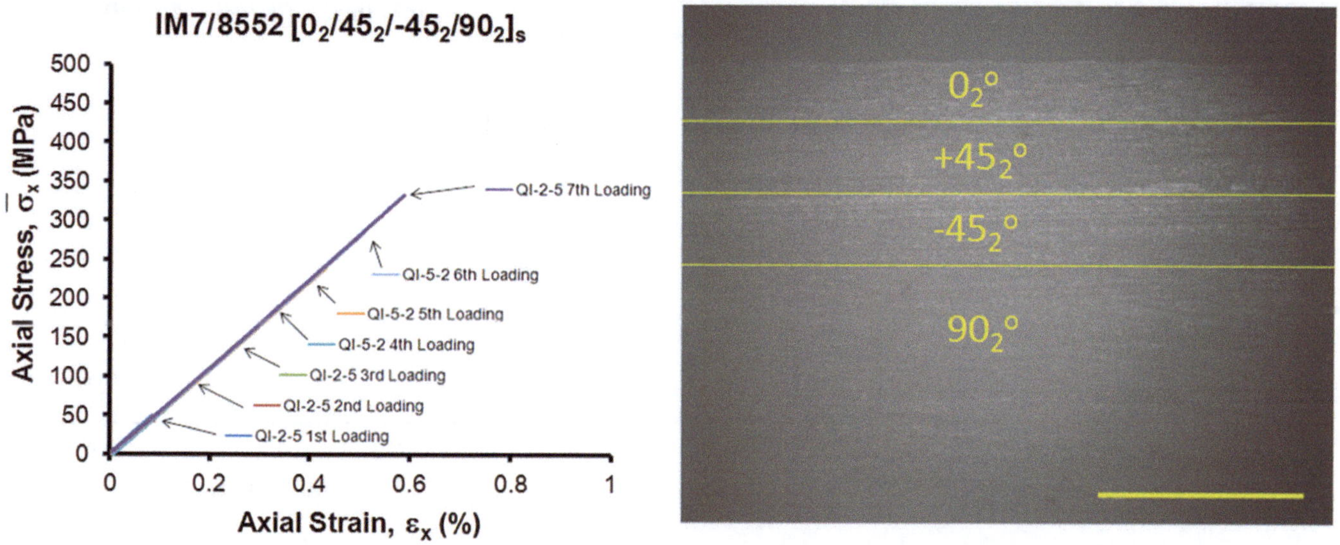

Fig. 20.10 Linear elastic stress-strain behavior of IM7/8552 quasi-isotropic laminate (*left*); laminate edge prior to FPF – scale bar = 0.84 mm (0.033 in.) (*right*)

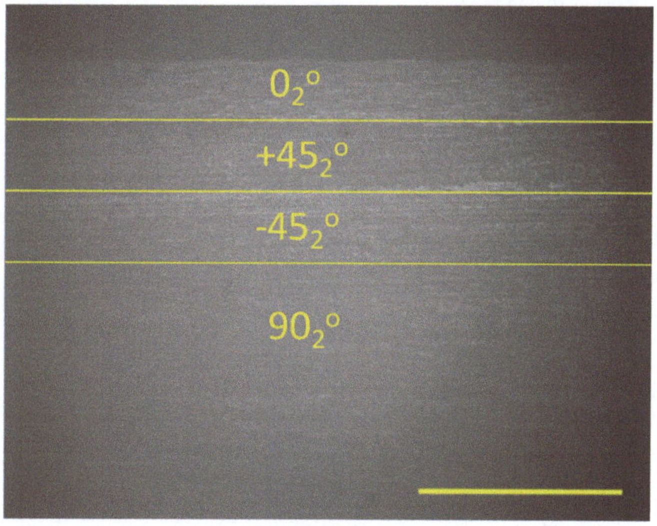

Fig. 20.11 Quasi-isotropic laminate edge prior to FPF; scale bar = 0.84 mm (0.033 in.)

investigate the effects of transverse cracking in the embedded 90° layers, the extensional stiffnesses [A] were calculated for the quasi-isotropic laminate once the 90° layers had reached FPF. The transverse modulus (E_2) and the shear modulus (G_{12}) were assumed to be 0, ¼, ½, ¾, and 1 times of their original values once FPF was reached, and the extensional stiffnesses were determined. The results are shown in Table 20.5.

Regarding the table, cracking in the embedded 90° layer is shown to have little influence on the overall laminate stiffnesses. This result is expected due to the relatively low stiffness of the 90° (9 GPa) layer compared to that of the 45° (20 GPa) and 0° (154 GPa) layers in the loading direction. Based on the analysis, it is assumed that no ply-discounting approach is required to determine the FPF of the embedded 45° layers. The Northwestern Failure Theory was applied to predict FPF of the 45° layers including residual stress and the results are shown in Table 20.6.

The NU Theory indicates that the 45° lamina will fail in shear. The experimental data for FPF of the 90° and 45° layers show a scatter of ~10%, which is reasonable given the approach. The axial stress for failure of the embedded 90° layer is predicted to occur at 274 MPa, while the 45° layer is expected to fail at 295 MPa. As expected, the values were quite close, and given the experimental scatter, the first ply failure of the embedded layers is assumed to occur at approximately the same stress as noted. The data was plotted in the τ_6 - σ_2 plane, and the results are shown in Fig. 20.18.

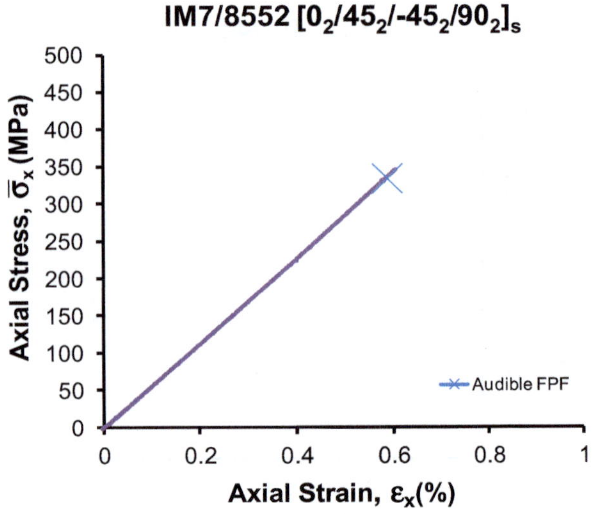

Fig. 20.12 First ply failure of IM7/8552 quasi-isotropic laminate (8th loading)

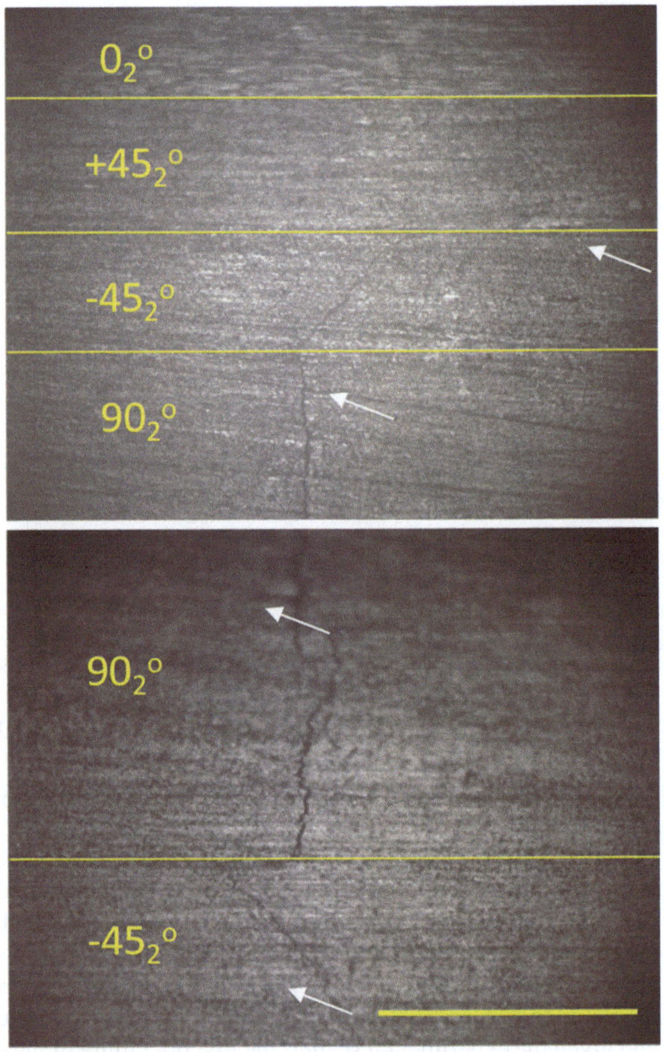

Fig. 20.13 Transverse crack in IM7/8552 quasi-isotropic laminate at FPF; scale bar is 0.56 mm (0.022 in.)

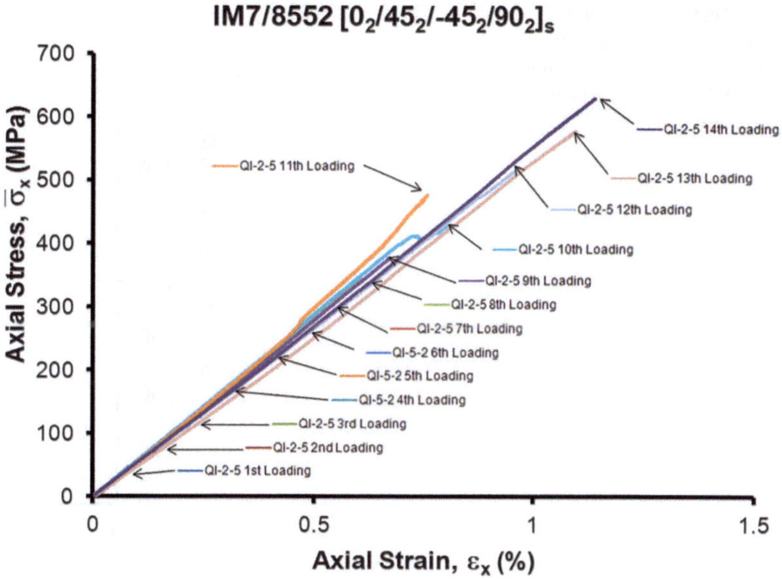

Fig. 20.14 Sequential loading of IM7/8552 quasi-isotropic laminate; all loadings shown

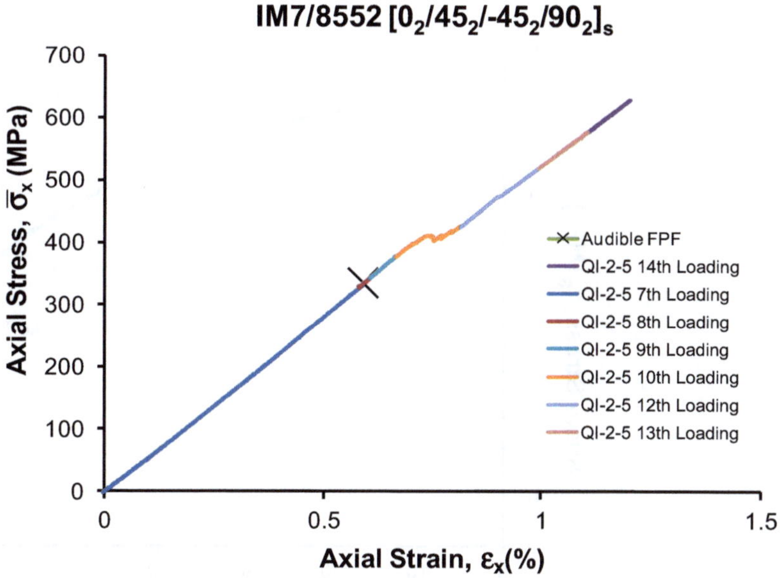

Fig. 20.15 Reconstituted loading curve for IM7/8552 quasi-isotropic sequentially loaded specimen

The Northwestern Failure Envelope is shown to fit the FPF of the 90° and 45° embedded layers quite well. The FPF points for the previously studied crossply laminate as well as the failure points for the tested lamina are included for comparison. Thus, the NU Theory is shown to predict matrix failure in a variety of increasingly complex composite layups.

In light of the present work, the Northwestern Failure Theory has been shown to be effective at modeling the material behavior of composite lamina and angle-ply laminates [80–83]. In terms of industry validation efforts, such failure mode based criteria provide an attractive capability for enabling the predictability of progressive damage and failure analysis methods [84, 85].

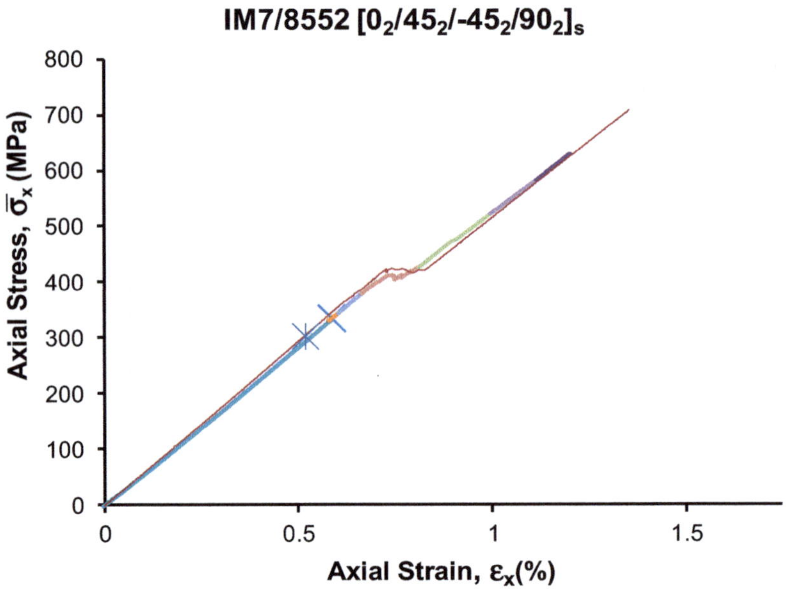

Fig. 20.16 Sequential loading vs. continuously loaded stress-strain behavior for IM7/8552 quasi-isotropic laminate

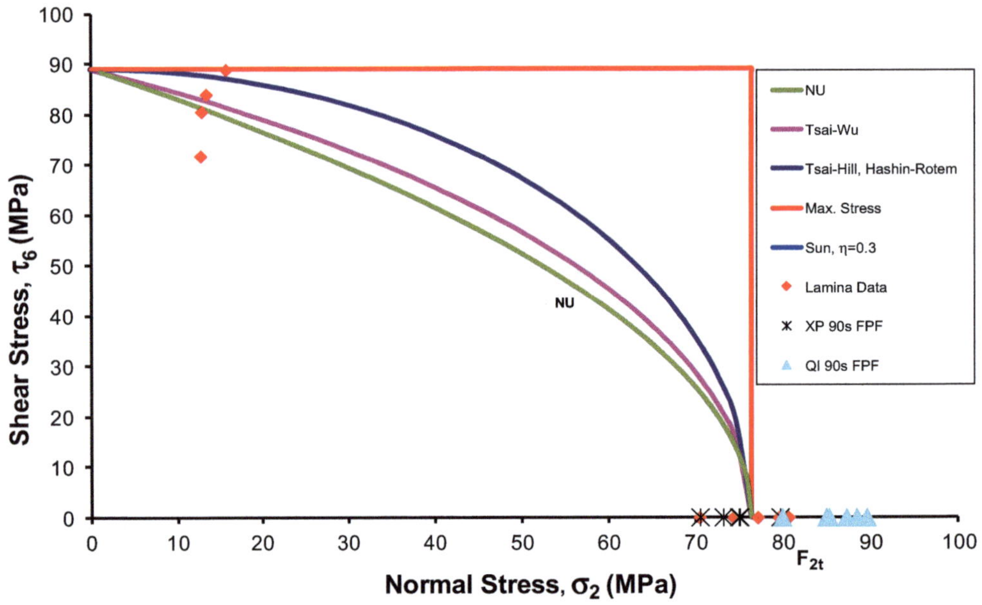

Fig. 20.17 Static failure envelopes for embedded 90° lamina FPF

20.11 Summary and Conclusions

The failure progression of a fiber-reinforced toughened-matrix composite (IM7/8552) was experimentally characterized at quasi-static (10^{-4} s^{-1}) strain rate using crossply and quasi-isotropic laminate specimens. The Northwestern Failure Theory (NU Theory) was used to provide a set of physics-based failure criteria for predicting the matrix-dominated failure of embedded plies using the lamina-based transverse tension, transverse compression, and shear failure strengths. The NU Theory was used to predict the first-ply-failure of embedded plies in $[0/90_4]_s$ and $[0_2/45_2/-45_2/90_2]_s$ laminates for the embedded 90° and 45° plies. The Northwestern Criteria were found to provide superior prediction of the matrix-dominated

Table 20.5 Extensional stiffnesses for quasi-isotropic laminate after 90° FPF

90° layer	N/m			
E_2&G_{12}	A_{xx}	A_{yy}	A_{xy}	A_{ss}
0	1.40E + 08	1.45E + 08	4.29E + 07	4.73E + 07
1/4	1.42E + 08	1.45E + 08	4.33E + 07	4.81E + 07
1/2	1.43E + 08	1.45E + 08	4.38E + 07	4.89E + 07
3/4	1.44E + 08	1.45E + 08	4.42E + 07	4.96E + 07
1	1.45E + 08	1.45E + 08	4.46E + 07	5.04E + 07

Table 20.6 NU Failure Theory prediction of 90° and 45° FPF in IM7/8552 quasi-isotropic laminate

ε_x(%)	σ_x(MPa)	NU Tension 90°(MPa)	NU Tension 45°(MPa)	NU Tension 45°(MPa)
0.59	334.7	274	332	295
0.56	321.6	274	332	295
0.55	331.6	274	332	295
0.52	301.3	274	332	295
0.52	303.7	274	332	295
0.46	269	274	332	295

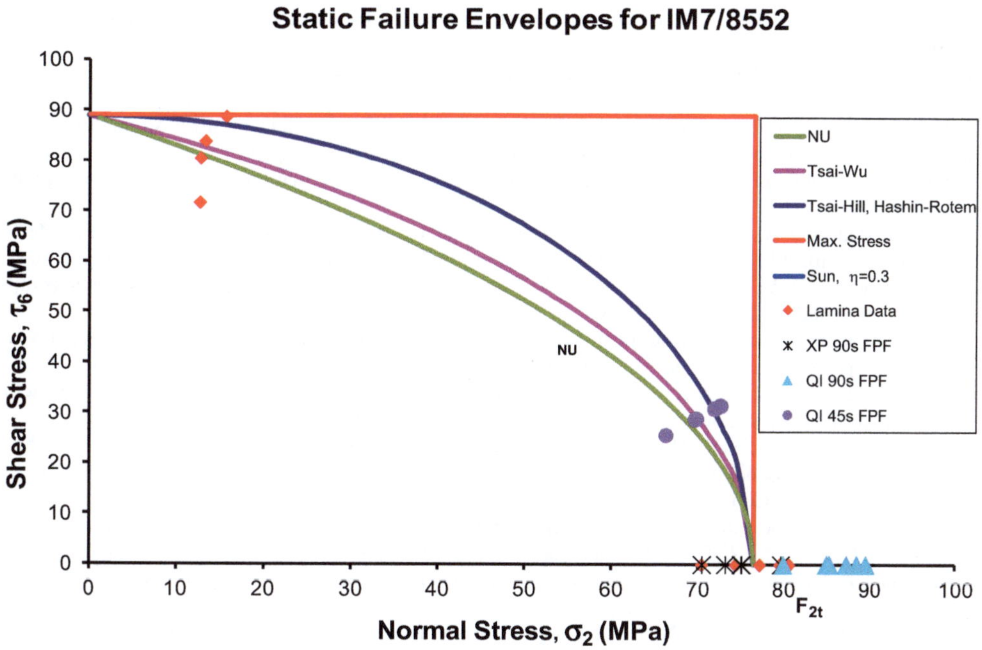

Fig. 20.18 Static failure envelopes for embedded 90° and 45° lamina FPF

embedded ply failure for all evaluated cases compared to the classical approaches. The results indicate the potential to use the Northwestern Criteria to provide the predictive baseline for damage propagation in composite laminates based on experimentally identified damage response on a length scale-relevant basis.

Acknowledgement Sandia National Laboratories is a multi-mission laboratory managed and operated by Sandia Corporation, a wholly owned subsidiary of Lockheed Martin Corporation, for the U.S. Department of Energy's National Nuclear Security Administration under contract DE-AC04-94AL85000.

References

1. Daniel, I.M., Ishai, O.: Engineering Mechanics of Composite Materials. Oxford University Press, Oxford (2006)
2. Reifsnider, K.L., Masters, J.E.: An investigation of cumulative damage development in quasi-isotropic graphite/epoxy laminates. In: Damage in Composite Materials ASTM STP 775, pp. 40–62. American Society for Testing and Materials, West Conshohocken (1982)
3. Henaff-Gardin, C., Lafarie-Frenot, M.C.: The use of a characteristic damage variable in the study of transverse cracking development under fatigue loading in cross-ply laminates. Int. J. Fatigue. **24**(2–4), 389–395 (2002)
4. Kashtalyan, M., Soutis, C.: Stiffness degradation in cross-ply laminates damaged by transverse cracking and splitting. Compos. A: Appl. Sci. Manuf. **31**(4), 335–351 (2000)
5. Lee, J.-W., Daniel, I.M.: Progressive transverse cracking of crossply composite laminates. J. Compos. Mater. **24**, 1225–1243 (1990)
6. Li, C., Ellyin, F., Wharmby, A.: On matrix crack saturation in composite laminates. Compos. Part B Eng. **34**(5), 473–480 (2003)
7. Li, S., Jiang, C., Han, S.: Modeling of the characteristics of fiber-reinforced composite materials damaged by matrix-cracking. Compos. Sci. Technol. **43**(2), 185–195 (1992)
8. Reifsnider, K.L., Highsmith, A.L.: Stiffness reduction mechanisms in composite laminates. In: Damage in Composite Materials ASTM STP 775, pp. 103–117. American Society for Testing Materials, West Conshohocken (1982)
9. Tagarielli, V.L., Minisgallo, G., McMillan, A.J., Petrinic, N.: The response of a multi-directional composite laminate to through-thickness loading. Compos. Sci. Technol. **70**(13), 1950–1957 (2010)
10. Chou, P.C., Wang, A.S.D.: A stochastic model for the growth of matrix cracks in composite laminates. J. Compos. Mater. **18**, 239–254 (1984)
11. Chou, T.W., Fukunaga, H.: Probabilistic failure strength analysis of graphite/epoxy cross-ply laminates. J. Compos. Mater. **18**, 339–356 (1984)
12. Gamby, D., Rebière, J.L.: A two-dimensional analysis of multiple matrix cracking in a laminated composite close to its characteristic damage state. Compos. Struct. **25**(1–4), 325–337 (1993)
13. Huang, Z.Q., Nie, G.H., Chan, C.K.: An exact solution for stresses in cracked composite laminates and evaluation of the characteristic damage state. Compos. Part B Eng. **42**(5), 1008–1014 (2011)
14. Joffe, R., Varna, J.: Analytical modeling of stiffness reduction in symmetric and balanced laminates due to cracks in 90° layers. Compos. Sci. Technol. **59**(11), 1641–1652 (1999)
15. Vaughan, T.J., McCarthy, C.T.: Micromechanical modelling of the transverse damage behaviour in fibre reinforced composites. Compos. Sci. Technol. **71**, 388–396 (2011)
16. Daniel, I.M., Lee, J.-W.: Damage development in composite laminates under monotonic loading. J. Compos. Technol. Res. **12**(2), 98–102 (1990)
17. Karthikeyan, K., Russell, B.P., Fleck, N.A., Wadley, H.N.G., Deshpande, V.S.: The effect of shear strength on the ballistic response of laminated composite plates. Eur. J. Mech. A. Solids. **42**(0), 35–53 (2013)
18. Pandya, K.S., Dharmane, L., Pothnis, J.R., Ravikumar, G., Naik, N.K.: Stress wave attenuation in composites during ballistic impact. Polym. Test. **31**(2), 261–266 (2012)
19. Gower, H.L., Cronin, D.S., Plumtree, A.: Ballistic impact response of laminated composite panels. Int. J. Impact Eng. **35**(9), 1000–1008 (2008)
20. Mohan, S., Velu, S.: Ballistic impact behaviour of unidirectional fibre reinforced composites. Int. J. Impact Eng. **63**(0), 164–176 (2014)
21. Morye, S.S., Hine, P.J., Duckett, R.A., Carr, D.J., Ward, I.M.: Modelling of the energy absorption by polymer composites upon ballistic impact. Compos. Sci. Technol. **60**(14), 2631–2642 (2000)
22. Naik, N.K., Doshi, A.V.: Ballistic impact behaviour of thick composites: Parametric studies. Compos. Struct. **82**(3), 447–464 (2008)
23. Naik, N.K., Shrirao, P.: Composite structures under ballistic impact. Compos. Struct. **66**(1–4), 579–590 (2004)
24. Sevkat, E., Liaw, B., Delale, F., Raju, B.B.: A combined experimental and numerical approach to study ballistic impact response of S2-glass fiber/toughened epoxy composite beams. Compos. Sci. Technol. **69**(7–8), 965–982 (2009)
25. Shaktivesh, N.N.S., Sesha Kumar, C.V., Naik, N.K.: Ballistic impact performance of composite targets. Mater. Des. **51**(0), 833–846 (2013)
26. Halabe, U.B.: 18 – Non-destructive evaluation (NDE) of composites: Techniques for civil structures. In: Karbhari, V.M. (ed.) Non-Destructive Evaluation (NDE) of Polymer Matrix Composites, pp. 483–517e. Woodhead Publishing (2013)
27. Karbhari, V.M.: 1 – Introduction: The future of non-destructive evaluation (NDE) and structural health monitoring (SHM). In: Karbhari, V.M. (ed.) Non-Destructive Evaluation (NDE) of Polymer Matrix Composites, pp. 3–11. Woodhead Publishing (2013)
28. Chang, R.R.: Experimental and theoretical analyses of first-ply failure of laminated composite pressure vessels. Compos. Struct. **49**(2), 237–243 (2000)
29. Huang, J.Q.: 2 – Non-destructive evaluation (NDE) of composites: Acoustic emission (AE). In: Karbhari, V.M. (ed.) Non-Destructive Evaluation (NDE) of Polymer Matrix Composites, pp. 12–32. Woodhead Publishing (2013)
30. Kim, R.J.-Y., Choi, N.-S., Ferracane, J., Lee, I.-B.: Acoustic emission analysis of the effect of simulated pulpal pressure and cavity type on the tooth–composite interfacial de-bonding. Dent. Mater. **30**, 876 (2014). (0)
31. Maimı´, P., Camanho, P.P., Mayugo, J.A., Turon, A.: Matrix cracking and delamination in laminated composites. Part I: Ply constitutive law, first ply failure and onset of delamination. Mech. Mater. **43**(4), 169–185 (2011)
32. Njuhovic, E., Bräu, M., Wolff-Fabris, F., Starzynski, K., Altstädt, V.: Identification of interface failure mechanisms of metallized glass fibre reinforced composites using acoustic emission analysis. Compos. Part B Eng. **66**, 443. (0)
33. Romanowicz, M.: Determination of the first ply failure load for a cross ply laminate subjected to uniaxial tension through computational micromechanics. Int. J. Solids Struct. **51**(13), 2549–2556 (2014)
34. Roozen, N.B., Tazelaar, K., Koussios, S., Beukers, A.: A new method to measure critical strain in composite materials – Combining the Euler–Fresnel spiral with acoustic emission to assess crack positions. Compos. Sci. Technol. **100**(0), 228–236 (2014)
35. Satish Kumar, Y.V., Srivastava, A.: First ply failure analysis of laminated stiffened plates. Compos. Struct. **60**(3), 307–315 (2003)
36. Woo, S.-C., Kim, T.-W.: High-strain-rate impact in Kevlar-woven composites and fracture analysis using acoustic emission. Compos. Part B Eng. **60**(0), 125–136 (2014)

37. Tittmann, B.R., Miyasaka, C., Guers, M., Kasano, H., Morita, H.: 16 – Non-destructive evaluation (NDE) of aerospace composites: Acoustic microscopy. In: Karbhari, V.M. (ed.) Non-Destructive Evaluation (NDE) of Polymer Matrix Composites, pp. 423–49e. Woodhead Publishing (2013)
38. Avdelidis, N.P., Gan, T.H.: 24 – Non-destructive evaluation (NDE) of Composites: Infrared (IR) thermography of wind turbine blades. In: Karbhari, V.M. (ed.) Non-Destructive Evaluation (NDE) of Polymer Matrix Composites, pp. 634–50e. Woodhead Publishing (2013)
39. Ley, O., Godinez, V.: 12 – Non-destructive evaluation (NDE) of aerospace composites: Application of infrared (IR) thermography. In: Karbhari, V.M. (ed.) Non-Destructive Evaluation (NDE) of Polymer Matrix Composites, pp. 309–36e. Woodhead Publishing (2013)
40. Shirazi, A., Karbhari, V.M.: 19 – Non-destructive evaluation (NDE) of composites: Application of thermography for defect detection in rehabilitated structures. In: Karbhari, V.M. (ed.) Non-Destructive Evaluation (NDE) of Polymer Matrix Composites, p. 515–541. Woodhead Publishing, 2013
41. Suratkar, A., Sajjadi, A.Y., Mitra, K.: 25 – Non-destructive evaluation (NDE) of composites for marine structures: Detecting flaws using infrared thermography (IRT). In: Karbhari, V.M. (ed.) Non-Destructive Evaluation (NDE) of Polymer Matrix Composites, pp. 649–68e. Woodhead Publishing (2013)
42. Feng, M.Q., Roqueta, G., Jofre, L.: 22 – Non-destructive evaluation (NDE) of composites: Microwave techniques. In: Karbhari, V.M. (ed.) Non-Destructive Evaluation (NDE) of Polymer Matrix Composites, pp. 574–616. Woodhead Publishing (2013)
43. Hsu, D.K.: 15 – Non-destructive evaluation (NDE) of aerospace composites: Ultrasonic techniques. In: Karbhari, V.M. (ed.) Non-Destructive Evaluation (NDE) of Polymer Matrix Composites, pp. 397–422. Woodhead Publishing (2013)
44. Dong, Y.: 23 – Non-destructive evaluation (NDE) of composites: Using fiber optic sensors. In: Karbhari, V.M. (ed.) Non-Destructive Evaluation (NDE) of Polymer Matrix Composites, pp. 617–633. Woodhead Publishing (2013)
45. Schaefer, J.D., Lee, J., Liguore, S.L., Richardson, T.D.: (2015, October 26–29). High Fidelity Test Database for Validation of Progressive Failure Analysis Methods. Composites and Advanced Materials Expo, Dallas, TX, USA
46. Francis, D.: 4 – Non-destructive evaluation (NDE) of composites: Introduction to shearography. In: Karbhari, V.M. (ed.) Non-Destructive Evaluation (NDE) of Polymer Matrix Composites, pp. 56–83. Woodhead Publishing (2013)
47. Schaefer, J.D., Justusson, B.P., Liguore, S.L., A comparison of emerging in-situ inspection techniques for validation of composite PDA methods, Society for the Advancement of Material and Process Engineering, Long Beach, CA, 23–26 May 2016
48. Daniel, I.M., Cho, J.-M., Werner, B.T., Fenner, J.S.: Characterization and constitutive modeling of composite materials under static and dynamic loading. AIAA J. **49**(8), 1658–1664 (2011. 2011/08/01)
49. Hart-Smith, L.J.: The role of biaxial stresses in discriminating between meaningful and illusory composite failure theories. Compos. Struct. **25** (1–4), 3–20 (1993)
50. Hart-Smith, L.J.: A re-examination of the analysis of in-plane matrix failures in fibrous composite laminates. Compos. Sci. Technol. **56**(2), 107–121 (1996)
51. Hart-Smith, L.J.: Predictions of a generalized maximum-shear-stress failure criterion for certain fibrous composite laminates. Compos. Sci. Technol. **58**(7), 1179–1208 (1998)
52. Hart-Smith, L.J.: Comparison between theories and test data concerning the strength of various fibre–polymer composites. Compos. Sci. Technol. **62**(12–13), 1591–1618 (2002)
53. Deng, S., Li, X., Lin, H., Weitsman, Y.J.: The non-linear response of quasi-isotropic composite laminates. Compos. Sci. Technol. **64**(10–11), 1577–1585 (2004)
54. Cândido, G.M., Costa, M.L., Rezende, M.C., Almeida, S.F.M.: Hygrothermal effects on quasi-isotropic carbon epoxy laminates with machined and molded edges. Compos. Part B Eng. **39**(3), 490–496 (2008)
55. Ogi, K., Kim, H.S., Maruyama, T., Takao, Y.: The influence of hygrothermal conditions on the damage processes in quasi-isotropic carbon/epoxy laminates. Compos. Sci. Technol. **59**(16), 2375–2382 (1999)
56. Tong, J., Guild, F.J., Ogin, S.L., Smith, P.A.: On matrix crack growth in quasi-isotropic laminates—I. Experimental investigation. Compos. Sci. Technol. **57**(11), 1527–1535 (1997)
57. Tong, J., Guild, F.J., Ogin, S.L., Smith, P.A.: On matrix crack growth in quasi-isotropic laminates—II. Finite element analysis. Compos. Sci. Technol. **57**(11), 1537–1545 (1997)
58. Swanson, S.R., Trask, B.C.: Strength of quasi-isotropic laminates under off-axis loading. Compos. Sci. Technol. **34**(1), 19–34 (1989)
59. Hallett, S.R., Jiang, W.-G., Khan, B., Wisnom, M.R.: Modelling the interaction between matrix cracks and delamination damage in scaled quasi-isotropic specimens. Compos. Sci. Technol. **68**(1), 80–89 (2008)
60. Chen, J.-F., Morozov, E.V., Shankar, K.: Simulating progressive failure of composite laminates including in-ply and delamination damage effects. Compos. A: Appl. Sci. Manuf. **61**(0), 185–200 (2014)
61. Herakovich, C.T.: Failure modes and damage accumulation in laminated composites with free edges. Compos. Sci. Technol. **36**(2), 105–119 (1989)
62. Zhou, G., Sim, L.M., Brewster, P.A., Giles, A.R.: Through-the-thickness mechanical properties of smart quasi-isotropic carbon/epoxy laminates. Compos. A: Appl. Sci. Manuf. **35**(7–8), 797–815 (2004)
63. Paradies, R.: Designing quasi-isotropic laminates with respect to bending. Compos. Sci. Technol. **56**(4), 461–472 (1996)
64. Edgren, F., Asp, L.E., Joffe, R.: Failure of NCF composites subjected to combined compression and shear loading. Compos. Sci. Technol. **66** (15), 2865–2877 (2006)
65. Esrail, F., Kassapoglou, C.: An efficient approach to determine compression after impact strength of quasi-isotropic composite laminates. Compos. Sci. Technol. **98**(0), 28–35 (2014)
66. Garg, A.C.: The fracture mechanics of some graphite fibre-reinforced epoxy laminates, part 1: Quasi-isotropic laminates. Composites. **17**(2), 141–149 (1986)
67. Guedes, R.M., de Moura, M.F.S.F., Ferreira, F.J.: Failure analysis of quasi-isotropic CFRP laminates under high strain rate compression loading. Compos. Struct. **84**(4), 362–368 (2008)
68. Kaddour, A.S., Hinton, M.J., Soden, P.D.: A comparison of the predictive capabilities of current failure theories for composite laminates: Additional contributions. Compos. Sci. Technol. **64**(3–4), 449–476 (2004)

69. Park, I.K., Park, K.J., Kim, S.J.: Rate-dependent damage model for polymeric composites under in-plane shear dynamic loading. Comput. Mater. Sci. **96**, 506. (0)
70. Schultheisz, C.R., Waas, A.M.: Compressive failure of composites, part I: Testing and micromechanical theories. Prog. Aerosp. Sci. **32**(1), 1–42 (1996)
71. Soden, P.D., Hinton, M.J., Kaddour, A.S.: A comparison of the predictive capabilities of current failure theories for composite laminates. Compos. Sci. Technol. **58**(7), 1225–1254 (1998)
72. Sun, C.T., Tao, J.: Prediction of failure envelopes and stress/strain behaviour of composite laminates. Compos. Sci. Technol. **58**(7), 1125–1136 (1998)
73. Wolfe, W.E., Butalia, T.S.: A strain-energy based failure criterion for non-linear analysis of composite laminates subjected to biaxial loading. Compos. Sci. Technol. **58**(7), 1107–1124 (1998)
74. Zubillaga, L., Turon, A., Maimí, P., Costa, J., Mahdi, S., Linde, P.: An energy based failure criterion for matrix crack induced delamination in laminated composite structures. Compos. Struct. **112**(0), 339–344 (2014)
75. Welsh, J.S., Mayes, J.S., Biskner, A.C.: 2-D biaxial testing and failure predictions of IM7/977-2 carbon/epoxy quasi-isotropic laminates. Compos. Struct. **75**(1–4), 60–66 (2006)
76. Tay, T.E., Lim, E.H.: Analysis of stiffness loss in cross-ply composite laminates. Compos. Struct. **25**(1–4), 419–425 (1993)
77. Bogetti, T.A., Hoppel, C.P.R., Harik, V.M., Newill, J.F., Burns, B.P.: Predicting the nonlinear response and progressive failure of composite laminates. Compos. Sci. Technol. **64**(3–4), 329–342 (2004)
78. Whitney, J.M.: On the 'ply discount method' for determining effective thermo-elastic constants of laminates containing transverse cracks. Compos. A: Appl. Sci. Manuf. **36**(10), 1347–1354 (2005)
79. Sun, C.T., Tao, J., Kaddour, A.S.: The prediction of failure envelopes and stress/strain behavior of composite laminates: Comparison with experimental results. Compos. Sci. Technol. **62**(12–13), 1673–1682 (2002)
80. Daniel, I.M., Schaefer, J.D., Werner, B.: Yield criteria for matrix and composite materials under static and dynamic loading, 20th international conference on composite materials, 19–24 July 2015
81. Schaefer, J.D., Daniel, I.M.: Strain-Rate-Dependent Yield Criteria for Composite Laminates, Fracture, Fatigue, Failure, and Damage Evolution, vol. 8, pp. 197–208. Springer International Publishing (2016)
82. Schaefer, J.D., Daniel, I.M.: Characterization and modeling of progressive damage of angle-ply composite laminates under varying strain rate loading, 31st ASC technical conference and ASTM D30 meeting 2016
83. Schaefer, J.D., Werner, B.T., Daniel, I.M.: Strain-rate-dependent failure of a toughened matrix composite. Exp. Mech. **54**(6), 1111–1120 (2014)
84. Schaefer, J.D. Justusson, B.P., Liguore, S.L., Renieri, G.D.: Assessment of predictive capabilities of progressive damage analysis methods using high fidelity experiments for validation, Society for the advancement of material and process engineering, Long Beach, 23–26 May 2016
85. Razi, H., Schaefer, J.D., Wanthal, S.: Rapid integration of new analysis methods in production, 31st ASC technical conference and ASTM D30 meeting 2016

Chapter 21
Experimental Mechanics for Multifunctional Composites and Next Generation UAVs

Jeffery W. Baur, Darren J. Hartl, Geoffrey J. Frank, Gregory Huff, Keith A. Slinker, Corey Kondash, W. Joshua Kennedy, and Gregory J. Ehlert

Abstract Low-cost unmanned aircraft that use affordable manufacturing and have limited service life can enable mission concepts in which there is a higher tolerance for aircraft loss, or attrition. Because of their higher risk tolerance, these low-cost attritable aircraft could also integrate emerging technology which may have previously been considered too risky for integration into expensive and long life aircraft. Light-weight multifunctional structural composites have the potential to integrate additional functions and enable mission agility without significantly adding weight or reducing payload capacity. However, design and development of these material systems are often difficult because of the traditional "building block" development approach used for traditional composites, the large option space available for structural and functional properties, and the potential complexity of the multiscale and multiphysics coupling. To realize integrated functionality, new multi-scales and multi-physical experimental mechanical characterization techniques should be merged with maturing integrated materials models. We discuss this need using examples of a reconfigurable liquid metal Structurally Embedded Vascular Antenna (SEVA), a plasmonic nanoparticle based method for measuring internal temperature gradients, and embedded micro-cantilever carbon-nanotube based sensors. The latter of these is also used to discuss the potential to accelerate development of multifunctional structural concepts by provide air flow measurement and structural feedback during testing of complex structures. This could, in turn, eliminate some testing of intermediate elements in the slow and expensive traditional "building block" approach.

Keywords Liquid metal antenna • Plasmonic nanoparticle • CNT sensor

21.1 Introduction

21.1.1 Multifunctional Concepts for Next Generation UAVs

Emerging unmanned aerial vehicle (UAV) concepts seek to balance mission agility, sustainment, and vehicle cost. Concepts like low-cost attritable aircraft technology (LCAAT) have limited life and even more limited sustainment cost [1]. The term "attritable" generally means that the user is tolerant of a limited life or the loss of vehicles during operations, due in part to its low-cost. Such a concept also affords the opportunity to consider emerging technology which does not have to be proven viable over the traditional life of an air vehicle. Structurally integrated multifunctional technology which does not

J.W. Baur (✉) • W. Joshua Kennedy • G.J. Ehlert
Materials and Manufacturing Directorate, Air Force Research Laboratory, 2941 Hobson Way, Wright-Patterson Air Force Base, OH, USA
e-mail: Jeffery.Baur@us.af.mil

D.J. Hartl
Materials and Manufacturing Directorate, Air Force Research Laboratory, 2941 Hobson Way, Wright-Patterson Air Force Base, OH, USA

Universal Technology Corporation, 1270 N. Fairfield Road, Beavercreek, OH, USA

Texas A&M University, H.R. Bright Building, 701 Ross St, College Station, TX 77843, USA

G.J. Frank
Materials and Manufacturing Directorate, Air Force Research Laboratory, 2941 Hobson Way, Wright-Patterson Air Force Base, OH, USA

University of Dayton Research Institute, 300 College Park, Dayton, OH 45469, USA

G. Huff
Texas A&M University, H.R. Bright Building, 701 Ross St, College Station, TX 77843, USA

K.A. Slinker • C. Kondash
Materials and Manufacturing Directorate, Air Force Research Laboratory, 2941 Hobson Way, Wright-Patterson Air Force Base, OH, USA

Universal Technology Corporation, 1270 N. Fairfield Road, Beavercreek, OH, USA

P.R. Thakre et al. (eds.), *Mechanics of Composite and Multi-functional Materials, Volume 6*,
Conference Proceedings of the Society for Experimental Mechanics Series, DOI 10.1007/978-3-319-63408-1_21

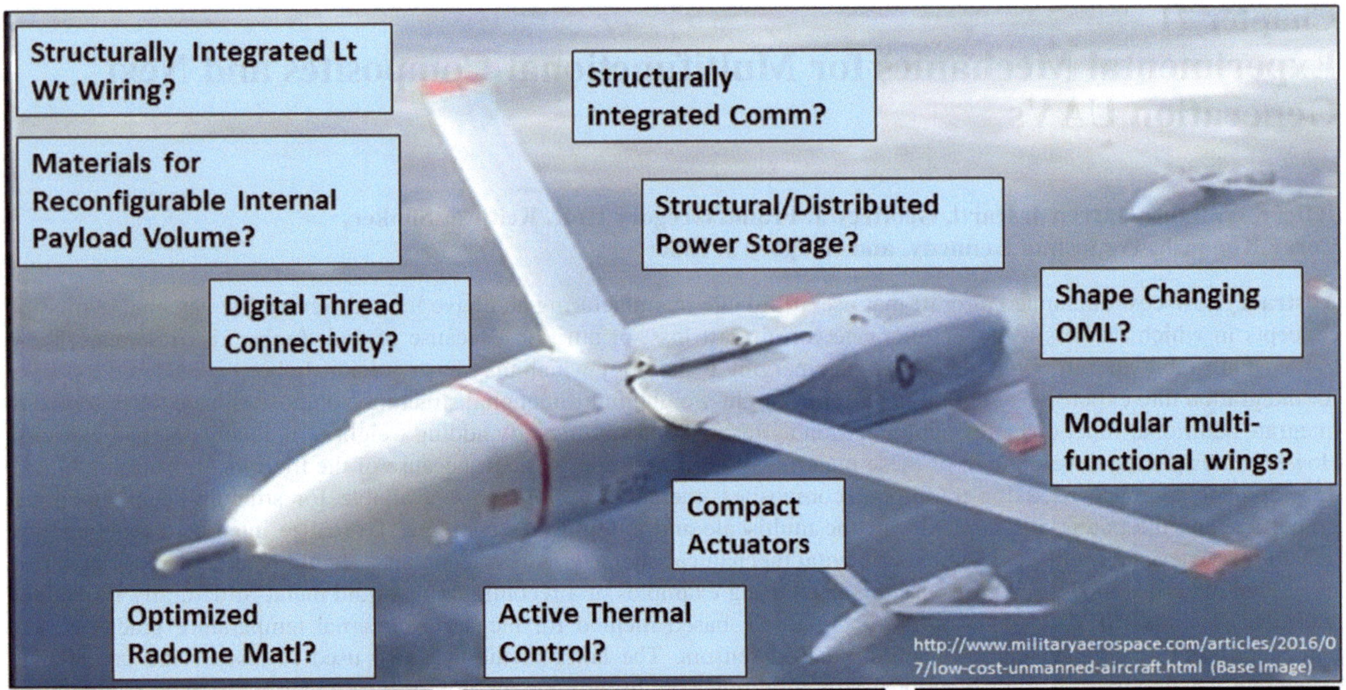

Range Improvement	Improved Comm, Comp, or Ctrl	New Performance
Conformal Systems for Increased Payload Volume (Fuel/other)	Structurally Integrated Antenna for Improved Bandwidth	Structurally Integrated Survivability to Extreme Environments
Low Drag, Seamless Control Surfaces for Increased Range	Structural Integrated Power Storage for Improved On-Board Computation	SHM or Rapidly Repair Structures for Enhanced Availability & Awareness
Integrated Wiring for Decreased Weight and Enhance Internal Volume	Materials for Compact Actuation for Control Surfaces	Shape-Changing OML Materials for Improve Range/Dash
Adaptive Inlet Ducts for Efficient Propulsion	Active Thermal Materials for Internal Cooling	Reconfigurable Materials for Internal Payload Redistribution

Fig. 21.1 Notional technical options for Multifunctional Composite Attritable Aircraft Technologies (Multi-CAAT) (Photo taken from reference in lower right Ref. [1])

significantly add size, weight or power requirements is one such promising technology. Specifically, light-weight structural composites which have functional systems integrated within, and/or on-top of, structural elements is an attractive option. Whichever multifunctional capability is considered, it must be sufficiently impactful and affordable to remain "attritable". For example, light-weight structural composites clearly enable greater range for the envisioned vehicles despite their traditionally higher cost. Reducing the development and manufacturing cost of such composites decreases the vehicle cost and makes the vehicle's attrition more acceptable. Similarly, affordable multifunctional composite technologies have the potential to impact future attritable aircraft and create Multifunctional Composite Attritable Aircraft Technologies (Multi-CAAT).

An important question to be answered is, "Which of the many potential multifunctional concepts can both significantly impact an attritable aircraft's mission effectiveness and remain significantly affordable to maintain the user's acceptance of limited life?" As shown in Fig. 21.1, one can consider many options in regards to enhancing the range of a vehicle, improving the communications, providing more efficient on-board electrical power storage, and decreasing the size, weight and power of aerodynamic controls or thermal management systems. One could also consider enabling new capability in survivability, awareness of vehicle status, external reconfigurability for aerodynamic performance, or internal reconfigurability for efficient payload distribution.

One example is the structurally integrated antenna with improved agility. Today's military air platforms require an ever-increasing range of frequencies for communication on platforms that are constrained by size, weight, power, aerodynamic efficiency, cost, and thermal conditions. Recently, the first ever reconfigurable liquid metal structurally embedded antenna was demonstrated. This light-weight Structurally Embedded Vascular Antenna (SEVA) has better than a 10:1 (from ~250 MHz to 4 GHz) frequency tuning range, gains of up to 2.5 dBi, 30–60° steering, and can potentially outperform copper at high power [2–4]. Heat created by either RF transmission or in the normal aero environment can essentially be wicked away by using the flowing liquid metal as a thermal transporting fluid. The embedded network of channels are created in a scalable, low-cost and agile manner by inserting an additive printed sacrificial polymer between uncured composite layers, curing the composite, and then removing the sacrificial polymer at a higher temperature [5, 6]. Thus, a wide range of antenna patterns are available simply by additively printing different patterns of sacrificial material. The volume of the channels can be small (<2%), so the mechanical properties, processibility, and structural weight are not significantly impacted. This technology could impact missions that require RF communication agility and don't wish to integrate separate high fidelity antennas.

Ultimately, the answer of which multifunctional technology to consider will depend on the user's need, the effectiveness with which a system can be designed, the cost of the technology, and the technical maturation of the concept. Since materials and processing are so impactful to realize the envisioned concept, new methods that manage risk and cost while also accelerating development are needed.

21.1.2 Tailoring the Building Block Approach to Accelerate Development

The current approach to manage risk for aerospace materials and structures is to use a "building block" approach with progressive fabrication and testing of elements with increasing complexity up to the full envisioned component which is the most extensively tested. This approach has been largely successful in mitigating risk for composite structures, but is rather lengthy and expensive [7]. Integrated Computational Science and Engineering (ICMSE) have sought to accelerate this process by integrating validated modeling methods with predictive capability in order to either reduce testing or to better assess which testing should be done [8, 9]. Experimental mechanical techniques which either validate or parameterize these tools is vital to their employment. Some believe that the greatest savings with predictive modeling over experimental testing will occur for the immediate elements within the building block pyramid as shown in Fig. 21.2. These integrated models would be adequately parameterized and validated at the materials level and can be used to predict performance of elements of increasing structural complexity. Yet, new approaches imply increased risk. Without a corresponding acceptance of additional risk from embracing new methods (like that taken for an attritable aircraft) the impact of this approach may not reach full fruition.

The problem of using the building block approach is compounded with multifunctional structures because the added dimension of the functional performance suggest a multiplication of testing to ensure both mechanical and functional

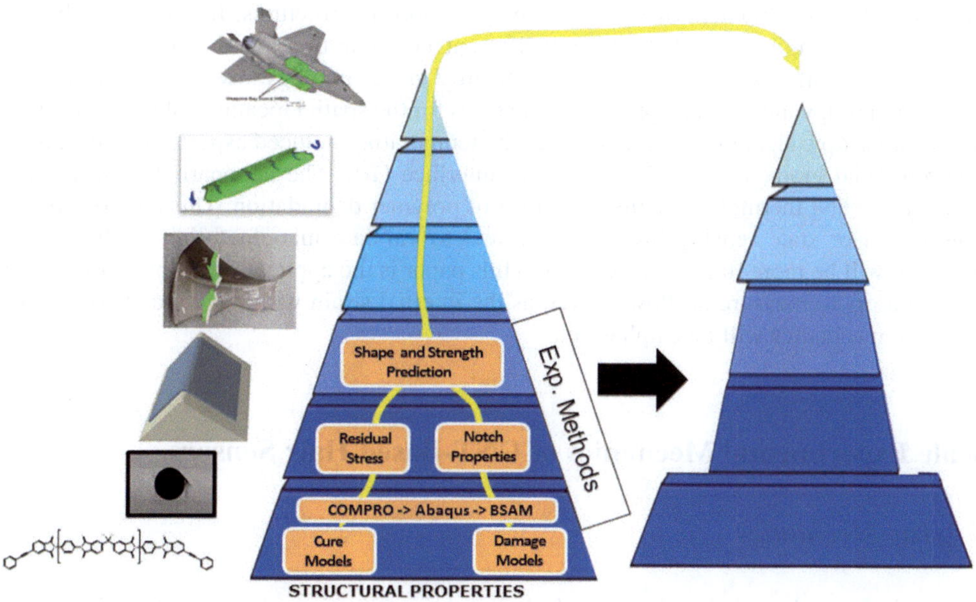

Fig. 21.2 Traditional building block approach to composite testing with overlay of integrated modeling and experimental methods to accelerate development. With the use of improved experimental methods, the development of components with moderate complexity may be accelerated as represented by the narrower mid-section stages

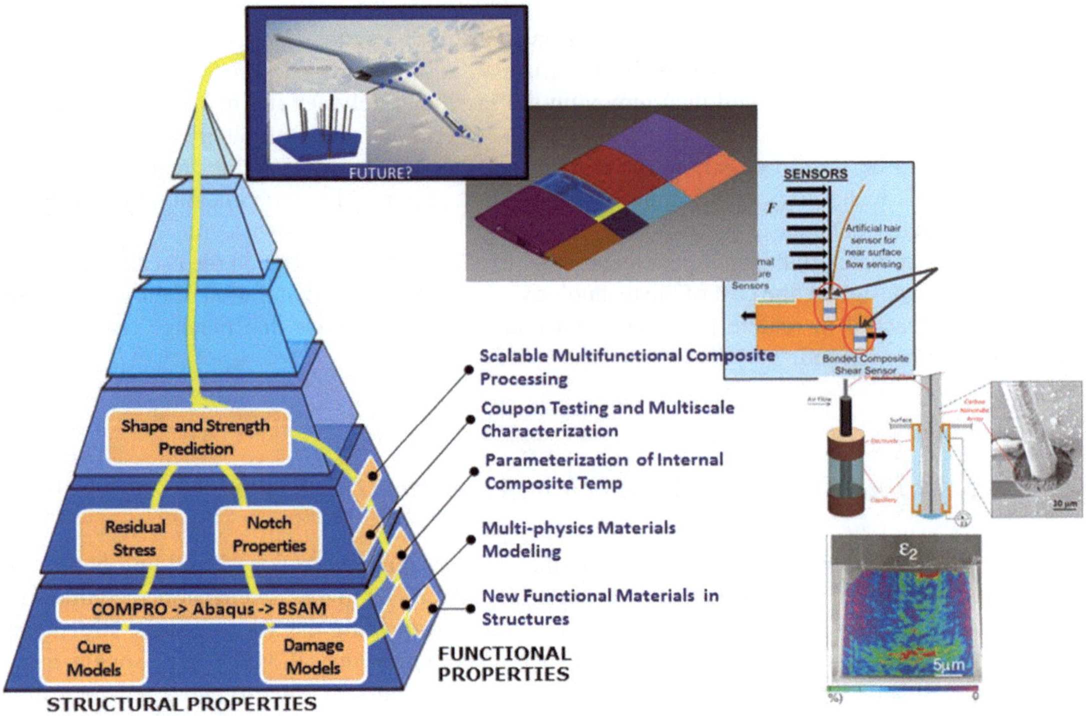

Fig. 21.3 Multi-dimensional building block approach to multifunctional composite with integrated multi-physics tools (*rounded boxes*) and multiscale experimental testing of embedded CNT-based hair sensor (*pictures outsider pyramid*) needed to accelerate development

performance is achieved. The importance of multi-physics modeling, coupled experimental techniques, and acceptance of risk then becomes even more valued in accelerating development as is suggested in Fig. 21.3. As an example, the previously mentioned liquid metal antenna required a multi-physics modeling framework that coupled RF, thermal, fluidic and structural properties to accelerate development. Optimization methods were also applied over this framework to determine the optimal design for a given antenna range, maximum surface temperature, and mechanical knock-down.

In general, technologies that can both give an additional function and feedback about the state of the structural element are ideal for accelerating the development of multifunctional composite structures. It would also be advantageous if the technology was low-cost and agile enough to accept different configurations in a scalable manner. Several of these technologies are currently being investigated within our team. One technology uses small concentrations of plasmon-resonant metal nanoparticles to validate multi-physics predictions for the spatial locations of the maximum temperature. The technique uses the spectroscopic signature associated with the temperature induced aspect ratio decrease and has been used to validate the sharp thermal gradients near the fiber-resin interface [10]. The nanoparticle sensing method shows great promise for passively recording thermal exposure and on-set of polymer degradation. This method could also be useful for non-destructive inspection or state sensing for operation, and to validate manufacturing methods during development. Another technology that will be more fully explored within this paper is the application of low-cost carbon nanotube micro-cantilever deflection sensors to measure air flow as well as the internal strain within a bonded composite. In this case, the multiscale experimental mechanics will be emphasized.

21.2 Multiscale Experimental Mechanics of CNT-Based Hair Sensors

21.2.1 CNT Array Mechanics

When coupled to the integrated modeling mentioned above, the application of low-cost strain sensors throughout the structure of an air vehicle has the potential to accelerate development by rapidly validating the predicted structural performance. At the extreme, one could envision that the final fabricated component or vehicle is structurally testing with feedback from appropriately distributed, trusted sensors both on the ground and in the air. This would assume high

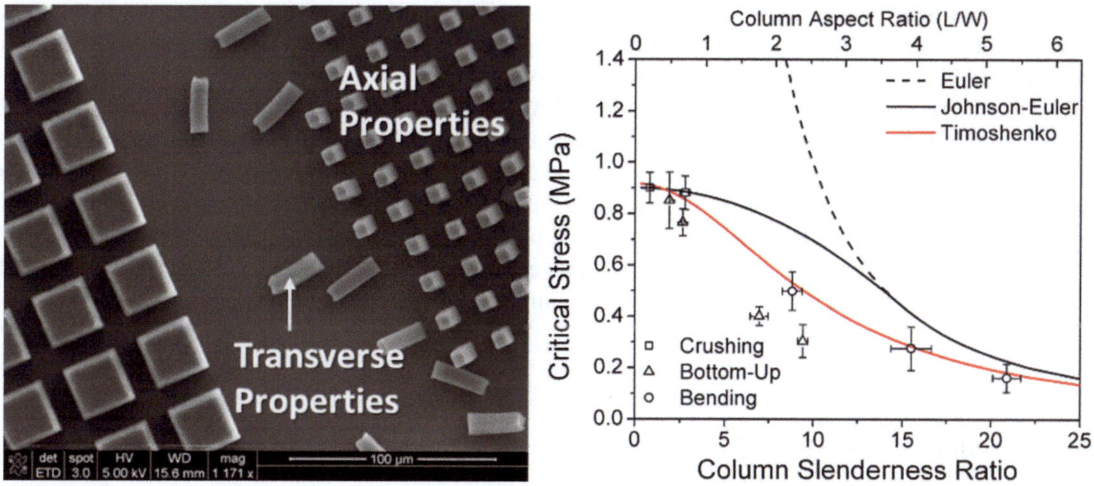

Fig. 21.4 Vertical aligned arrays of CNTs (*left*) characterized by nanoindenting, in-situ testing and digital image correlation to achieve a validated mechanical model [11]

confidence in the sensor, low fabrication and sensor cost, and a trustworthy multi-scale structural model against which to compare. If realized, it would significantly reduce the fabrication and testing of the intermediate elements and provide a "digital thread" by which a history of aerodynamic loading could be captured to inform decisions. The high confidence in the sensor would be a result of both the individual sensor, as well as the sensor network and the accompanying algorithms.

Within this paper, we discuss the challenges of characterizing, via experimental mechanics, the performance of the CNT arrays which form the mechanoresistive element in a microscale, low-cost artificial hair sensor. As shown in Fig. 21.3 above, carbon nanotube (CNT) based systems as inherently multi-scaled. We consider the examples of in-situ mechanical characterization of micro-scaled arrays of aligned CNTs, the internal mechanics of a millimeter-sized CNT-based hair-like sensor, and the application of the sensor to the internal mechanics of a macro-scale adhesively bonded joint.

The mechanical behaviour of vertically aligned CNT forests is governed by complex, multi-scale mechanics related to the geometrical arrangement and properties of individual CNTs. CNT forests deform in a cooperative manner as a persistent network rather than as fields of isolated and non-interacting CNTs. Predictive design of CNT forest mechanical properties requires knowledge of the pertinent elastic constants including moduli, Poisson's ratio, and strength values. Yet, consistency of these properties via controlled synthesis remains a challenge. While the mechanical properties parallel to the CNT growth direction have been thoroughly examined, the transverse modulus, shear modulus, and Poisson's ratios have received relatively less attention. Despite their anisotropic organization (i.e. vertical alignment), CNT forests have been modelled as isotropic materials or as arrays of non-interacting elastic Euler beams because of the lack of understanding of these parameters. The result is that the buckling load predicted by the elastic Euler buckling model may be in error by orders of magnitude when considering the entire CNT forest height [11].

Using in-situ measurements, we previously found that the strain distribution in compressed CNT columns is highly non-uniform during compression and varies greatly with column dimensions [11]. Recently, we independently measured the five independent elastic constants of the CNT forest using uniaxial compression and out of plane shear tests facilitated by nanoindentation, in situ SEM compression, and digital image correlation (DIC) to predict the critical buckling stress of CNT columns of various aspect ratios using Timoshenko and Johnson-Euler column models, as shown in Fig. 21.4. Overall, the in-situ experimental mechanical methods provided the understanding to fabricate micro-cantilever, hair-like sensor could be fabricated [11].

21.2.2 *Internal Mechanics of Hair-Like CNT-Based Sensor*

The understanding of planar arrays was applied to the radial CNT arrays grown on a single structural fiber within an electrode glass capillary. More specifically, this artificial hair sensor (AHS) consists of an array of carbon nanotubes (CNTs) radially grown on a high stiffness structural glass microfiber within an electrode capillary which is integrated using MEMS-free processes. The CNTs in these devices form self-aligned, radial arrays during synthesis that have been shown to exhibit elastic, foam-like behaviour and large piezoresistance in response to compressive strain against a conductive surface. Using in-situ microscale deflection experiments, the moment sensitivity of the hair is shown to scale inversely with the CNT length and stiffness to a maximum of 1.3 ± 0.4% resistance change nN^{-1} m^{-1} and to have a consistent normalized CNT piezoresistivity (1.1 ± 0.2) for a majority of the more than two dozen sensors examined [12].

In comparison to other hair sensor devices, the CNT-based AHS have been shown to have high sensitivity and capable of detecting the separation and stagnation points in air flow over a cylinder [13]. The sensor sensitivity and noise both change distinctly as the flow transitions from steady and laminar to turbulent, suggesting the sensor is capable of detecting flow transitions. When arrays of these sensors were incorporated into an air foil and tested within a wind-tunnel with an feed-forward neural net, they were shown to be useful for determining the angle of attack, lift coefficient, drag coefficient, and wind speed [14]. Considering they are small, low-cost and easily distributed, they are expected to find utility in gust alleviation or alternative navigation for small or limited-life air vehicles.

21.2.3 Application of Hair-Like Micro-Cantilever Sensor to Bonded Joints

Alternatively, we consider the application of the hair sensor to a bonded composite joint to provide additional structural awareness. With the growing use of polymer matrix composites in aerospace structural applications, adhesively bonded joints with assured mechanical properties are of increasing interest. These joints have the potential to lower structural weight, decrease assembly time, decrease component cost, and provide design flexibility. Yet, their performance is variable and there exists limited methods for assessing the bond strength between composites. Often an additional mechanical fastener is used to assure performance, but this adds weight, complexity, and leads to stress concentrations in the composite. A minimally invasive sensor which measures the strain within a bonded joint could make a significant contribution to the vision of having a self-reporting structure which is tested when assembled and avoids the fabrication and testing of the intermediate elements in the traditional building block approach.

Recently, we inserted the above hair sensor into a bonded composite as a means to measure the relative shear strain within the bonded joint during cycling [15]. We compared the internal measurement response from the hair with external laboratory techniques such as an extensometer and load frame. The extensometers are precise and give good indications of strain throughout the bonded overlap length, but are not accurate at spatially measuring local strain at discrete points within bonded joint. An embedded hair sensor could measures relative displacements of opposing bonded adherends at specific locations. But, another measurement technique is required to verify the local measurement response of the hair sensor. To remedy this, we developed a new technique called dual-plane digital image correlation which was able to independently measure the relative shear displacement of a transparent single lap shear specimen. This new measurement technique offered excellent accuracy and provided a reference for the shear strain measurement to compare with the internally mounted hair sensor and the external load frame. The hair sensor was able to detect the cyclic loading within the bonded joint and give indications of precursor damage prior to initial disbond, which was later followed by complete joint failure [15]. Since strain measurements with the external laboratory devices (load frame/extensometer/dual plane DIC) are not viable for fielded applications, such a technology as the hair sensor could provide part of a low-cost system which informs the state of bonded composites within a low-cost attritable aircraft. In considering the information desired for an aircraft structure as shown in Fig. 21.5, it is possible to consider that the low-cost, artificial hair sensor described here may provide part of the desired solution.

21.3 Conclusion

Within this paper we have discussed the systematic aspects of inserting multifunctional composite concepts into emerging air vehicle concepts like the low-cost attritable aircraft. We conclude that this type of vehicle should be fruitful to exploring concepts related to the Multifunctional Composite Attritable Aircraft Technologies (Multi-CAAT). Yet, there are significant challenges with using the traditional "building block" approach to managing risk. Instead, we consider coupling the increased risk tolerance of the user with integrated computational and experimental methods which are multiscale and multiphysics to potentially accelerate development and assure agility.

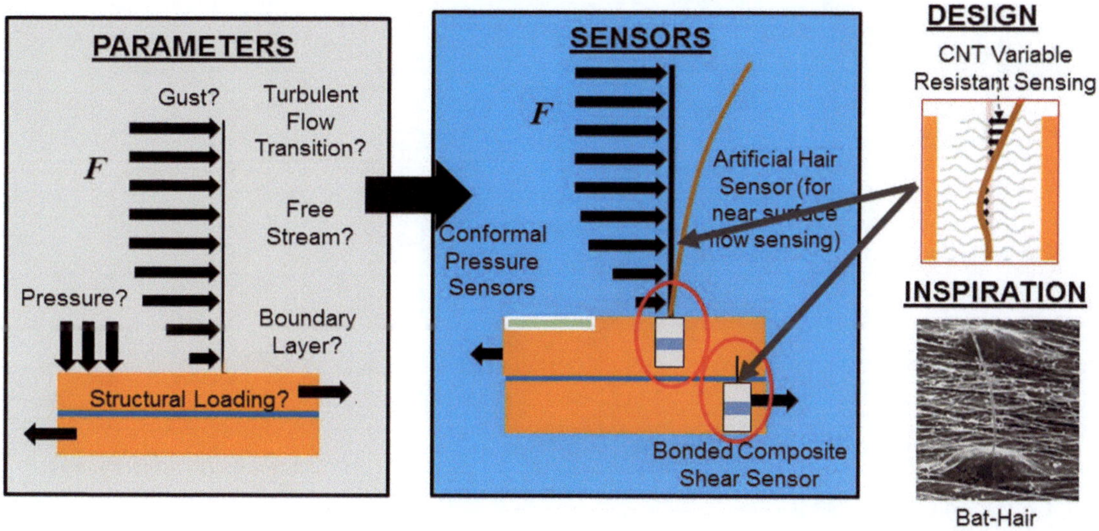

Fig. 21.5 Aero and structural parameters desired (*left*) and approach to providing information (*right*) with embedded micro-cantilever hair-like sensor (pressure sensor not discussed) [15]

References

1. Keller, J.: "Composite engineering to develop enabling technologies for low-cost unmanned attack aircraft" Military & Aerospace Electronics, July 11 2016. Accessed online at http://www.militaryaerospace.com/articles/2016/07/low-cost-unmanned-aircraft.html
2. Hartl, D.J., et al.: A liquid metal-based structurally embedded vascular antenna: I. Concept and multiphysical modeling. Smart Mater. Struct. **26**, 25001–25015 (2017)
3. Hartl, D.J., et al.: A liquid metal-based structurally embedded vascular antenna: II. Mulitobjective and parameterized design exploration. Smart Mater. Struct. **26**, 25002–25014 (2017)
4. Huff, G.H., et al.: A physically reconfigurable structurally embedded vascular antenna (SEVA). IEEE Trans. Antennas Propag., Accepted 2017
5. See http://www.cuaerospace.com/Products/VascTechMicrovascularComposites.aspx
6. Esser-Khan, A.P., et al.: Three-dimensional microvascular fiber-reinforced composite. Adv. Mater. **23**(32), 3654–3658 (2011)
7. Xie, J., et al.: Study on airworthiness requirements on composite aircraft structure for transport category aircraft in FAA. Procedia Eng. **17**, 270–278 (2011)
8. Wood, K.: Virtual testing of composites: Beyond make and break Composite World Nov, (2012)
9. Hoos, K., et al.: Static strength prediction in laminated composites by using discrete damage modeling. J. Compos. Mater. **51**(10), 1473–1492 (2016). See http://journals.sagepub.com/toc/jcma/51/10
10. Kennedy, W.J., et al.: High-resolution mapping of thermal history in polymer nanocomposites: gold nanorods as microscale temperature sensors. ACS Appl. Mater. Interfaces. **7**(50), 27624–27631 (2015)
11. Maschmann, M.R., et al.: Continuum analysis of carbon nanotube array buckling enabled by anisotropic elastic measurements and modeling. Carbon. **66**, 377–386 (2014)
12. Slinker, K.A. et al.: "CNT-based artificial hair senosrs for predictable boundary layer air flow". Advanced Materials Technologies, Accepted (2016)
13. Phillips, D.M., et al.: Detection of flow separation and stagnation points using artificial hair sensors. Smart Mater. Struct. **24**, 1115026 (2015)
14. Kaman, T.M., et al.: Aerodynamic parameters for distributed heterogeneous CNT hair sensors with feedforward neural network. Bioinspir. Biomim. **11**, 066006 (2016)
15. Slinker, K.A., et al.: Artificial hair sensors from structural microfibers and cnt arrays for sensing air flow or mechanical shear, 20th international conference on composite materials, Copenhagen, 19–24 July 2015

MIX
Papier aus verantwortungsvollen Quellen
Paper from responsible sources
FSC® C105338

If you have any concerns about our products,
you can contact us on
ProductSafety@springernature.com

In case Publisher is established outside the EU,
the EU authorized representative is:
Springer Nature Customer Service Center GmbH
Europaplatz 3, 69115 Heidelberg, Germany

Printed by Libri Plureos GmbH
in Hamburg, Germany